C. Christiansen, William Francis Magie

Elements of Theoretical Physics

C. Christiansen, William Francis Magie

Elements of Theoretical Physics

ISBN/EAN: 9783337277543

Printed in Europe, USA, Canada, Australia, Japan

Cover: Foto ©berggeist007 / pixelio.de

ELEMENTS

OF

THEORETICAL PHYSICS

BY

DR. C. CHRISTIANSEN

PROFESSOR OF PHYSICS IN THE UNIVERSITY OF COPENHAGEN

TRANSLATED INTO ENGLISH BY

W. F. MAGIE, PH.D.

PROFESSOR OF PHYSICS IN PRINCETON UNIVERSITY

London

MACMILLAN AND CO., LIMITED

NEW YORK: THE MACMILLAN COMPANY

1897

TRANSLATOR'S PREFACE.

The treatise of Professor Christiansen, of which a translation is here given, presents the fundamental principles of Theoretical Physics, and develops them so far as to bring the reader in touch with much of the new work that is being done in that subject. It is not in every respect exhaustive, but it is stimulating and informing, and furnishes a view of the whole field, which will facilitate the reader's subsequent progress in special parts of it. The need of such a book, in which the various branches of the subject are developed in connection with one another and in a consistent notation, has been long felt by both teachers and students.

The thanks of the translator are due to Professor Christiansen for his courtesy in permitting the use of his book. The translation was made from the German of Müller. The first draft of it was prepared by the translator's wife, without whose aid the task might never have been accomplished.

W. F. MAGIE.

PRINCETON UNIVERSITY, September, 1896.

TABLE OF CONTENTS.

PAGE

INTRODUCTION, - - - - - - - - - - - - 1

CHAPTER I.

GENERAL THEORY OF MOTION.

SECTION

I. Freely Falling Bodies, - - - - - - - - - - 5

II. The Motion of Projectiles, - - - - - - - - - 7

III. Equations of Motion for a Material Point, - - - 8

IV. The Tangential and Normal Forces, - - - - - - 13

V. Work and Kinetic Energy, - - - - - - - - - 14

VI. The Work Done on a Body during its Motion in a Closed Path, - - - - - - - - - - - - - 16

VII. The Potential, - - - - - - - - - - - 21

VIII. Constrained Motion, - - - - - - - - - - 24

IX. Kepler's Laws, - - - - - - - - - - - 27

X. Universal Attraction, - - - - - - - - - - 30

XI. Universal Attraction (*continued*), - - - - - - - 31

XII. The Potential of a System of Masses, - - - - - - 34

XIII. Examples. Calculation of Potentials, - - - - - - 36

XIV. Gauss's Theorem. The Equations of Laplace and Poisson, - 41

XV. Examples of the Application of Laplace's and Poisson's Equations, - - - - - - - - - - - - 46

XVI. Action and Reaction. On the Molecular and Atomic Structure of Bodies, - - - - - - - - - - - - - 48

XVII. The Centre of Gravity, - - - - - - - - - 50

SECTION PAGE
XVIII. A Material System, - 53
XIX. Moment of Momentum, - 55
XX. The Energy of a System of Masses, - 56
XXI. Conditions of Equilibrium. Rigid Bodies, - 58
XXII. Rotation of a Rigid Body. The Pendulum, - 60

CHAPTER II.

THE THEORY OF ELASTICITY.

XXIII. Internal Forces, - 62
XXIV. Components of Stress, - 64
XXV. Relations among the Components of Stress, - 67
XXVI. The Principal Stresses, - 69
XXVII. Faraday's Views on the Nature of Forces Acting at a Distance, - 72
XXVIII. Deformation, - 74
XXIX. Relations between Stresses and Deformations, - 79
XXX. Conditions of Equilibrium of an Elastic Body, - 82
XXXI. Stresses in a Spherical Shell, - 83
XXXII. Torsion, - 85
XXXIII. Flexure, - 87
XXXIV. Equations of Motion of an Elastic Body, - 89
XXXV. Plane Waves in an Infinitely Extended Body, - 90
XXXVI. Other Wave Motions, - 93
XXXVII. Vibrating Strings, - 95
XXXVIII. Potential Energy of an Elastic Body, - 96

CHAPTER III.

EQUILIBRIUM OF FLUIDS.

XXXIX. Conditions of Equilibrium, - 99
XL. Examples of the Equilibrium of Fluids, - 101

CHAPTER IV.

MOTION OF FLUIDS.

XLI. Euler's Equations of Motion, - 103
XLII. Transformation of Euler's Equations, - 106

SECTION PAGE
XLIII. Vortex Motions and Currents in a Fluid, - - - 107
XLIV. Steady Motion with Velocity-Potential, - - - 109
XLV. Lagrange's Equations of Motion, - - - - - - 111
XLVI. Wave Motions, - - - - - - - - - - 112

CHAPTER V.

INTERNAL FRICTION.

XLVII. Internal Forces, - - - - - - - - - - 115
XLVIII. Equations of Motion of a Viscous Fluid, - - - 118
XLIX. Flow through a Tube of Circular Cross Section, - - 119

CHAPTER VI.

CAPILLARITY.

L. Surface Energy, - - - - - - - - - - 121
LI. Conditions of Equilibrium, - - - - - - - 123
LII. Capillary Tubes, - - - - - - - - - 125

CHAPTER VII.

ELECTROSTATICS.

LIII. Fundamental Phenomena of Electricity, - - - 127
LIV. Electrical Potential, - - - - - - - - 128
LV. The Distribution of Electricity on a Good Conductor, - 130
LVI. The Distribution of Electricity on a Sphere and on an Ellipsoid, - - - - - - - - - - 132
LVII. Electrical Distribution, - - - - - - - - 135
LVIII. Complete Distribution, - - - - - - - - 139
LIX. Mechanical Force Acting on a Charged Body, - - 141
LX. Lines of Electrical Force, - - - - - - - 143
LXI. Electrical Energy, - - - - - - - - - 145
LXII. A System of Conductors, - - - - - - - 147
LXIII. Mechanical Forces, - - - - - - - - 150
LXIV. The Condenser and Electrometer, - - - - - 151
LXV. The Dielectric, - - - - - - - - - - 155

SECTION		PAGE
LXVI.	Conditions of Equilibrium,	157
LXVII.	Mechanical Force and Electrical Energy in the Dielectric,	158

CHAPTER VIII.

MAGNETISM.

LXVIII.	General Properties of Magnets,	163
LXIX.	The Magnetic Potential,	166
LXX.	The Potential of a Magnetized Sphere,	168
LXXI.	The Forces which Act on a Magnet,	169
LXXII.	Potential Energy of a Magnet,	171
LXXIII.	Magnetic Distribution,	173
LXXIV.	Lines of Magnetic Force,	174
LXXV.	The Equation of Lines of Force,	178
LXXVI.	Magnetic Induction,	179
LXXVII.	Magnetic Shells,	180

CHAPTER IX.

ELECTRO-MAGNETISM.

LXXVIII.	Biot and Savart's Law,	184
LXXIX.	Systems of Currents,	186
LXXX.	The Fundamental Equations of Electro-Magnetism,	188
LXXXI.	Systems of Currents in General,	190
LXXXII.	The Action of Electrical Currents on each other,	192
LXXXIII.	The Measurement of Current-Strength on the Quantity of Electricity,	194
LXXXIV.	Ohm's Law and Joule's Law,	197

CHAPTER X.

INDUCTION.

LXXXV.	Induction,	199
LXXXVI.	Coefficients of Induction,	202
LXXXVII.	Measurement of Resistance,	205
LXXXVIII.	Fundamental Equations of Induction,	208
LXXXIX.	Electro-Kinetic Energy,	210
XC.	Absolute Units,	211

CHAPTER XI.

ELECTRICAL OSCILLATIONS.

SECTION		PAGE
XCI.	Oscillations in a Conductor,	215
XCII.	Calculation of the Period,	217
XCIII.	The Fundamental Equations for Electrical Insulators or Dielectrics,	219
XCIV.	Plane Waves in the Dielectric,	221
XCV.	The Hertzian Oscillations,	223
XCVI.	Poynting's Theorem,	224

CHAPTER XII.

REFRACTION OF LIGHT IN ISOTROPIC AND TRANSPARENT BODIES.

XCVII.	Introduction,	229
XCVIII.	Fresnel's Formulas,	231
XCIX.	The Electro-Magnetic Theory of Light,	235
C.	Equations of the Electro-Magnetic Theory of Light,	237
CI.	Refraction in a Plate,	242
CII.	Double Refraction,	246
CIII.	Discussion of the Velocities of Propagation,	249
CIV.	The Wave Surface,	251
CV.	The Wave Surface (*continued*),	254
CVI.	The Direction of the Rays,	256
CVII.	Uniaxial Crystals,	259
CVIII.	Double Refraction at the Surface of a Crystal,	261
CIX.	Double Refraction in Uniaxial Crystals,	264

CHAPTER XIII.

THERMODYNAMICS.

CX.	The State of a Body,	266
CXI.	Ideal Gases,	270
CXII.	Cyclic Processes,	272
CXIII.	Carnot's and Clausius' Theorem,	274
CXIV.	Application of the Second Law,	279

SECTION		PAGE
CXV.	The Differential Coefficients,	280
CXVI.	Liquids and Solids,	281
CXVII.	The Development of Heat by Change of Length,	282
CXVIII.	Van der Waal's Equation of State,	283
CXIX.	Saturated Vapours,	290
CXX.	The Entropy,	292
CXXI.	Dissociation,	295

CHAPTER XIV.

CONDUCTION OF HEAT.

CXXII.	Fourier's Equation,	298
CXXIII.	Steady State,	300
CXXIV.	The Periodic Flow of Heat in a given Direction,	301
CXXV.	A Heated Surface,	303
CXXVI.	The Flow of Heat from a Point,	304
CXXVII.	The Flow of Heat in an Infinitely Extended Body,	305
CXXVIII.	The Formation of Ice,	307
CXXIX.	The Flow of Heat in a Plate whose Surface is kept at a Constant Temperature,	308
CXXX.	The Development of Functions in Series of Sines and Cosines,	312
CXXXI.	The Application of Fourier's Theorem to the Conduction of Heat,	315
CXXXII.	The Cooling of a Sphere,	318
CXXXIII.	The Motion of Heat in an Infinitely Long Cylinder,	322
CXXXIV.	On the Conduction of Heat in Fluids,	325
CXXXV.	The Influence of the Conduction of Heat on the Intensity and Velocity of Sound in Gases,	330

INTRODUCTION.

In the Science of Physics it is assumed that all phenomena are capable of ultimate representation by motions, that is, by changes of place considered with reference to the time required for their accomplishment. We therefore begin with a brief discussion of the theory of pure motion (Kinematics). We will treat first the motion of a point. The continuous line traced out by the successive positions which a moving point occupies in space is called its *path*. The symbol s represents the distance which the point traverses along its path in the time t. In measuring these quantities the second is used as the unit of time; the centimetre, as the unit of length. The measures of all the magnitudes which occur in the discussion of motions may be stated in terms of these two units.

Motions are distinguished by the form of the path, as rectilinear, curvilinear, or periodic. Rectilinear and curvilinear motions are sufficiently defined by their names. A periodic motion is one in which the same condition of motion recurs after a definite interval of time; that is, one in which the moving point returns after a definite time to the same position with the same velocity and direction of motion.

Rectilinear motion may be either *uniform* or *variable*. It is uniform if the moving point traverses equal distances in equal times. In this case the point traverses the same distance in each unit of time, and the distance traversed in the unit of time measures its velocity. If the point traverses the distance s in the time t with a uniform motion, the velocity c is the ratio of s to t, or (a) $c = s/t$. A velocity is therefore a length divided by a time.

If a point moves on the circumference of a circle with a constant velocity, the radius vector drawn to this point sweeps out equal sectors in equal times. In this case the angle which is swept out by this radius vector in the unit of time measures the angular velocity.

If s_1, s_2, s_3, etc., represent the distances which are successively traversed in the corresponding times t_1, t_2, t_3, etc., a uniform motion is defined by $s_1/t_1 = s_2/t_2 = s_3/t_3 = \ldots$. On the other hand, in the case of a variable motion, $s_1/t_1 \gtrless s_2/t_2 \gtrless s_3/t_3 \gtrless \ldots$, that is, a motion is variable if it is uniform in no part. The mean velocity during the time t_1 is s_1/t_1, that is, the velocity with which the point must move uniformly, in order to traverse the distance s_1 in the time t_1. The mean velocity depends on the length of path that is taken into consideration. The ratio $\Delta s/\Delta t$ has, however, a finite limit, if Δs and Δt vanish simultaneously. This limit is expressed by ds/dt, and represents the velocity at the instant t when the motion is variable. The velocity of a point, whose motion is variable, is therefore given by the first differential coefficient of length with respect to time. The increment ds of the distance, during the time-element dt, divided by dt, represents the distance that would be traversed at this rate in a unit of time.

We may assume arbitrarily a special unit for velocity, as also for all the other physical magnitudes which we wish to measure. But all phenomena depend on the motion of masses, and any one of them can therefore be measured in terms of the *absolute units* of *mass*, *length*, and *time*. As unit of mass we take the mass of a cubic centimetre of water at the temperature 4° C., or the *gram*; as unit of length, the *centimetre*; as unit of time, the *second*. In contrast with the absolute units all others are called *derived* or *composite* units. These may always be expressed in terms of the absolute units. The *dimensions* of a derived unit express the manner in which it involves the absolute units. If we express length by $[L]$, mass by $[M]$, and time by $[T]$, we have for the dimensions of velocity $[LT^{-1}]$.

We will follow the notation introduced by Newton, and write for ds/dt the abbreviation $\dot{s}$. Similarly $\dot{x}$ is used to represent dx/dt, etc.

In general, the velocity of a moving body varies with the time; the velocity is then a function of the time. If the moving point has the velocity v' at the time t', and the velocity v'' at the time t'', then $v'' - v'$ is the increment of velocity in the time interval from t' to t''. Let this time interval be infinitely small and equal to Δt; the gain in velocity during this time interval divided by Δt will express the velocity p that would be gained at this rate in a unit of time, that is, the *acceleration*. The equation defining p is

(b) $$p = (v'' - v')/(t'' - t') = \Delta v/\Delta t \quad \text{or} \quad p = dv/dt = \dot{v}.$$

Since $v = ds/dt = \dot{s}$, we have $p = d\dot{s}/dt = d^2s/dt^2 = \ddot{s}$. The *acceleration* is therefore the second differential coefficient of length with respect to time. Since the difference between two velocities is itself a velocity, and the difference between two times a time interval, we obtain from (b) for the dimensions of acceleration $[LT^{-2}]$.

If the velocity increases by the same amount in each equal interval of time, the acceleration is constant and the motion is said to be *uniformly accelerated.* As may be seen from the foregoing discussion, it is in this case only that the acceleration is measured by the increment of velocity in the unit of time. If the acceleration is variable and a function of the time, the motion is said to be *variably accelerated.*

We will consider next the *curvilinear motion* of a point. Let ds be an element of the curved path. The direction of motion of the point changes. It coincides at each point on the curve with the direction of the element of the curve at that point, or with the tangent to the curve. If the element ds makes the angles α, β, γ, with the axes of a system of rectangular coördinates, and if dx, dy, dz are the projections of ds on the axes, then

$$dx = ds \cos \alpha\,; \quad dy = ds \cos \beta\,; \quad dz = ds \cos \gamma.$$

The elements dx, dy, dz are the edges of an infinitely small rectangular parallelepiped, whose diagonal is ds. Forming the expressions

$$dx/dt = ds/dt \,.\, \cos \alpha, \quad dy/dt = ds/dt \,.\, \cos \beta, \quad dz/dt = ds/dt \,.\, \cos \gamma$$

or

$$\dot{x} = \dot{s} \cos \alpha, \quad \dot{y} = \dot{s} \cos \beta, \quad \dot{z} = \dot{s} \cos \gamma,$$

we see that $\dot{x}$, $\dot{y}$, $\dot{z}$ are the projections of the velocity of the moving point on the axes. In this way a velocity may be resolved into three components in three directions at right angles to one another. This resolution of velocity corresponds to the representation of a curve in rectangular coordinates. The velocity $\dot{x}$ is that with which the moving point departs from the yz-plane. Instead of dealing with the motion of the point expressed in terms of elements of the curve, we introduce three other motions which produce the same result, namely, a motion of the point in the x-axis with the velocity $\dot{x}$, a motion of the x-axis with the velocity $\dot{y}$ in the direction of the y-axis, during which the x-axis remains parallel to its original position, and a motion of the xy-plane in the direction of the z-axis, with the velocity $\dot{z}$, during which the xy-plane remains parallel to its original position.

If a, b, and c represent the coordinates of the initial position of a point, and if this point is considered as affected simultaneously by two motions, whose projections on the x-axis are x_1 and x_2, the whole distance traversed by the point in the direction of the x-axis is $x_1 + x_2$. The component of velocity in the direction of the x-axis is $\dot{x} = \dot{x}_1 + \dot{x}_2$.

Similar expressions hold for motions in the directions of the other axes. The resultant velocity is represented by the diagonal of the parallelepiped, whose edges are $\dot{x}$, $\dot{y}$, $\dot{z}$, or by $\dot{s} = \sqrt{\dot{x}^2 + \dot{y}^2 + \dot{z}^2}$.

Since an acceleration is the increment of a velocity, the resultant acceleration will be determined in a similar manner. Let $\ddot{x}_1$, $\ddot{x}_2$ represent the x-components of the increments of velocity due to the two motions. We then have for the total acceleration in the direction of the x-axis, $\ddot{x} = \ddot{x}_1 + \ddot{x}_2$, and for the acceleration of the point, $\ddot{s} = \sqrt{(\ddot{x}_1 + \ddot{x}_2)^2 + (\ddot{y}_1 + \ddot{y}_2)^2 + (\ddot{z}_1 + \ddot{z}_2)^2}$, by which $\ddot{s}$ is expressed as the diagonal of the parallelepiped whose edges are $\ddot{x}$, $\ddot{y}$, $\ddot{z}$.

If the coordinates of the moving point are given as functions of the time, the equation of the path is obtained by determining the values of x and y which hold for the same time t. If, for example, $x = f_1(t)$ and $y = f_2(t)$, the relation between x and y is found by eliminating the variable t from the equations by any appropriate method.

From this brief discussion of these purely kinematic questions we turn to the consideration of the causes of motion, taking as our starting point the researches of Galileo on freely falling bodies.

CHAPTER I.

GENERAL THEORY OF MOTION.

SECTION I. FREELY FALLING BODIES.

THE investigation by Galileo of the motion of freely falling bodies was the first step in the development of modern physics. It is advantageous to start from the same point in our study of the subject. Galileo concluded from his experiments *that all bodies falling freely in vacuo will fall at the same rate.* This is one of the most important discoveries in natural science, since it shows that all bodies, independent of their condition in other respects, have one property in common. No parallel to this has been found in Nature. It points to a unity in the constitution of matter, of which we certainly do not as yet appreciate the full significance.

Galileo's conclusions have been confirmed by the careful experiments of Newton, Bessel, and others. Galileo concluded further, that *the distance s traversed by a falling body in the time t is proportional to the square of the time,* so that (a) $s=\frac{1}{2}gt^2$, where g is a constant. The constant g is called the *acceleration of gravity.* The falling body has a uniformly accelerated motion, since

$$ds/dt=\dot{s}=gt \quad \text{and} \quad d^2s/dt^2=\ddot{s}=g.$$

Its acceleration is therefore constant. This second law of falling bodies is not to be considered a fundamental law in the sense in which the first is.* In the time τ immediately following the time t, the body traverses the distance σ, which is determined from the equation $s+\sigma=\frac{1}{2}(t+\tau)^2g$. By the use of equation (a) we obtain

* Later researches have shown that the value of the force of gravity depends on the distance of the falling body from the centre of the earth, and therefore g varies during the fall. However, the variation of g is so slight that it has not yet been detected by direct experiment on falling bodies.

(b) $\sigma = gt\tau + \frac{1}{2}g\tau^2$. During the time τ the velocity is variable, but if v is the mean velocity during that time, we will have $v = \sigma/\tau = gt + \frac{1}{2}g\tau$. If τ is infinitely small and equal to dt, we have $\sigma = ds$, and neglecting $\frac{1}{2}gdt$ in comparison with gt, (c) $v = ds/dt = \dot{s} = gt$. *The velocity therefore increases proportionally to the time, and g represents the increment of velocity in the unit of time.*

The body falls through the space s in the time t, which, from (a), is determined by (d) $t = \sqrt{2s/g}$.

The velocity at the time t is obtained by substituting this value of t in (c); making this substitution, we have (e) $v = \sqrt{2sg}$. We reach the same result by eliminating t between equations (a) and (c).

From the laws of falling bodies we deduce the *law of inertia.*

In order to explain the fact that the velocity of a falling body increases uniformly with the time, we make the assumption, that *a body retains a velocity once imparted to it unchanged in magnitude and direction; any change of its velocity is due to external causes.* This law is called the *principle of inertia.*

At the time $(t + \tau)$ the velocity v' is $v' = gt + g\tau$. The initial velocity is here gt, to which, in consequence of an external cause, namely, the force of gravity, the velocity $g\tau$ is added. Under the action of gravity the body traverses the space $\sigma = gt\tau + \frac{1}{2}g\tau^2$ in the time τ, immediately following the time t. The falling body traverses the space $gt\tau$ during the time τ, with the velocity gt attained at the end of the time t; the additional distance $\frac{1}{2}g\tau^2$ traversed by the falling body is due to the action of gravity during the time τ.

The principle of inertia holds not only when the increment of velocity is in the same direction as the original velocity, but also when it makes any angle whatever with the original velocity. This principle justifies the application of the methods of geometrical addition to the motions and accelerations of bodies.

The laws of falling bodies lead also to an answer to the question, how forces are to be measured. It is evident that the gain in velocity of a body, and its pressure on a support, that is, its weight, are properly regarded as actions of one and the same force. The increasing velocity of a body in its fall is an evidence of that force, and the increment of velocity in the unit of time gives a new measure of it. This definition shows what before Galileo's time was not clearly understood, how the combined action of several forces may be measured. The increments of velocity corresponding to the separate forces are combined by the method previously described, and the resulting increment gives a measure of the combined action of the forces.

SECTION II. THE MOTION OF PROJECTILES.

We will apply the foregoing principles to the motion of projectiles, which is closely connected with that of freely falling bodies. We consider

1. Vertical projection, both downward and upward.

Galileo, in his study of the motion of projectiles, proceeded on the assumption that a body which is given an initial motion in any direction retains this motion, which is combined with that imparted to it by gravity in accordance with the laws of freely falling bodies. If, at the time $t=0$, the velocity u is given to a body, directed vertically downward, its velocity v, after the lapse of the time t, is (a) $v=u+gt$, and the distance traversed is (b) $s=ut+\frac{1}{2}gt^2$. If the body is given the initial velocity u, directed vertically upward, the corresponding formulas are (c) (d) $v=u-gt$ and $s=ut-\frac{1}{2}gt^2$.

2. Projection in a direction inclined to the vertical.

Let a body be projected in the direction OA, making an angle α with the horizontal. Let OA represent the initial velocity u (Fig. 1). The space which the body would traverse in the time t if gravity did not act on it is $OB=ut$. The body, however, does not reach B, but, at the end of the time t, is beneath B at the point C, so that $BC=\frac{1}{2}gt^2$. Let the x-axis Ox and the y-axis Oy lie in the vertical plane containing OB; then the coordinates of the point C at the time t are

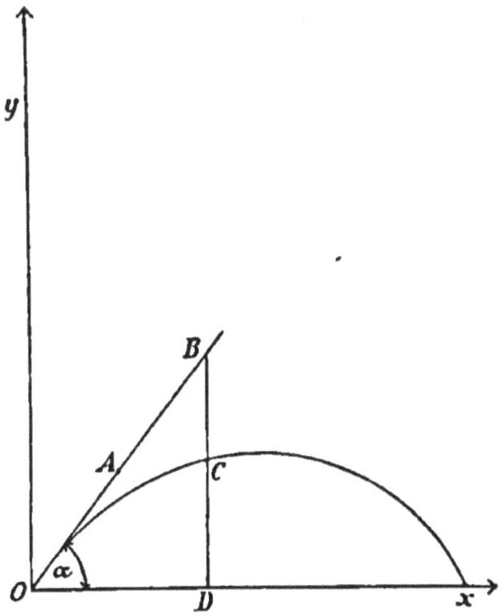

FIG. 1.

(e) $$x=OD=ut\cos\alpha,$$
$$y=CD=ut\sin\alpha-\tfrac{1}{2}gt^2.$$

By these equations the position of the body at any time is determined. During the time-element dt the coordinates x and y increase by

(f) $$dx=u\cos\alpha\,dt \text{ and } dy=u\sin\alpha\,dt-gtdt.$$

The distance ds traversed in the time dt is determined by

$$ds^2=dx^2+dy^2=[(u\cos\alpha)^2+(u\sin\alpha-gt)^2]dt^2.$$

The velocity v is given by

(g) (h) $v = \dot{s}$ and $v^2 = \dot{s}^2 = \dot{x}^2 + \dot{y}^2 = u^2 - 2\,ugt \sin\alpha + g^2t^2$.

The components of the velocity parallel to the axes Ox and Oy are respectively $v\,.\,dx/ds$ and $v\,.\,dy/ds$, or, by (g), are equal to $\dot{x}$ and $\dot{y}$. From equation (f) we have (i) $\dot{x} = u\cos\alpha$; $\dot{y} = u\sin\alpha - gt$. Hence the horizontal velocity is constant, while the vertical velocity diminishes uniformly; this follows because the only force acting is directed vertically downward.

If t is eliminated from the equations (e) we obtain for the equation of the path (k) $y = x\tan\alpha - x^2/4h\,.\,(1 + \tan^2\alpha)$, where $h = \frac{1}{2}u^2/g$ is the distance through which the body must fall under the action of gravity to attain the velocity u. Equation (k) shows that *the path is a parabola.* The range, or the distance reckoned from O, at which the path cuts the x-axis, is given by (k) if we set $y = 0$. We have $0 = \tan\alpha - \frac{1}{2}gx(1 + \tan^2\alpha)/u^2$. The range W, or the particular value of x given by this equation, is $W = u^2 \sin 2\alpha/g$; the maximum range is attained when $\alpha = \frac{1}{4}\pi$.

If the velocity u is given, we may determine from equation (k) the direction in which a body must be projected in order to reach a prescribed point. Transposing, we obtain

$$\tan\alpha = (2h \pm \sqrt{4h^2 - 4hy - x^2})/x.$$

This equation shows that there are in general two directions in which the body may be projected with the initial velocity u so as to reach a prescribed point. If the expression under the radical is zero, there is only one possible direction. If the point to be reached by the body is so situated that $4h^2 - 4hy - x^2 < 0$, $\tan\alpha$ will be imaginary, and the body will not reach the prescribed point.

SECTION III. EQUATIONS OF MOTION FOR A MATERIAL POINT.

In the theory of motion we use the word *force* to designate the causes, known or unknown, of a change in the motion of a body. If a body at rest is set in motion, or if a moving body comes to rest, these changes are ascribed to the action of a force. If the change is sudden, the force acting on the body is called *an instantaneous force* or *impulse.* Close examination shows, however, that finite changes in the motion of a body are never instantaneous, but occur only in a finite time. This time may, in many cases, be very small. The motion of a body, which is measured by its velocity, may vary both in amount and in direction. The velocity of a freely falling body varies only in amount; the velocity of a body revolving round a centre varies in direction, and sometimes also in amount. Experiment shows that all changes in the direction, as well as in the amount of velocity, are due to external causes, which act during a longer or shorter time, but never instantaneously.

We may set aside all questions as to the origin of force, and measure the amount of a force by its action. We may take as a measure of a force either the space which a body, starting from rest, traverses under the action of the force, or the velocity which the force imparts to the body in a given time. There is no essential difference between these two modes of measurement, but generally the velocity produced, or better, the change in velocity, is used for the purpose. *We measure the amount of an impulse by the change in velocity imparted to the body by the impulse, and the amount of a constant force acting continually upon the body, by the change in velocity which occurs in a second. Newton assumed further, that the force is proportional to the quantity of that which is set in motion, that is, to the mass* m *of the body.* He therefore set $F=f \, . \, m \, . \, b$ where b is the acceleration of the body, and f is a factor dependent on the units of force, mass, and acceleration, or on the units of mass, time, and length. If we set $f=1$, then $F=m \, . \, b$, and we obtain the following definition for the unit of force: *The unit of force is that force which imparts the unit of acceleration to the unit of mass, or which imparts to a body in a second the unit of momentum* (cf. XVI.). This unit of force, called *a dyne*, is therefore that force which, acting for one second, imparts to a mass of one gram the velocity of one centimetre per second. Hence the dimensions of force are MLT^{-2} (cf. Introduction).

The force with which a body is attracted by the earth is called its weight, and is measured by the product of its mass and the acceleration which it would have if it were falling freely. If a body is prevented from falling by a support, it exerts a pressure on the support which is equal to its weight. Conversely, the support exerts the same pressure on the body, in accordance with the *law of action and reaction.* This pressure may be determined by the balance, by the elasticity of a spring, etc.

Since the velocity which is caused by a force F may be resolved into components in the directions of the three axes of a system of rectangular coordinates, so, in the same way, the force F may be resolved into components along the three coordinate axes. If these components are represented by X, Y, and Z, we have $F^2=X^2+Y^2+Z^2$.

We may also resolve forces into their components in other ways. These will be treated later.

If a body moves with the velocity v in the direction AB (Fig. 2), and if a force acts on it in the direction AC, the path of the body may be determined by the method used by Galileo to obtain the law of the motion of projectiles. Consider the motion in the time τ. In that time the body will traverse the distance $AM = v\tau$, in consequence of its initial velocity; in the same time it will traverse the distance $AN = \frac{1}{2}\gamma\tau^2$ under the action of the force F, if γ represents the acceleration due to the force F. If the parallelogram $AMDN$ is constructed, D will be the position of the body at the end of the time τ.

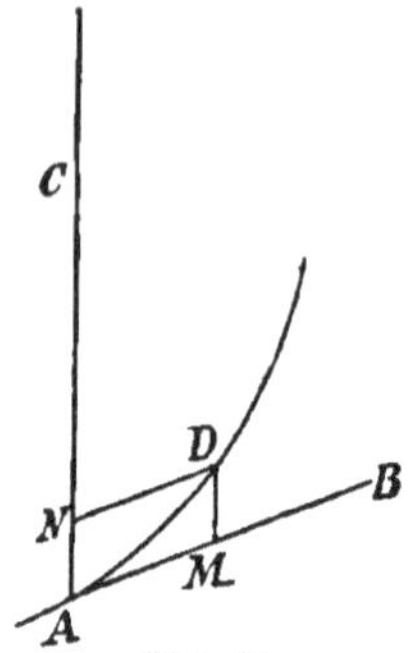

FIG. 2.

Let the direction of motion make the angles α, β, γ with the axes OX, OY, OZ of a system of rectangular coordinates; let the direction of the force make the angles λ, μ, ν with the same axes. If the coordinates of the point A are x, y, and z, the x coordinate of the point D is (b) $x + AM\cos\alpha + DM\cos\lambda = x + v\tau\cos\alpha + \frac{1}{2}\gamma\tau^2\cos\lambda$. Since the coordinates are functions of the time, we may obtain an expression for the x-coordinate of D by the use of Taylor's theorem. Applying this theorem, we obtain (c) $x + \dot{x}\tau + \frac{1}{2}\ddot{x}\tau^2 + \dots$. By comparing (b) and (c) it follows that (d) (e) $v\cos\alpha = \dot{x}$ and $\gamma\cos\lambda = \ddot{x}$. In a similar way we obtain $v\cos\beta = \dot{y}$, $\gamma\cos\mu = \ddot{y}$; $v\cos\gamma = \dot{z}$, $\gamma\cos\nu = \ddot{z}$. The symbols $\dot{x}$, $\dot{y}$, $\dot{z}$ represent the velocities along the coordinate axes; this appears also if we write $v = ds/dt$, and notice that $\cos\alpha = dx/ds$, etc., so that $v\cos\alpha = dx/ds \cdot \dot{s} = \dot{x}$. From (e) it follows further that $m\gamma\cos\lambda = m\ddot{x}$. Since $m\gamma$ is the force $F = \sqrt{X^2 + Y^2 + Z^2}$ and $m\gamma\cos\lambda$ represents the x component X of the force F, we have (f) $X = m\ddot{x}$; similarly $Y = m\ddot{y}$, $Z = m\ddot{z}$. These equations (f) are the equations of motion of the particle m. If X, Y, and Z are given functions of the coordinates, of the time, and sometimes of the velocity, equations (f) will determine the motion of the mass m, if its position and velocity are given at the beginning of the motion. To determine the motion, however, it is necessary to integrate equations (f), which can be done in only a very few cases. If the motion is known, that is, if x, y, and z are given as functions of the time t, these equations may be more easily applied to find the force which causes the motion.

We will now consider some examples.

1. *Motion in a Circle.*

Let a body of mass m move with constant velocity in the circle ABC, whose centre lies at the origin of coordinates, and whose radius is R (Fig. 3). Let T represent the time of revolution, or period. If ω represents the angular velocity of the body, and if the x-axis is drawn through the point occupied by the body at the time $t=0$, we have $x=R\cos(\omega t)$, $y=R\sin(\omega t)$. It then follows from (f) that $X=m\ddot{x}=-m\omega^2R\cos(\omega t)$, $Y=m\ddot{y}=-m\omega^2R\sin(\omega t)$ or $X=-m\omega^2x$, $Y=-m\omega^2y$. The force acting on the body is, therefore, $F=\sqrt{X^2+Y^2}=m\omega^2R$. The cosines of the angles made by the direction

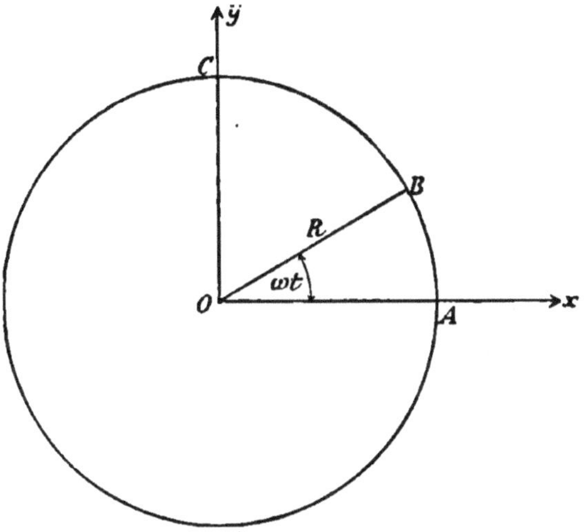

FIG. 3.

of the force with the x- and y-axes are respectively $-x/R$ and $-y/R$. The force is therefore directed toward the centre of the circle. If v represents the velocity of the body in the circle, we have $v=R\omega=2\pi R/T$ and $F=mv^2/R=4\pi^2mR/T^2$. *The acceleration directed toward the centre, the so-called centripetal acceleration, is equal to* $v^2/R=R\omega^2$. F is called the *centripetal force.* This result was first obtained by Huygens.

2. *The Motion of Projectiles.*

Let a body be projected from the origin of coordinates with the velocity u in a direction which makes the angle α with the horizontal x-axis; let the positive y-axis be directed upward. Then

$$X=0, \quad Y=-mg.$$

The equations of motion are $m\ddot{x}=0$, $m\ddot{y}=-mg$. By integration we have $x=a+a_1t$, $y=b+b_1t-\frac{1}{2}gt^2$, where a, a_1, b, b_1 are constants. Since the body is at the origin at the time $t=0$, we have $a=0$ and $b=0$. The components of velocity at the time t are $\dot{x}=a_1$, $\dot{y}=b_1-gt$ From the value of the velocity at the time $t=0$ we have

$$a_1=u\cos\alpha,\quad b_1=u\sin\alpha.$$

We thus obtain again the equations given in II. (e).

3. *Oscillatory Motion.*

If an elastic cylindrical rod, whose weight is so small as to be negligible, and which carries on one end a heavy sphere, is clamped firmly by the other end, and if it is then bent, it will be urged back toward its position of equilibrium by a force which is proportional to its displacement from that position. If r (Fig. 4) is the distance

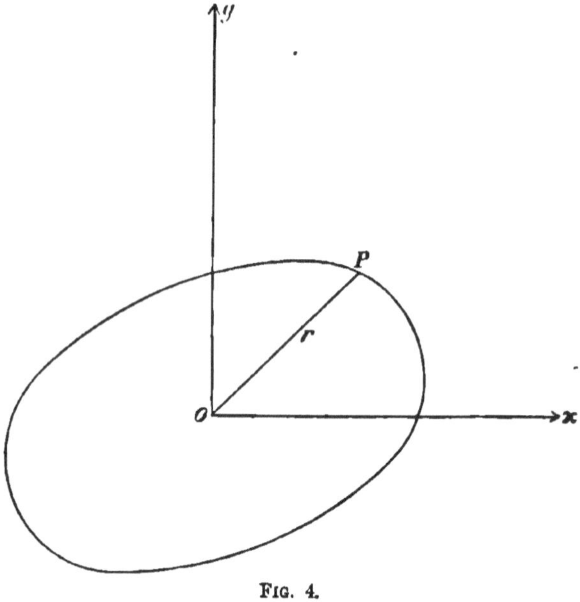

FIG. 4.

from the position of equilibrium O to the point P, which the body occupies at the time t, the force which urges it toward O may be set equal to $-mk^2r$, where k is a constant. The components of this force are $X=-mk^2x$, $Y=-mk^2y$, and the equations of motion are

$$\ddot{x}=-k^2x,\quad \ddot{y}=-k^2y.$$

The integrals of these equations are

$$x = a_1 \cos kt + b_1 \sin kt, \quad y = a_2 \cos kt + b_2 \sin kt,$$

where a_1, b_1, a_2, b_2 are constants. If the coordinates of the point P at the time $t=0$ are x_0, y_0, and if the components of the velocity v at the same time are u_0 and v_0, we have

$$x_0 = a_1, \ y_0 = a_2; \quad u_0 = b_1 k, \ v_0 = b_2 k.$$

With these values of the constants, the equations become

$$x = x_0 \cos kt + u_0/k \, . \sin kt, \quad y = y_0 \cos kt + v_0/k \, . \sin kt.$$

The components of the velocity are

$$\dot{x} = -kx_0 \sin kt + u_0 \cos kt, \quad \dot{y} = -ky_0 \sin kt + v_0 \cos kt.$$

If, when $t=0$, the point P lies on the axis Oy, and if its initial velocity v is parallel to the x-axis, $v = u_0$ and $v_0 = 0$; and

$$x = u_0/k \, . \sin kt, \quad y = y_0 \cos kt.$$

For two values of t which differ by $2\pi/k$, the values of x and y are the same. The motion is therefore *periodic.* The period is $T = 2\pi/k$. If we divide the first equation by u_0/k, the second by y_0, and add the squares of the right and left sides of both equations, we eliminate t, and obtain the equation of an ellipse as the equation of the path of the body.

SECTION IV. THE TANGENTIAL AND NORMAL FORCES.

Let MAD (Fig. 5) be a part of the path of a body whose mass

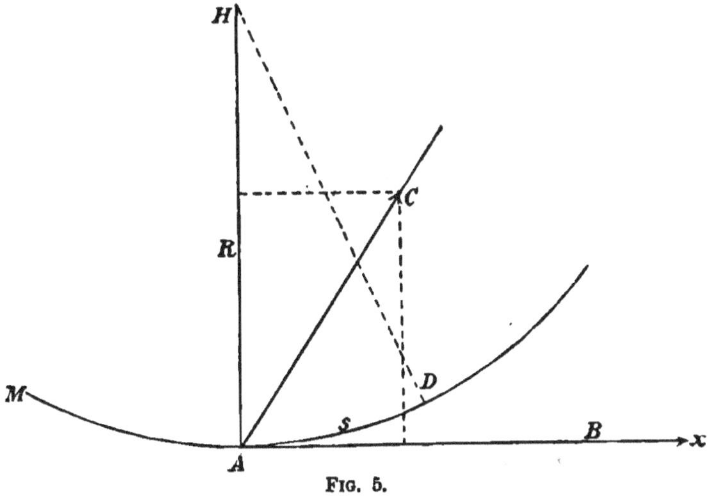

FIG. 5.

is m, let AB be the tangent to the path at the point A, AC, the direction of the force F acting on the body. The directions of the motion and of the force lie in the plane of the path. We choose this plane for the xy-plane of a system of rectangular coordinates whose x axis lies in the direction AB. The normal AH, drawn to the same side as the force AC, is taken as the positive y-axis. The equations of motion are $m\ddot{x}=T$, $m\ddot{y}=N$.

T and N are the components of the force in the direction of the tangent and of the normal, and are called in consequence *tangential* and *normal forces*. If the small arc AD is represented by s, if H is the centre of curvature of the curve at the point A, and if the radius of curvature AH is represented by R, the coordinates of D are

$$x=R\,.\,\sin(s/R),\quad y=R-R\cos(s/R).$$

Hence

$$\ddot{x}=\ddot{s}\,.\,\cos(s/R)-\sin(s/R)\,.\,\dot{s}^2/R,$$

$$\ddot{y}=\ddot{s}\,.\,\sin(s/R)+\cos(s/R)\,.\,\dot{s}^2/R.$$

If s is very small, we may set $\cos(s/R)=1$ and $\sin(s/R)=0$. We have then

$$\ddot{x}=\ddot{s},\quad \ddot{y}=\dot{s}^2/R=v^2/R,$$

and therefore

$$T=m\ddot{s}\ \text{and}\ N=mv^2/R,$$

that is, *the tangential force is proportional to the acceleration in the path. The normal force is proportional directly to the square of the velocity, and inversely to the radius of curvature.*

SECTION V. WORK AND KINETIC ENERGY.*

If a particle, under the action of a force S, moves along a path ds, whose direction is that of the force S, the force is said to do work equal to Sds. If the direction of motion and the direction of the force make the angle θ with each other, we must use the component of the force in the direction of motion, instead of the total force S; the work done is $Sds\cos\theta$. If the body moves in a given path s_0s under the action of the tangential force T, the work done by motion through the element ds is Tds, and the work done in the path s_0s is given by the integral $\int_{s_0}^{s}Tds$. If the velocity of the particle is represented by v, $v=ds/dt$ and $T=m\ddot{s}=m\dot{v}$. Hence

(a) $$\int_{s_0}^{s}Tds=\int m\dot{v}vdt=\tfrac{1}{2}mv^2-\tfrac{1}{2}mv_0^2,$$

where v_0 represents the velocity of the body in its initial position s_0. The quantity $\frac{1}{2}mv^2$, or the product of one half the mass and the

* Kinetic energy is also called actual energy or *vis viva*.

square of the velocity, is called the kinetic energy of the body. From equation (a) *the gain in kinetic energy is equal to the work done by the tangential force,* or is equal to the work done by the total force, since in the calculation of the work as it has been defined it is necessary to consider only the component of the total force which acts in the direction of the path. If α, β, γ are the angles which ds makes with the coordinate axes, and X, Y, Z the components of the force T, then the following equations hold:

$$T = X\cos\alpha + Y\cos\beta + Z\cos\gamma,$$

$$ds\cos\alpha = dx, \quad ds\cos\beta = dy, \quad ds\cos\gamma = dz.$$

The work done by the force T in the infinitely small distance ds is

$$Xdx + Ydy + Zdz.$$

Equation (a) then takes the form

(b) $$\int(Xdx + Ydy + Zdz) = \tfrac{1}{2}mv^2 - \tfrac{1}{2}mv_0^2.$$

This equation may be applied to advantage in many cases, especially if the force is a function of the coordinates only. If the path is also given, we may use this equation to determine the velocity at any point in the path.

1. *Example.*—Let the xz-plane of a system of rectangular coordinates be horizontal, and let the y-axis be directed vertically upward. Let a body of mass m be situated on the y-axis, and let the only force acting on it be gravity. Its components are

$$X = 0, \quad Y = -mg, \quad Z = 0.$$

We have therefore $\int(Xdx + Ydy + Zdz) = -mg(y - b)$, if the body begins to move at the point $y = b$. From (b) we obtain

(c) $$v^2 = v_0^2 - 2g(y - b).$$

Hence the velocity is determined by the y-coordinate alone. This example is discussed in II.

2. *Example.*—The force is a function of the distance of the particle from a fixed point. Let the force be a repulsion and a central force, that is, one whose direction passes always through a fixed point O. We take this point as the origin. The components of the force which acts at the point (x, y, z) are

$$X = f(r)\,.\,x/r, \quad Y = f(r)\,.\,y/r, \quad Z = f(r)\,.\,z/r.$$

Using these values, we have

$$\int(Xdx + Ydy + Zdz) = \int\frac{f(r)}{r}(xdx + ydy + zdz).$$

Since $r^2 = x^2 + y^2 + z^2$, and therefore $rdr = xdx + ydy + zdz$, the work which is done by the force during the movement of the body from the point A to the point B is $\int_{r_0}^{r} f(r)dr$, if r_0 and r are respectively the distances from the point O to the points A and B. Let the velocities at the points A and B be respectively v_0 and v, then

$$\tfrac{1}{2}mv^2 - \tfrac{1}{2}mv_0^2 = \int_{r_0}^{r} f(r)dr.$$

The gain in kinetic energy depends only on r_0 and r, and is consequently independent of the form of the path. The general condition which must be fulfilled that the work done may depend only on the initial and final positions of the body, and be independent of the path traversed, will be examined in the next section.

SECTION VI. THE WORK DONE ON A BODY DURING ITS MOTION IN A CLOSED PATH.

If a body describes a closed path $ABCD$ (Fig. 6) under the action of a force whose components are X, Y, Z, the work done upon it is determined by taking the integral (a) $\int(Xdx + Ydy + Zdz)$ over the whole path. If the body, moving from A with the velocity v_0,

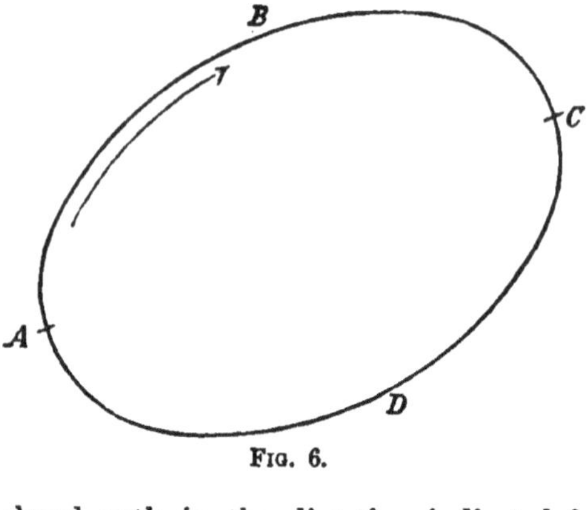

FIG. 6.

traverses the closed path in the direction indicated by the arrow, and returns to A with the velocity v, the work done is equal to

$$\tfrac{1}{2}mv^2 - \tfrac{1}{2}mv_0^2.$$

Supposing $v > v_0$, kinetic energy is produced during the motion, and will increase continuously if the motion is continued. On the other hand, supposing $v < v_0$, kinetic energy will be produced if the body traverses the path $ABCD$ in the opposite direction. Now, we know by experience that a body, under the action of forces proceeding from fixed points, after traversing a closed path, returns to the starting point with the same kinetic energy which it had when it started. It is therefore important to investigate to what conditions the components of the force must conform in order that the integral (a), taken over a closed path, shall be zero; that is, the conditions which must hold in order that a body, moving through a closed path, shall return to its original position with the same kinetic energy with which it started.

If the integral taken over the closed path $ABCD$ equals zero, that is, if

$$\int^{ABC}(Xdx+Ydy+Zdz)+\int^{CDA}(Xdx+Ydy+Zdz)=0,$$

where the letters connected with the integral signs indicate that the first integral is to be taken over the line ABC, the second, over CDA, we have

$$\int^{ABC}(Xdx+Ydy+Zdz)=\int^{ADC}(Xdx+Ydy+Zdz).$$

If the work done during the passage of the body from one point to another is independent of the path and dependent only on the initial and final points of its path, the components X, Y, Z are single valued and continuous functions of the position of the point.

Before we deduce the general conditions which must hold in order that the work performed by a force shall be dependent only on the initial and final points of the path, we will determine the work done in the case in which the area enclosed by the path is infinitely small. Through the point O (Fig. 7), whose coordinates are x, y, z, we draw the lines Ox, Oy, Oz parallel to the coordinate axes, whose positive directions are determined in the following way. If the right hand is stretched out in the direction of the positive x-axis, a line drawn from the palm will give the direction of the positive y-axis, and the thumb that of the positive z-axis. A positive rotation around the x-axis is that by which the $+y$-axis is brought by a rotation through a right angle into coincidence with the $+z$-axis. This rule, by cyclic* interchange of the letters x, y, z, gives the directions of the positive rotations about the y- and z-axes. If $OBDC$ is a rectangle in the yz-plane, and if its perimeter is

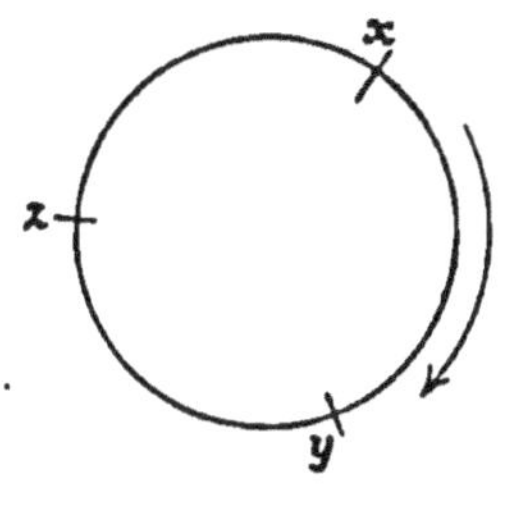

* That is, if y is replaced by x, z will be replaced by y, and x by z.

traversed in the direction $OBDCO$, the motion by which it is traversed is said to be in the positive direction. This convention as to the sign of the direction of rotation shall hold in all our subsequent work. If we set $OB = dy$, the work done by the transfer of the body from O to B equals Ydy. If the body moves from B to D, the work done is $(Z + \partial Z/\partial y \,.\, dy)dz$. The work done in the path DC is $-(Y + \partial Y/\partial z \,.\, dz)dy$, and that done in the path CO is $-Zdz$.

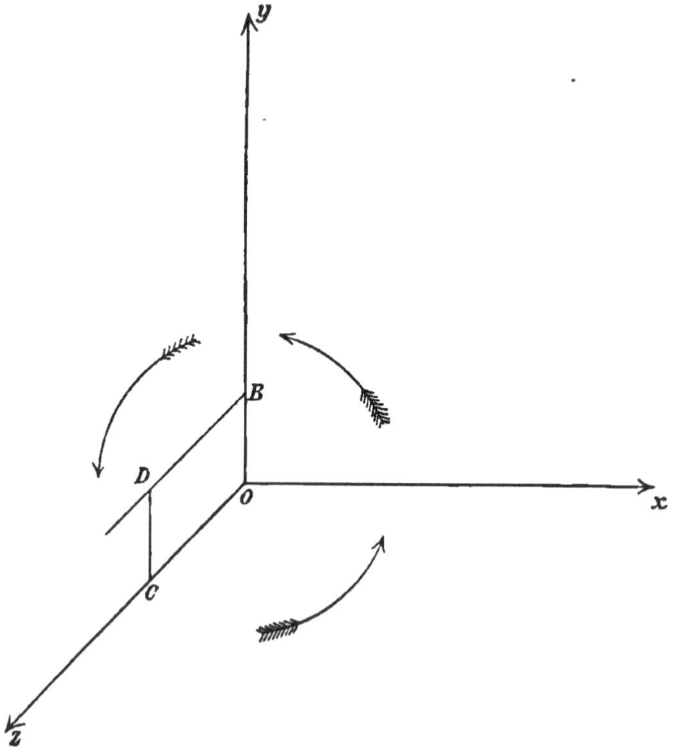

FIG. 7.

Hence the total work done is $(\partial Z/\partial y - \partial Y/\partial z)dydz$. In general, the work done by a force during the movement of a body around a surface element dS_x, which is parallel to the yz-plane, is

(b) $$F \,.\, dS_x = (\partial Z/\partial y - \partial Y/dz)dS_x.$$

In the same way we obtain

$$G \,.\, dS_y = (\partial X/\partial z - \partial Z/\partial x)dS_y\,;$$

$$H \,.\, dS_z = (\partial Y/\partial x - \partial X/\partial y)dS_z.$$

F, G, and H are the quantities of work done during the movement of the body around a unit area at the point O, when perpendicular to the x-, y-, and z-axes respectively.

If $OABC$ (Fig. 8) is an infinitely small tetrahedron, whose three edges OA, OB, OC are parallel to the coordinate axes, and if the body moves on the boundary of the surface ABC in the direction given by the order of the letters, the work done is equal to that which is done by moving the body in succession about OAB, OBC, and OCA. By this set of motions, the distances AB, BC, CA will each be traversed once in the positive direction, while the distances

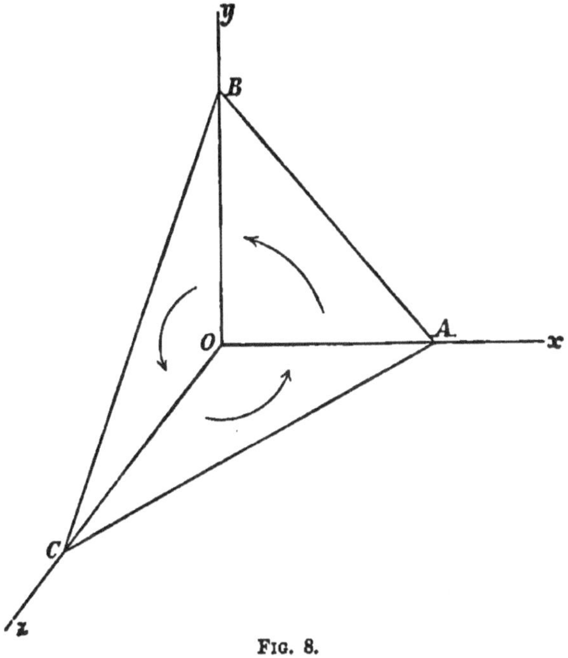

FIG. 8.

OA, OB, OC will each be traversed twice and in opposite directions, so that the work done in them is zero. The work done during the movement of the body about the surface $ABC = ds$ is, therefore, (c) $J \,.\, dS = F \,.\, dS \,.\, l + G \,.\, dS \,.\, m + H \,.\, dS \,.\, n$, where l, m, n are the cosines of the angles which the normal to the surface dS drawn outward from the tetrahedron makes with the coordinate axes. Hence the work J done during the movement of the body around unit area is (d) $J = Fl + Gm + Hn$, where l, m, and n determine the position of the unit area.

If the work done during the movement of a body about an infinitely small surface is zero, we must have $J=0$ for all positions of the surface, or $F=G=H=0$,

$$\text{(e)} \qquad \begin{cases} \partial Z/\partial y - \partial Y/\partial z = 0, \quad \partial X/\partial z - \partial Z/\partial x = 0, \\ \partial Y/\partial x - \partial X/\partial y = 0. \end{cases}$$

When the equations of condition (e) are satisfied, the expression under the integral sign in (a) is the complete differential of a function V of x, y, z, whence $X=\partial V/\partial x$, $Y=\partial V/\partial y$, $Z=\partial V/\partial z$. The equations of condition (e) are satisfied by this assumption. The function V is *the potential* of the acting forces; we here obtain for the first time the mathematical definition of this function, whose differential coefficients with respect to x, y, z are the components of force X, Y, Z. If V is a value of the potential of the acting forces, V' is also a value, if $V'=V+C$, where C is a constant; since

$$X=\partial V/\partial x=\partial V'/\partial x, \text{ etc.}$$

The value of the potential therefore involves an unknown or arbitrary constant. We will return to the consideration of this point in the next section.

If the equations of condition (e) are everywhere satisfied, the work done during the movement of the body about a surface is also zero when the surface is finite. The surface may be divided into surface-elements, as shown in Fig. 9. If the body moves about these elements one after another in the same direction, the total work done will equal zero. It is here assumed that the forces X, Y, Z are continuous and single valued functions of the coordinates. Every line-element thus introduced will be traversed twice in opposite directions, with the exception of those which form the boundary of the finite surface.

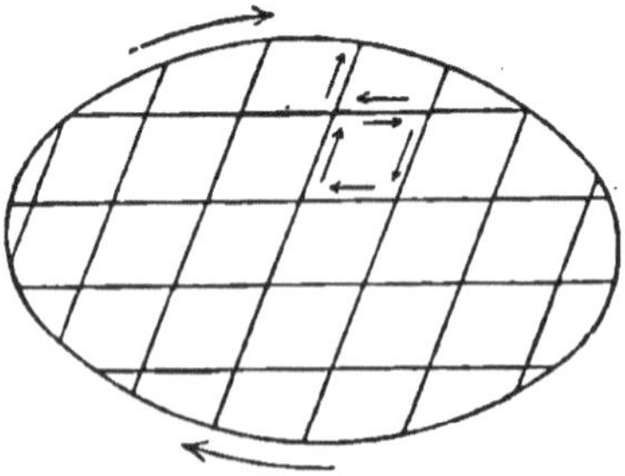

FIG. 9.

Those forces or systems of forces, which are such that the work done by them is independent of the path in which the body is transferred from its initial to its final position, are called *conservative forces.* The most important examples of such forces are those which act from a fixed point, and have values which depend only on their distance from it. If the force acting at the point P depends only on the distance of that point from the origin of coordinates O, that is, if it equals $f(r)$, then $X = f(r) . x/r$, since x/r is the cosine of the angle which the line OP makes with the x-axis. We have similarly

$$X = f(r) . x/r, \quad Y = f(r) . y/r, \quad Z = f(r) . z/r.$$

If we set $f(r)/r = R$, then $X = Rx$, $Y = Ry$, $Z = Rz$. We have then

$$\partial Z/\partial y = dR/dr . yz/r, \quad \partial Y/\partial z = dR/dr . yz/r.$$

The equation of condition $\partial Z/\partial y - \partial Y/\partial z = 0$ is therefore satisfied. The same is true of the other equations of condition (e).

The work done during the movement of the body about a surface is given by the integral $\int(Xdx + Ydy + Zdz)$. This work is also done if the body moves in succession about all the surface-elements into which the finite surface is divided (Fig. 9). In this process the motion must be uniformly carried out in the same sense. From (c) this work is equal to $\int(Fl + Gm + Hn)dS$. If we substitute the expressions for F, G, H formerly obtained, we have, by the use of (a) and (b),

$$\text{(f)} \quad \left\{ \begin{aligned} &\int(X . dx/ds + Y . dy/ds + Z . dz/ds)ds \\ &\qquad = \iint[(\partial Z/\partial y - \partial Y/\partial z)l + (\partial X/\partial z - \partial Z/\partial x)m \\ &\qquad\qquad + (\partial Y/\partial x - \partial X/\partial y)n]dS, \end{aligned} \right.$$

where s is the perimeter of the surface S, and l, m, n are the direction cosines of the normal to each surface-element. Equation (f) shows that the line integral along a closed curve may be replaced by a surface-integral over a surface bounded by this curve. The only conditions which the surface S must fulfil are that it shall be bounded by the curve and have no singular points. The theorem contained in (f) was discovered by Stokes.

SECTION VII. THE POTENTIAL.

The only applications of the potential that we will discuss are those like the foregoing, in which the work done during the motion is completely determined by the initial and final points of the path. That this may be the case, we must have

$$\partial Z/\partial y = \partial Y/\partial z, \quad \partial X/\partial z = \partial Z/\partial x, \quad \partial Y/\partial x = \partial X/\partial y.$$

We exclude from the discussion all cases in which these equations do not hold.

Let the components of the force in the field be X, Y, Z. Let there be a unit of mass at the point O (Fig. 10), whose rectangular coordinates are a, b, c, and let it move from O to P along the path s. The work V done by the force during this motion is

$$\text{(a)} \qquad V = \int_O^P (Xdx + Ydy + Zdz) = V_P - V_O,$$

it being assumed that X, Y, Z are the partial derivatives of a single function V, which itself is a function only of x, y, z. The work required to transfer the unit of mass from any point O to P is equal to the difference of the potentials V_P and V_0 at those points, or is equal to the *difference of potential. Such differences of potential are all that can be directly measured.* The value of the potential itself

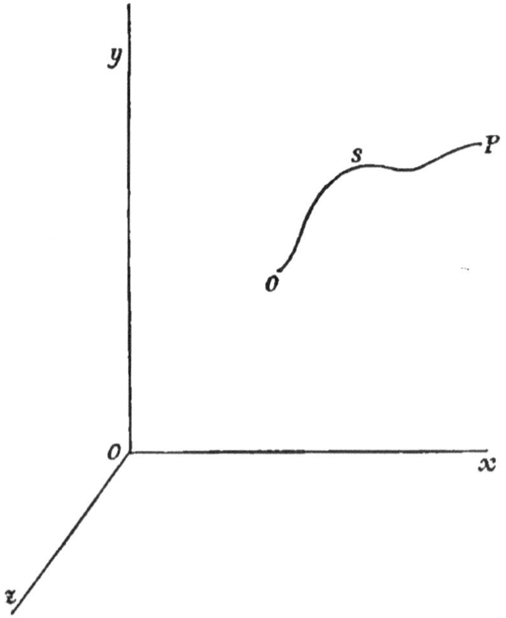

FIG. 10.

involves an unknown constant, and therefore cannot be completely determined. If we assume that the potential is zero at the point O, then V_P is the potential at P. Hence *the potential at any point is the work required to transfer the unit of mass to that point from a point where the potential is zero.*

The potential V is a function of the coordinates. The equation (b) $V(x, y, z) = C$, when C is constant, represents a surface which is the locus of points, such that the amount of work required to transfer the unit of mass from a point where the potential is zero to any one of them is the same. If different values of C are taken, we obtain a system of surfaces, which are called *level* or *equipotential surfaces.* Let PP' and QQ' (Fig. 11) be two infinitely near surfaces of this system; let the potential on PP' be V, and on QQ' be $V+dV$. Let ds be the element of an arbitrary curve crossing these surfaces, which is cut off by them. If the force acting in the direction of ds is T, the quantity of work Tds will be done by the transfer of the unit of mass from P to Q; this work is also equal to $V_Q - V_P = dV$. We have, therefore, (c) $T.ds = dV$ or $T = dV/ds$. Hence the force in any direction at a point is determined from the potential; the relation between force and potential being given by equation (c). Since the direction of the element ds is arbitrary, we may substitute for ds the elements dx, dy, dz, and obtain for the components of the force $X = \partial V/\partial x$, $Y = \partial V/\partial y$, $Z = \partial V/\partial z$. From equation (c) the force is inversely proportional to the element ds drawn between the two equipotential surfaces V and $V+dV$. If the direction of ds is that of the normal to the surface PP', the force has its greatest value. If a series of lines is drawn which cut the equipotential surfaces orthogonally, their directions are the directions of the force at the points of intersection. Such lines are consequently called *lines of force.* The tangent to the line of force at a point gives the direction of the force at that point.

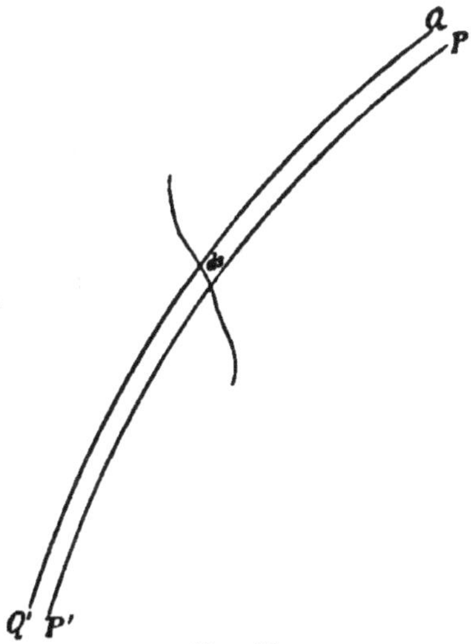

FIG. 11.

If P_1 and P_2 are two infinitely near points in an equipotential surface, no work need be done to transfer a body from P_1 to P_2, for $V_{P_1} - V_{P_2} = 0$; the force acting on the body is perpendicular to the direction of motion.

1. *Example.—Gravity.*—If at a place near the earth's surface we set up a system of rectangular coordinates, so that the xz-plane is horizontal, and the positive y-axis directed vertically upward, then $X=0$, $Y=-mg$, $Z=0$. Hence we have $V=-mgy$, that is, the equipotential surfaces are horizontal planes.

2. *Example.*—In the case discussed in V., Ex. 2, the work

$$V=\int_{r_0}^{r} f(r)dr$$

is needed to move the body from its position at the distance r_0 from a fixed point to another position at the distance r. Hence we have $V=F(r)-F(r_0)$, and the equipotential surfaces are spheres whose centres are at the centre of attraction O.

SECTION VIII. CONSTRAINED MOTION.

Galileo investigated not only freely falling bodies and the motion of projectiles, but also motion on an inclined plane and the motion of a pendulum, and thus made the first step in the investigation of constrained motion.

If a body is compelled by any cause to move in a given path, which is not that which it would follow if free to yield to the action of the forces applied to it, its motion is said to be *constrained.*

1. *Example.—The Inclined Plane.*—Let the body D (Fig. 12), acted on by gravity, slide down an inclined plane AB, which makes the

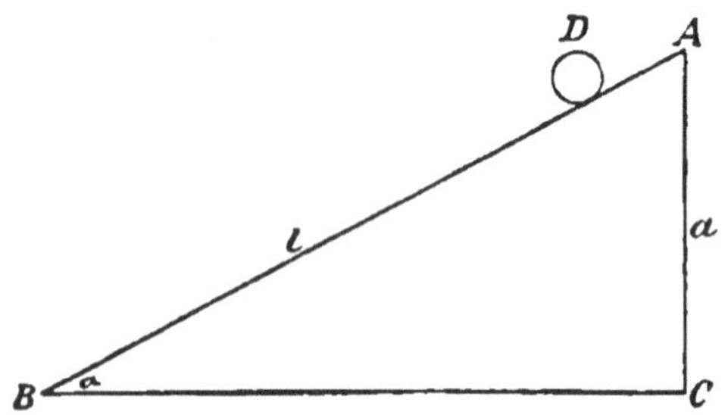

FIG. 12.

angle α with the horizontal plane BC. We neglect any resistance which may arise from friction. The force or reaction exerted by the inclined plane acts perpendicularly to the plane AB, and does not affect the motion of the body. The expression sought may be best obtained by using the relation between kinetic energy and work. If m represents the mass of the body, v the velocity acquired at B, g the acceleration of gravity, and l the length AB of the inclined plane, we have $\frac{1}{2}mv^2=mg\sin\alpha\,.\,l$, assuming that the motion begins at A, so that the initial velocity is zero. If a represents the height AC of the inclined plane, we have $l\sin\alpha=a$, and the work alone equals mga. Hence the velocity of the body at the foot of the inclined plane B is $v=\sqrt{2ga}$, and is the same as that which it would have at C, if it were to fall freely through the distance AC. If a body moves on the curve AB (Fig. 13) under the action of gravity, we determine the velocity at B in a similar way, from the initial velocity v_0 at A and the distance of the fall AC. That is, we have

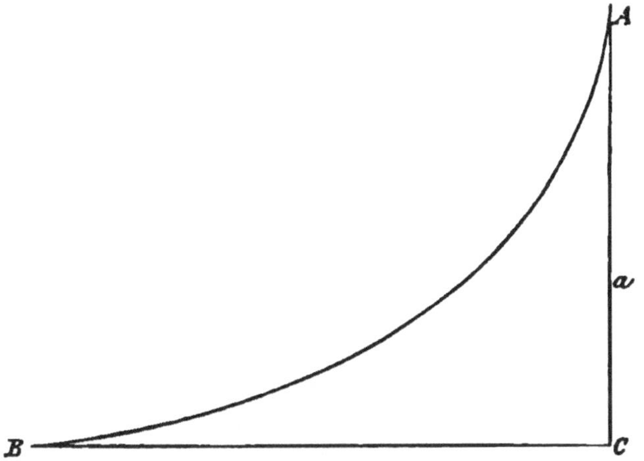

FIG. 13.

$\frac{1}{2}mv^2 - \frac{1}{2}mv_0{}^2 = mga$, and therefore (a) $v^2 = v_0{}^2 + 2ga$. The time t required for the movement of the body from A to B is

(b) $$t=\int_A^B \frac{ds}{v},$$

where ds is an element of the path AB.

2. *Example.—The Pendulum.*—If we suspend a body A (Fig. 14) at the end of a weightless rod which can swing freely about the point O, it is compelled to move on the surface of a sphere whose radius is equal to the length l of the rod. We will treat only the simple case in which the departure of the pendulum from its position of equilibrium is small. If, at the time $t=0$, the body starts from rest at A, it will move in the arc $ABCD$ through the point C lying perpendicularly under O. If we set $OA=l$, $\angle AOC=\alpha$, $\angle BOC=\theta$, and if AA' and BB' are drawn perpendicular to OC, then the velocity which the body gains in moving from A to B equals that which would be gained if it were to fall freely from A' to B'. The distance from A' to B' is $A'B'=l(\cos\theta-\cos\alpha)$, and therefore the velocity at B is

$$v=\sqrt{2gl(\cos\theta-\cos\alpha)}.$$

For $v=0$, or for $\theta=\pm\alpha$, the pendulum bob will be at rest, and is then at A or D, if $\angle DOC=\angle AOC$. The time t taken by the body to move from A to B is found by substituting this value of v in (b). We thus obtain

(c) $$t=-\int_{\alpha}^{\theta} l d\theta/\sqrt{2gl(\cos\theta-\cos\alpha)}.$$

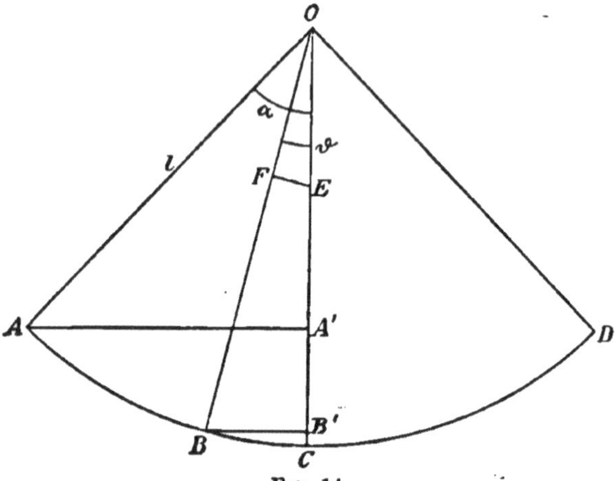

FIG. 14.

This expression is easily integrated if α, and therefore θ, are so small that we may set $\cos\theta=1-\frac{1}{2}\theta^2$ and $\cos\alpha=1-\frac{1}{2}\alpha^2$. The expansion of the cosine in a series is of the form

$$\cos x=1-x^2/2!+x^4/4!-\ldots,$$

and if x is very small we may neglect terms of higher orders than the second. Making this restriction, we have

$$t=\sqrt{l/g}\,.\int_{\theta}^{\alpha} d\theta/\sqrt{\alpha^2-\theta^2},$$

and by integration (d) $\theta=\alpha\cos(t\sqrt{g/l})$. If $t\sqrt{g/l}=\frac{1}{2}\pi$, we have $\theta=0$; the body moving from A reaches the lowest point of its path in the time $t=\frac{1}{2}\pi\,.\sqrt{l/g}$. The time T required for the movement of the body from A to D is twice this, or (e) $T=\pi\sqrt{l/g}$. T is called the period of oscillation.* *The period of oscillation is directly proportional*

* In the case of the pendulum here treated, which swings in a plane, it must be clearly understood that by the *period of oscillation* T only one advancing or returning beat is meant. In other periodic motions, the period of oscillation is the time between two instants, at which the motion of the body is precisely similar, that is, at which the body has the same velocity and direction of motion; or, it is the time required for both the advancing and returning beats.

to the square root of the length of the pendulum, and is inversely proportional to the square root of the acceleration of gravity.

The equation (e) holds only for very small arcs. In the case of finite values of a, we use the formula

$$\text{(f)} \qquad T = \pi \cdot \sqrt{\frac{l}{g}} \cdot \left(1 + \left(\frac{1}{1 \, . \, 2}\right)^2 \sin^2(\tfrac{1}{2} a) + \left(\frac{1 \, . \, 3}{2 \, . \, 4}\right)^2 \sin^4(\tfrac{1}{2} a) + \ldots\right).$$

It is only for very small arcs that the oscillations of the pendulum are *isochronous*, that is, independent of the size of the arcs. If the arc is not infinitesimal, the period will increase rather rapidly with the length of the arc.

The pendulum may also be studied in the following way. Let an oscillating body of mass m be at the point B, and be acted on by the force mg. We may represent this force by the line OE (Fig. 14). Draw EF perpendicular to OB; then OF and FE are components of the force OE. The magnitude of the tangential force is $mg \sin \theta$. If we set $BC = s$, and reckon the tangential force positive, when it tends to increase s, we have $P = -mg \sin(s/l)$, and if we assume s to be very small, $P = -mgs/l$. Hence the equation of motion is (g) $m\ddot{s} = P$ or $\ddot{s} = -gs/l$. By integration we obtain, by a suitable choice of constants, (h) $s = a \cos(t\sqrt{g/l})$. This equation corresponds to (d).

If a body is compelled to move on a given surface, the determination of its motion is in general very difficult. We will not enter into the discussion of the general case, but will consider only the motion of an infinitely small body on a spherical surface, when the body during the motion always remains near the lowest point C of this surface, and when gravity is the only force acting on it. We may then assume that the component of gravity which moves the body is directed toward the point C, and that it is equal to mgs/l, when l represents the radius of the sphere. This assumption gives the motion treated in III., Ex. 3. The path is an ellipse and the time of oscillation is $T = 2\pi\sqrt{l/g}$. The time of oscillation is therefore independent of the form and dimensions of the path.

SECTION IX. KEPLER'S LAWS.

In our deduction of the principal theorems of the general theory of motion, we proceeded from Galileo's laws of falling bodies. We turn now to that force of which gravity is a special example, and from whose properties the laws of planetary motion may be deduced. Starting with the hypothesis of Copernicus, that the sun is stationary and that the earth rotates on its own axis and also revolves round the sun, Kepler announced the following laws:

1. *A radius vector drawn from the sun to a planet describes equal sectors in equal times.*

2. *The orbits of the planets are ellipses with the sun at one of the foci.*

3. *The squares of the periodic times of two planets are proportional to the cubes of the semi-major axes of their orbits.*

These laws may be expressed analytically in the following way. Let S be the centre of the sun (Fig. 15) and APQ a part of the orbit of a planet. Let the planet be at A at the time $t=0$, and at P at the time t. In the next time-element dt the planet moves from P to Q, and its radius vector drawn from the sun describes the sector PSQ. Let $\angle ASP=\Theta$, $\angle PSQ=d\Theta$, and $SP=r$. The surface PSQ is equal to $\frac{1}{2}r^2d\Theta$. Since by Kepler's first law the surface described by the radius vector increases proportionally to the time, we have $r^2d\Theta=kdt$, where k is constant, or, writing the equation in another form,

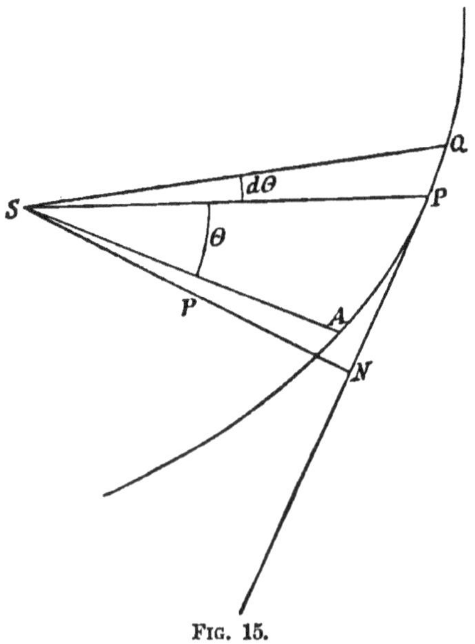

FIG. 15.

(a) $$r^2 . \dot{\Theta}=k.$$

Kepler's first law is a special case of a general law which is called *the law of areas.* This law is: *If the force which acts upon a moving body proceeds from a fixed point, the surface described by the radius vector drawn from that point to the body increases at a constant rate.* Hence Kepler's first law holds for all central forces.

From (a) it appears that the angular velocity $\dot{\Theta}$ is inversely proportional to the square of the distance of the planet from the sun.

We represent the velocity of the planet at P by v, and the perpendicular from S upon the tangent to the orbit at the point P by $SN=p$.

If we set $PQ=ds$, the area of the sector PSQ is equal to $\frac{1}{2}pds=\frac{1}{2}pvdt$. But it is also equal to $\frac{1}{2}r^2d\Theta=\frac{1}{2}k\,.\,dt$. Hence we have $pvdt=kdt$, or $pv=k$, that is, *the velocities of the planet at different points in its orbit are inversely proportional to the distances of the tangents at those points from the sun, the centre of attraction.*

Let BPC be the elliptical orbit of the planet (Fig. 16), with the sun situated at the focus S. Let the major axis be $BC=2a$, and let SA be a fixed radius vector which makes the angle a with the major axis. We set $SP=r$, $\angle ASP=\Theta$. If F and S are the foci, we have

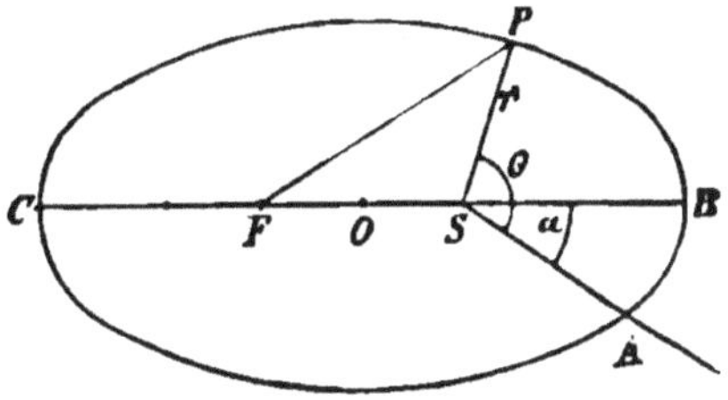

FIG. 16.

$$PF+PS=2a, \quad PF=2a-r,$$

and hence $$(2a-r)^2=4a^2e^2+r^2+4aer\cos(\Theta-a),$$

if e is the *eccentricity*, and if, therefore, $FS=2ae$. From this equation we obtain for the equation of the path in polar coordinates

(b) $$1/r=[1+e\cos(\Theta-a)]/a(1-e^2).$$

From equation (a) we have $\int\frac{1}{2}r^2\,.\,d\Theta=\frac{1}{2}k\,.\,T$, if the integration is taken over the whole orbit and if T is the periodic time. The integral is equal to the area of the ellipse, or to $a\,.\,b\,.\,\pi$, if b represents the minor axis. Hence we have $2\pi ab=k\,.\,T$. If we notice that $a^2=b^2+a^2e^2$, we obtain (c) $2\pi a^2\sqrt{1-e^2}=kT$, and squaring,

$$4\pi^2a(1-e^2)/k^2=T^2/a^3.$$

By Kepler's third law T^2/a^3 is constant for all the planets. We must therefore have (d) $\mu=k^2/a(1-e^2)=4\pi^2a^3/T^2$, a constant.

The velocity v may be determined in the following way. Let S (Fig. 16) be the origin of a system of rectangular coordinates, and let SA be the x-axis. We have $x=r\cos\Theta$ and $y=r\sin\Theta$, and $v^2=\dot{x}^2+\dot{y}^2$. From the equations

(e) $$\dot{x}=\dot{r}\cos\Theta-r\sin\Theta\,.\,\dot{\Theta}; \quad \dot{y}=\dot{r}\sin\Theta+r\cos\Theta\,.\,\dot{\Theta},$$

we obtain (f) $v^2=\dot{r}^2+r^2\dot{\Theta}^2$.

If we substitute for $r\dot{\Theta}$ its value given by equations (a) and (b), and for $\dot{r}$ the value got by differentiating equation (b), we have

$$v^2=\left(1+2e\cos(\Theta-a)+e^2\right).\,k^2/a^2(1-e^2)^2.$$

Noticing that $1+2e\cos(\Theta-a)+e^2=2\left(1+e\cos(\Theta-a)\right)-(1-e^2)$, we obtain, by the help of equation (b), $v^2=(2/r-1/a)\,.\,k^2/a(1-e^2)$, or, introducing the quantity μ defined by (d), (g) $v^2=2\mu/r-\mu/a$.

SECTION X. UNIVERSAL ATTRACTION.

We owe to Newton the determination of the law of the force which must act on a planet in order that its motions may conform to Kepler's laws. To determine this force, we use equation (g) IX. Let the centre of the sun be the origin of a system of rectangular coordinates, and let the planet be situated at the point (x, y) Represent the components of the unknown force by X and Y. From the law of kinetic energy (V.) we have $\frac{1}{2}v^2 - \frac{1}{2}v_0^2 = \int Xdx + Ydy$. If v_0 is the velocity at the distance r_0, we obtain, by the help of equation (g) IX., $\mu/r - \mu/r_0 = \int Xdx + Ydy$. If $Xdx + Ydy$ is a complete differential $d\phi$, we will have

$$X = \partial\phi/\partial x = \partial(\mu/r)\partial x \text{ and } Y = \partial\phi/\partial y = \partial(\mu/r)/\partial y,$$

or $$X = -\mu/r^2 \,.\, \partial r/\partial x = -\mu x/r^3, \quad Y = -\mu/r^2 \,.\, \partial r/\partial y = -\mu y/r^3.$$

The force R with which the sun acts on the planet is $R = -\mu/r^2$, that is, is a force which is *inversely proportional to the square of the distance of the planet from the sun.* It is evident from equation (d) IX. that the quantity μ has the same value for all the planets.

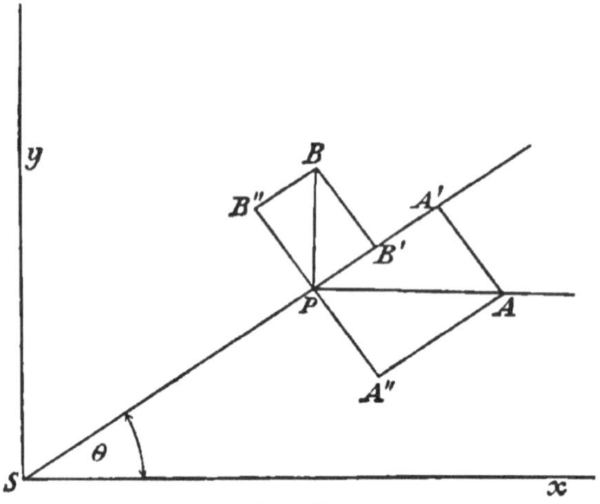

FIG. 17.

We may also obtain these results from the general equations $\ddot{x} = X$ and $\ddot{y} = Y$. The unknown components of force X and Y may be represented by the lines PA and PB (Fig. 17), and resolved into a component R in the direction $SP = r$ and a component T perpendicular to SP. Setting $\angle PSX = \Theta$, we have

$$R = X\cos\Theta + Y\sin\Theta, \quad T = -X\sin\Theta + Y\cos\Theta.$$

By the help of equation (e) IX., this becomes

(d) $R = \ddot{r} - r\dot{\theta}^2$ and $T = 2\dot{r}\dot{\theta} + r\ddot{\theta} = 1/r \, . \, d(r^2\dot{\theta})/dt.$

But since $r^2\dot{\theta} =$ constant by Kepler's first law, we have $T = 0$. *The attractive force is therefore directed toward the sun.* Using equations (a) and (b) IX., we obtain (e) $R = -k^2/a(1-e^2)r^2 = -\mu/r^2$.

We apply this result to the motion of the moon. By reference to (d) IX., where the value of μ is given, we find that

$$R = -4\pi^2a^3/T^2r^2.$$

The orbit of the moon is approximately a circle with a radius 60,27 times as great as that of the earth. Setting

$$r = a = 4 \, . \, 10^9 \, . \, 60{,}27/2\pi \text{ cm},$$

we have the acceleration γ of the moon toward the earth,

$$\gamma = 4\pi^2a/T^2 = 8\pi \, . \, 60{,}27 \, . \, 10^9/2\,360\,600^2 \text{ cm},$$

since the period of the moon's rotation is 27,322 days or 2 360 600 seconds. Hence we have $\gamma = 0{,}27183$ cm. If the centre of the moon were situated at the distance of the radius of the earth from the earth's centre, it would have an acceleration equal to

$$0{,}27183 \, . \, \overline{60{,}27}^2 \text{ cm} = 987 \text{ cm},$$

assuming that the force is inversely proportional to the square of the distance. This value accords so well with that of the acceleration at the surface of the earth, that we are justified in assuming that the motion of a falling body is an action of the same force as that which keeps the moon and the planets in their orbits. The final proof of the validity of Newton's law of mass attraction is obtained from the complete agreement of the theoretical conclusions drawn from it with the results of observations on the heavenly bodies.

SECTION XI. UNIVERSAL ATTRACTION (*continued*).

We will now use a method precisely the opposite of our former one. We will assume the law of attraction known, and determine the path of a planet whose position and velocity at the time $t = 0$ are given. Let S be the centre of the sun (Fig. 18), let the attracted body be situated at A at the time $t = 0$, and let AC represent the velocity v_0, whose direction makes the angle $CAD = \phi$ with $SA = r_0$ produced. If the acceleration which the sun imparts to the planet is set equal to μ/r^2, then, using a system of polar coordinates whose origin is at S, and writing the force with the minus sign because it is directed toward the sun, we obtain

(a) (b) $$\ddot{r} - r\dot{\theta}^2 = -\mu/r^2 \text{ and } 1/r \,.\, d(r^2\dot{\theta})/dt = 0.$$

It follows from (b) that (c) $r^2\dot{\theta} = k$, where k is a constant. This formula was obtained from Kepler's first law, (a) IX. Since $r\dot{\theta}$ is the component of velocity perpendicular to the direction of r, we obtain for $t = 0$, (d) $k/r_0 = v_0 \sin\phi$. By the help of equation (c), (a) takes the form $\ddot{r} - k^2/r^3 = -\mu/r^2$. If this equation is multiplied by $2\dot{r}dt$, we have $d(\dot{r}^2) + d(k^2/r^2) = 2d(\mu/r)$, and by integration

$$\dot{r}^2 + k^2/r^2 = 2\mu/r + \text{Const.}$$

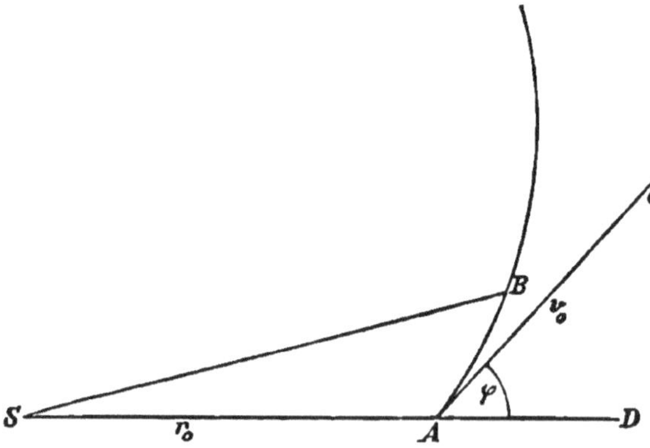

FIG. 18.

In the initial point A, we have $v = v_0$ and $\dot{r} = v_0 \cos\phi$, therefore for $t = 0$, we obtain $v_0^2 \cos^2\phi + k^2/r_0^2 = 2\mu/r_0 + \text{Const.}$, from which by the help of (d) it follows that $v_0^2 = 2\mu/r_0 + \text{Const.}$ Hence we have

(e) $$\dot{r}^2 = v_0^2 - 2\mu/r_0 + 2\mu/r - k^2/r^2.$$

Since the velocity v, from (f) IX., may be expressed generally by $v^2 = \dot{r}^2 + r^2\dot{\theta}^2$, we obtain by the use of (c) and (e)

(f) $$\tfrac{1}{2}v^2 - \tfrac{1}{2}v_0^2 = \mu/r - \mu/r_0.$$

This equation agrees with (g) IX.

The same result may be derived from the theorem connecting kinetic energy and work. From (e) we have

(g) $$\dot{r} = \pm\sqrt{v_0^2 - 2\mu/r_0 + 2\mu/r - k^2/r^2},$$

where the upper sign is to be taken, if r and t increase or diminish together. It follows from equation (c) that $\dot{\theta} = k/r^2$. Since r and θ depend only on t, we obtain from (c) and (g)

$$k \,.\, d(1/r)/d\theta = \mp\sqrt{v_0^2 - 2\mu/r_0 + 2\mu/r - k^2/r^2}.$$

This is the differential equation of the orbit. By adding and subtracting μ^2/k^2 under the radical sign, and by noticing that μ/k is a constant, and that therefore its differential is zero, this equation may be written

(h) $$d\Theta = d(k/r - \mu/k)/ \mp \sqrt{v_0^2 - 2\mu/r_0 + \mu^2/k^2 - (k/r - \mu/k)^2}.$$

If we set $u^2 = v_0^2 - 2\mu/r_0 + \mu^2/k^2$, we get by integration

$$\Theta = \text{arc}\cos(k/ur - \mu/uk) + a,$$

where a is a constant.

Hence the equation of the path is

(i) $$1/r = \big(1 + ku/\mu \cos(\Theta - a)\big)/(k^2/\mu).$$

In this equation u may always be considered positive, since a is arbitrary.

The polar equation of a conic is (k) $1/r = \big(1 + e\cos(\Theta - a)\big)/a(1 - e^2)$, which represents an ellipse, a parabola, or a branch of an hyperbola respectively, according as $e<1$, $e=1$ or $e>1$. If $e=0$ we have the equation of the circle. From the equation $e = ku/\mu$, by introducing the value of u, we obtain (l) $1 - e^2 = (2\mu/r_0 - v_0^2) \,.\, k^2/\mu^2$. If a body approaches the sun from infinity to the distance r_0, its velocity will be v_1, determining by the following equation

$$\tfrac{1}{2}v_1^2 = -\int_\infty^{r_0} \frac{\mu \,.\, dr}{r^2} = \frac{\mu}{r_0}.$$

Therefore we have (m) $e^2 = 1 - (v_1^2 - v_0^2) \,.\, k^2/\mu^2$. Hence the path is either an ellipse, a parabola, or an hyperbola, according as

$$v_0 < v_1,\quad v_0 = v_1 \quad \text{or} \quad v_0 > v_1;$$

that is, the path of the body is an ellipse, parabola, or hyperbola, according as the *vis viva* imparted to the planet at the first instant is too small to send it to infinity against the attraction of the sun, or exactly sufficient, or more than is sufficient, to accomplish that result.

By comparison of the formulas (i) and (k), we obtain

(n) $$a(1 - e^2) = k^2/\mu.$$

This corresponds to (d) IX. From (m) and (n) it follows, moreover, that (o) $\mu = \pm(v_1^2 - v_0^2) \,.\, a$. The upper sign is used when $v_1 > v_0$, the lower when $v_1 < v_0$.

In the first case, if the value for v_1^2 is substituted in (o), we have $v_0^2 = 2\mu/r_0 - \mu/a$. In conjunction with (f) this equation becomes $v^2 = 2\mu/r - \mu/a$, which corresponds to (g) IX.

SECTION XII. THE POTENTIAL OF A SYSTEM OF MASSES.

In the previous discussion Newton's law of gravitation was derived from Kepler's laws by the assumption that the attraction proceeds from the centre of the sun, or, what is the same thing, that the whole mass of the sun is concentrated at its centre. A similar assumption was made in the case of the planets. These assumptions might be made without further demonstration if the radius of the sun were infinitely small in comparison with the orbits of the planets; since this is not the case, it is necessary to investigate with what force a mass distributed throughout a given space acts on a body. This problem in the simplest cases was solved by Newton. His researches, and those of other distinguished mathematicians, have led to results of the greatest importance, both in physics and mathematics. The method by which such problems are treated is due to Laplace, and the theory was developed by Poisson, Green, Gauss, and others.

Let the masses m_1, m_2, m_3 be situated at the points A, B, C (Fig. 19), and let a unit of mass be concentrated at the point whose coordinates are x, y, z. The force with which the unit of mass is attracted by m_1 is $-fm_1/r_1^2$ where r_1 is the distance AP, and f a constant dependent on the units of mass, force, and length. Call the coordinates of A, ξ_1, η_1, ζ_1. The components X_1, Y_1, Z_1 of the force by which P is attracted to A are evidently $X_1 = -fm_1/r_1^2 . (x-\xi_1)/r_1$, etc. In the same way we calculate the components of the forces which originate at the other points, B, C, etc. If the sum of all the X components is represented by X, we obtain

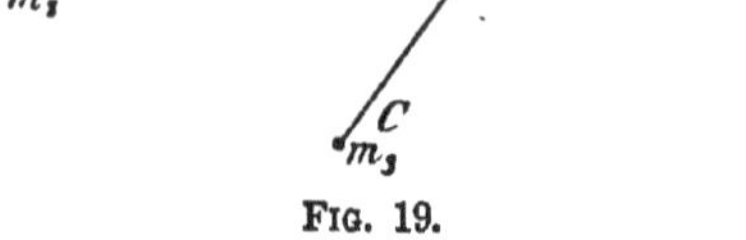

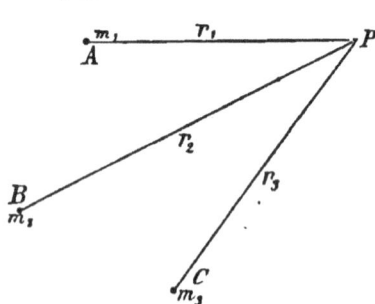

FIG. 19.

(a) $$X = -f\{m_1(x-\xi_1)/r_1^3 + m_2(x-\xi_2)/r_2^3 + \dots\}.$$

We set (b) $$V = m_1/r_1 + m_2/r_2 + m_3/r_3 + \dots .$$

Now $$r_1^2 = (x-\xi_1)^2 + (y-\eta_1)^2 + (z-\zeta_1)^2,$$

and therefore $$r_1 \partial r_1/\partial x = (x-\xi_1), \text{ etc.}$$

Hence we have

$$\partial V/\partial x = -m_1(x-\xi_1)/r_1^3 - m_2(x-\xi_2)/r_2^3 - \dots$$

and (c) $$X = f . \partial V/\partial x.$$

In a similar way we derive the equations

(d) (e) $$Y = f \cdot \partial V/\partial y \text{ and } Z = f \cdot \partial V/\partial z.$$

The quantity V defined by equation (b) is (VII.) *the potential at the point P of the given system of masses.* If the potential is given, the equations (c) (d) and (e) determine the forces acting in the directions of the coordinate axes. Since the position of the system of coordinates is arbitrary, the force acting in any direction may be derived from V. This has been already shown in VII. The force acting in the direction s is therefore

$$dV/ds = \partial V/\partial x \,.\, dx/ds + \partial V/\partial y \,.\, dy/ds + \partial V/\partial z \,.\, dz/ds.$$

The work A performed by the force in moving a unit of mass along an arbitrarily chosen path is given by

$$A = \int_o^s (Xdx + Ydy + Zdz),$$

where o and s are respectively the initial and final points of the path. If the values given in the formulas (c) (d) (e) are substituted for X, Y, and Z, and the element of the path whose projections on the coordinate axes are dx, dy, dz is designated by ds, then

$$A = f\int_o^s (\partial V/\partial x \,.\, dx/ds + \partial V/\partial y \,.\, dy/ds + \partial V/\partial z \,.\, dz/ds)ds = f\int_o^s dV,$$

and hence we have $A = f(V_s - V_o)$. If the body traverses a closed path, the work done by the forces equals zero (cf. VI., VII.). Therefore, if we let a body traverse a closed path under the action of gravity, the work which gravity performs in moving the body forward is equal in absolute value to the work which must be performed against gravity in order to bring the body back to the starting point. There is no surplus work performed. Hence it is evidently impossible to produce a *perpetuum mobile*, that is, an arrangement which continuously creates work out of nothing.

We have assumed that the masses considered are concentrated at points; this, however, does not occur in nature. Matter is more or less continuously distributed throughout space or on surfaces. If it is uniformly distributed in space, the mass ρ contained in the unit of volume is called the density. If it is not uniformly distributed, let a sphere of infinitely small radius be constructed about the point P; the ratio of the mass contained in the sphere to its volume is the volume density ρ at the point P. If the mass is distributed over a surface, the surface density σ at the point P is defined by the ratio of the mass contained within a circle of infinitely small radius drawn about the point P as centre to the area of the circle.

If the mass contained in the unit of volume is ρ, the element of volume $d\omega$ will contain the mass $\rho d\omega$. The potential of a mass which is distributed in space is therefore, from equation (b),

(g) $$V=\iiint \rho d\omega/r.$$

This integral is extended over the whole volume occupied by the mass. r is the distance between $d\omega$ and the point for which the potential is to be determined.

It is sometimes necessary to consider the mass as distributed in an infinitely thin sheet over a surface. If the mass on the unit of surface is σ, the surface-element dS will contain the mass σdS. The potential takes the form (h) $V=\iint \sigma dS/r$.

The potential cannot be determined without some further information; in the next paragraph we will discuss some of the simplest cases.

SECTION XIII. EXAMPLES. CALCULATION OF POTENTIALS.

The sun and planets are approximately spherical. On the supposition that they are spheres, their potential can be easily calculated, if the density ρ is given, and if we assume that it is a function of the radius, and therefore has the same value for all parts of the concentric spherical layers which compose the sphere.

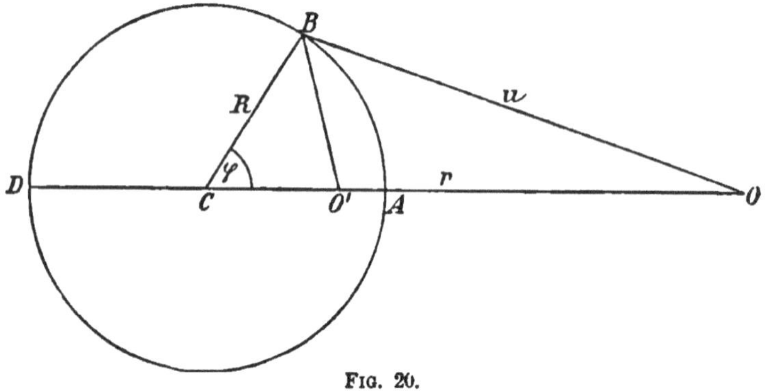

FIG. 20.

1. *The Potential of an Infinitely Thin Spherical Shell of Constant Surface Density* σ.

Let ABD (Fig. 20) be a sphere, whose centre is the point C and whose radius is R. The potential at the point O is to be determined. If we set $OC=r$, $\angle OCB=\phi$ and $OB=u$, we have

$$V=\int_0^\pi 2\pi R\sin\phi\,.\,Rd\phi\,.\,\sigma/u.$$

Since $u^2 = r^2 + R^2 - 2Rr\cos\phi$ and $udu = Rr\sin\phi d\phi$, the integral takes the form $V = \int 2\pi R/r \,.\, udu/u \,.\, \sigma = 2\pi R\sigma/r \,.\, \int du$. If O lies outside the sphere, we have $\int du = (r+R) - (r-R) = 2R$; while if it lies within the sphere, we have $\int du = (R+r) - (R-r) = 2r$. If we designate the potential outside the sphere by V_a and that within the sphere by V_i, we have (a) $V_i = 4\pi R\sigma$; $V_a = 4\pi R^2\sigma/r$. *The potential is therefore constant inside the spherical shell; and at points outside the spherical shell it is inversely proportional to the distance from its centre.* Hence the potential for the whole region is given by the two different expressions V_a and V_i. It is not discontinuous at the surface, since for $r = R$ we have $V_a = V_i = 4\pi R\sigma$. On the other hand, its differential coefficient is discontinuous at the surface. For we have

$$dV_a/dr = -4\pi R^2\sigma/r^2 \text{ and } dV_i/dr = 0,$$

and therefore

(b) $$[dV_a/dr]_{r=R} = -4\pi\sigma \text{ and } [dV_i/dr]_{r=R} = 0.$$

Hence an infinitely thin spherical shell exerts no force at a point lying within it. The sphere acts only on points outside of it, as if its whole mass were concentrated at its centre. Hence a solid sphere made up of homogeneous concentric spherical shells acts on outside points in a similar manner. If the attracted point is situated within the mass of a spherical shell, it will be attracted to the centre by the portion of the mass which lies within a sphere described about the centre with the distance of the point from the centre as radius. The portion lying outside this surface exerts no action.

2. *The Potential of a Solid Sphere.*

We will now calculate the potential of a solid sphere of constant density ρ. We have for points outside the sphere

(c) $$V_a = \int_0^R \frac{4\pi R^2 \,.\, dR \,.\, \rho}{r} = \frac{4\pi R^3\rho}{3r},$$

and for points inside the sphere

$$V_i = \int_0^r 4\pi R^2 \,.\, dR \,.\, \rho/r + \int_r^R 4\pi R \,.\, dR \,.\, \rho = \frac{4\pi}{3}r^2\rho + 2\pi\rho(R^2 - r^2),$$

or (d) $V_i = 2\pi\rho(R^2 - \frac{1}{3}r^2)$. In this case also, the potentials within and without the sphere are represented by two different expressions V_i and V_a. Both values, however, coincide at the surface, since for $r = R$ we have for the potential $V_i = V_a = \frac{4}{3}\pi R^2\rho$.

The function V, which represents the potential of a mass distributed through space, is everywhere continuous. Its differential coefficients with respect to r are

(e) $$dV_i/dr = -\tfrac{4}{3}\pi r\rho,\ dV_a/dr = -\tfrac{4}{3}\pi R^3\rho/r^2,$$

and these values at the surface, where $r=R$, are also equal, that is,

$$[dV_i/dr]_{r=R}=[dV_a/dr]_{r=R}=-\tfrac{4}{3}\pi\rho R.$$

Hence the first derivatives of the potential of a mass distributed in space are nowhere discontinuous, but are continuous throughout all space. On the other hand, its second derivatives vary continuously in the interior and the exterior regions, but on passage through the spherical surface a discontinuity occurs. That is, d^2V/dr^2 at the surface has two values, since

$$[d^2V_i/dr^2]_{r=R}=-\tfrac{4}{3}\pi\rho \text{ and } [d^2V_a/dr^2]_{r=R}=+\tfrac{8}{3}\pi\rho.$$

From equation (e) the force outside the sphere is inversely proportional to the square of the distance of the unit of mass from the centre of the sphere. We thus justify the assumption that the planets and the sun may be treated as points in which their respective masses are concentrated. In the interior of the sphere the force is proportional to the distance of the attracted point from the centre. If we transform equation (e) to $dV_i/dr=-\tfrac{4}{3}\pi r^3\rho\,.\,1/r^2$, we see that the force proceeds from that portion of the sphere whose distance from the centre is less than r. This only holds on the assumption made about ρ. The earth's density very probably increases toward the centre; hence the force of gravity will not have its greatest value at the surface, but at some point beneath it. This corresponds with the results of experiments on the time of vibration of a pendulum in a deep mine.

3. *The Potential of a Circular Plate.*

Let AB (Fig. 21) be a circular plate of surface density σ; the centre of the plate is O and the axis OP. The point P, for which the potential is to be determined, lies on the axis at the distance x from the plate. The potential V is then

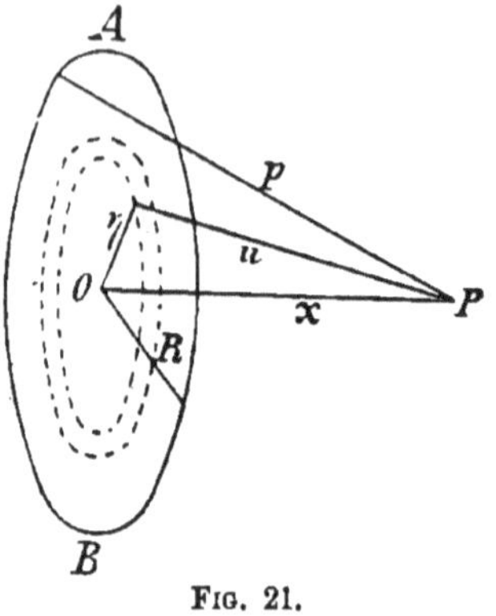

FIG. 21.

$$V=\int_0^R 2\pi\eta\,.\,d\eta\,.\,\sigma/u,$$

where R is the radius of the plate and η and u are the distances of a point on the plate from O and P respectively. We have $u^2=\eta^2+x^2$, therefore $udu=\eta d\eta$, and hence

$$V=\int 2\pi du\,.\,\sigma=2\pi\sigma(p-x),$$

if p is the distance of the point P from the edge of the plate. If the x drawn from one face of the plate is considered positive, the potential for negative values of x is $V = 2\pi\sigma(p+x)$. Hence

$$\text{(f)} \qquad \begin{cases} \text{for } x > 0, & V_1 = 2\pi\sigma(p-x) \\ \text{for } x < 0, & V_2 = 2\pi\sigma(p+x). \end{cases}$$

If the radius of the plate is infinitely great in comparison with x, we may set $V_1 = C - 2\pi\sigma x$ and $V_2 = C + 2\pi\sigma x$, where C is an infinitely great constant, since p is infinitely great and σ remains finite. We have

$$\text{for } x>0, \quad dV_1/dx = -2\pi\sigma,$$

and

$$\text{for } x<0, \quad dV_2/dx = +2\pi\sigma.$$

By passage through the surface, dV/dx, that is, the force, changes discontinuously by $4\pi\sigma$.

4. *The Potential of an infinitely long straight line.*

Suppose each unit of length of the line AB (Fig. 22) to have the mass μ. Let C be a point at the distance a from AB, and CD the perpendicular let fall from C upon AB. The potential V, at the point C, is

$$V = 2\int_0^\infty (\mu/r)dz = 2\mu\int_0^\infty dz/\sqrt{a^2+z^2},$$

$$V = 2\mu \log(z'/a + \sqrt{1+z'^2/a^2}).$$

Since z' is infinitely great in comparison with a, we may neglect 1 under the radical and write

$$\text{(k)} \quad V = 2\mu\log(2z'/a) = C - \mu\log a^2,$$

where C is an infinitely great constant, if z' is infinitely great. Further, we obtain

$$dV/da = -2\mu/a,$$

that is, *the force is inversely proportional to the distance of the point from the straight line.*

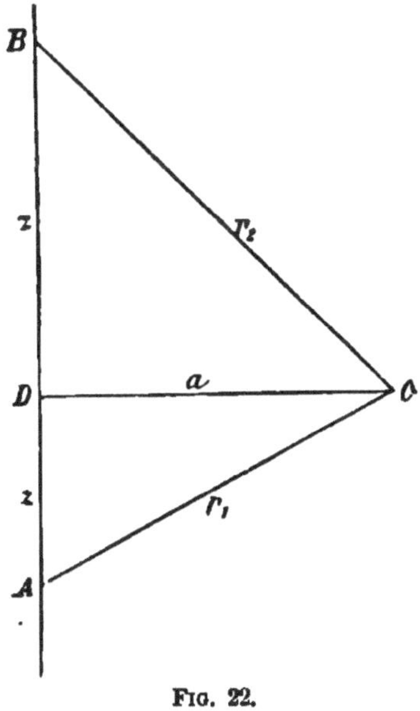

FIG. 22.

5. *The Potential of a Circular Cylinder.*

Represent the surface density of a circular cylinder by σ. Through the point P, for which the potential is to be determined, pass a plane perpendicular to the axis of the cylinder (Fig. 23). Let R be the radius of the cross section of the cylinder and r the distance of the point P from its centre.

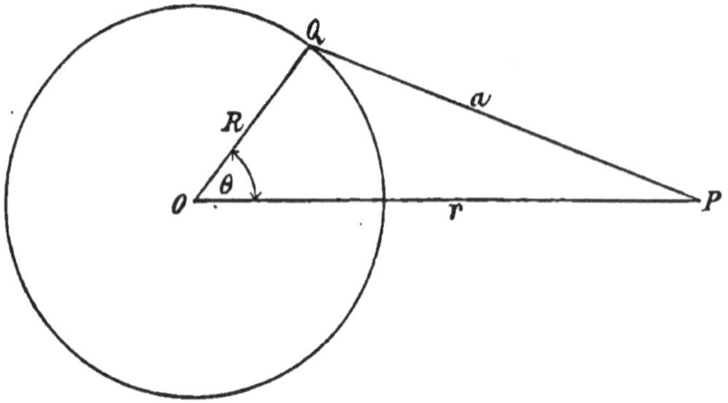

FIG. 23.

We then have from (k)

$$V = C - 2\int_0^\pi R d\theta \, . \, \sigma \, . \log a^2.$$

We now find the value of the integral

$$A = \int_0^\pi R d\theta \log . \, a,$$

in which $a = \sqrt{r^2 - 2rR\cos\theta + R^2}$. First consider the case in which $r > R$. The integral may then be written, if we set $\alpha = R/r$,

$$A = 2\int_0^\pi R d\theta (\log r + \log\sqrt{1 - 2\alpha\cos\theta + \alpha^2}).$$

Since $\cos\theta = \frac{1}{2}(e^{i\theta} + e^{-i\theta})$, we have $1 - 2\alpha\cos\theta + \alpha^2 = (1 - \alpha e^{i\theta})(1 - \alpha e^{-i\theta})$, and $\int_0^\pi d\theta \log\sqrt{1 - 2\alpha\cos\theta + \alpha^2} = \frac{1}{2}\int_0^\pi d\theta\big(\log(1 - \alpha e^{i\theta}) + \log(1 - \alpha e^{-i\theta})\big)$.

Now developing the terms in this integral in series, and carrying out the integration, we find that the integral is equal to zero, and hence that $A = 2\pi R \log r$. Thus the mean value of $\log a$ for all points of the circumference of the circle is equal to $\log r$ or to the logarithm of the mean distance from P to the circumference of the circle.

If now $r < R$, that is, if P lies between O and the circumference, and if we set $\alpha = r/R$, we have

$$A = 2\int_0^\pi R d\theta (\log R + \log\sqrt{1 - 2\alpha\cos\theta + \alpha^2}),$$

and hence $A = 2\pi R \log R$. In this case also the mean value of $\log a$

is equal to the logarithm of the mean distance from P to the circumference of the circle.

Now setting V_a and V_i for the potentials of points outside and inside the cylinder respectively, we have from these values,

$$V_a = C - 4\pi R\sigma \log r, \quad V_i = C - 4\pi R\sigma \log R.$$

The potential is therefore constant, and the force zero within the cylinder. Outside the cylinder the force is given by

(n) $$dV_a/dr = -4\pi R\sigma/r,$$

that is, the force is inversely proportional to the distance of the point from the axis of the cylinder.

SECTION XIV. GAUSS'S THEOREM. THE EQUATIONS OF LAPLACE AND POISSON.

Let ABF (Fig. 24) be a closed surface, of which $AB = dS$ is a surface-element, and at the point O within the surface, let the mass m be concentrated. On the element ds at C, construct the normal

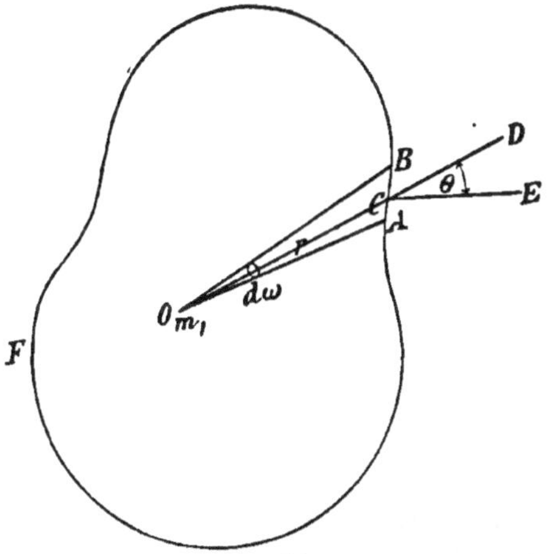

FIG. 24.

CE. Let the length of the line connecting O and C be r, and let the normal CE make the angle $DCE = \Theta$ with OC produced. If the potential at C due to m_1 is V_1, then $V_1 = m_1/r$, and the force N_1 acting at the point C in the direction CE is $N_1 = \partial V_1/\partial n$, while the total force in the direction of CO is m_1/r^2. We have

(a) $$N_1 = m_1/r^2 \,.\, \cos(\pi - \Theta) = -m_1/r^2 \,.\, \cos\Theta.$$

If a sphere of unit radius is described about the point O as centre, the straight lines drawn to the contour of dS mark out on this unit sphere a surface-element whose magnitude is equal to

(b) $$d\omega = dS \, . \cos\Theta / r^2.$$

From (a) and (b) we obtain

$$N_1 dS = -m_1/r^2 \, . \cos\Theta dS = -m_1 d\omega \text{ and } \partial V_1/\partial n \, . \, dS = -m_1 d\omega.$$

If there are still other masses, m_2, m_3, etc., within the closed surface, we obtain similarly

$$\partial V_2/\partial n \, . \, dS = -m_2 d\omega, \quad \partial V_3/\partial n \, . \, dS = -m_3 d\omega, \ldots .$$

V_1, V_2, V_3 are the potentials at the point C due to the masses m_1, m_2, m_3 respectively. For the total potential at the point C we have $V = V_1 + V_2 + V_3 + \ldots$, and therefore

$$\partial V/\partial n \, . \, dS = -(m_1 + m_2 + m_3 + \ldots) d\omega.$$

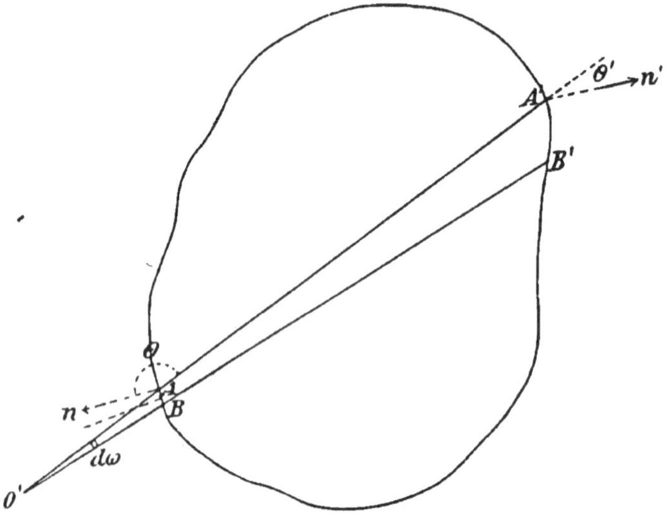

FIG. 25.

If we designate the mass enclosed by the surface by Σm, integration over the whole surface gives (c) $\int \partial V/\partial n \, . \, dS = -4\pi \Sigma m$. The force acting in the direction of the normal to the surface S is $\partial V/\partial n$; we call $\partial V/\partial n \, . \, dS$ the *flux of force* which passes through the element dS. *Hence the total flux of force passing through a finite closed surface equals the sum of the acting masses contained within the surface multiplied by* -4π. Hence, if the entire acting mass is enclosed by the surface, and $\partial V/\partial n$ is given for all points on the surface, the sum of the masses may be determined by the help of equation (c).

The theorem expressed in (c) also holds in case the acting mass lies outside the closed surface. Let the mass m' be situated at the point O' (Fig. 25) outside the surface $ABB'A$. If the surface-element $d\omega$ is taken on the surface of the unit sphere described about O' as centre, the straight lines drawn from O' through the boundary of this surface-element mark out on the closed surface the surface-elements $AB = dS$ and $A'B' = dS'$. Let the normals to AB and $A'B'$ directed outward from the closed surface be n and n' respectively, and let the forces $\partial V'/\partial n$ and $\partial V'/\partial n'$ act in the direction of these normals. The V' in these expressions represents the potential due to m'. If the angles made by the normals directed outward and the straight line drawn from O' are designated by Θ and Θ' respectively, and if we set $O'A = r$, $O'A' = r'$, we then obtain

$$\partial V'/\partial n = m'/r^2 . \cos(\pi - \Theta); \quad \partial V'/\partial n' = m'/r'^2 . \cos(\pi - \Theta'),$$
$$dS . \cos(\pi - \Theta) = r^2 d\omega; \quad dS' \cos\Theta' = r'^2 . d\omega,$$

and therefore $\partial V'/\partial n . dS + \partial V'/\partial n' . dS' = 0$. We therefore have (d) $\int \partial V'/\partial n . dS = 0$, if the integral is extended over the whole surface. *The flux of force proceeding from a point outside a closed surface, and passing through the surface, is equal to zero.* Therefore the value of the integral is independent of the mass outside the surface. We have then, generally, (e) $\int \partial V/\partial n . dS = -4\pi M$, where S is a closed surface, V the potential, n the normal directed outward, and M the sum of all the masses within the surface. This theorem is due to Gauss.

Equation (e) may be put into another form. We have

$$\partial V/\partial n = \partial V/\partial x . dx/dn + \partial V/\partial y . dy/dn + \partial V/\partial z . dz/dn,$$

and $dx/dn = \lambda$, $dy/dn = \mu$, $dz/dn = \nu$, where λ, μ, and ν are the cosines of the angles which the normal to the surface makes with the axes. We have then $\partial V/\partial n = \lambda X + \mu Y + \nu Z$. X, Y, and Z are the components of the force, and f is set equal to 1. We then obtain from Gauss's theorem (f) $\int (X\lambda + Y\mu + Z\nu) dS = -4\pi M$.

Let x, y, z (Fig. 26) be the coordinates of the point O, Ox, Oy, and Oz be parallel to the coordinate axes, and OO' be a parallelepiped whose edges are parallel to these axes. Let X, Y, Z be the components of the force acting at O. The components of the force at the point A, whose coordinates are $x + dx$, y, z, will be

$$X + \partial X/\partial x . dx, \quad Y + \partial Y/\partial x . dx, \quad Z + \partial Z/\partial x . dx.$$

We apply Gauss's theorem to the surface of this parallelepiped. The force acting normal to the surface OA' is $-X$, that acting normal to AO' is $+X + \partial X/\partial x . dx$. In the same way the force acting normal to OB' is $-Y$, and that acting normal to BO' is $Y+\partial Y/\partial y\,.\,dy$. A similar statement holds for the z coordinate. We have therefore

$$\begin{aligned}\int\partial V/\partial n\,.\,dS=\iiint[-Xdydz+(X+\partial X/\partial x\,.\,dx)dydz]&\\ +[-Ydxdz+(Y+\partial Y/\partial y\,.\,dy)dzdx]&\\ +[-Zdxdy+(Z+\partial Z/\partial z\,.\,dz)dxdy],&\end{aligned}$$

or
$$\int\partial V/\partial n\,.\,dS=\iiint(\partial X/\partial x+\partial Y/\partial y+\partial Z/\partial z)dxdydz.$$

We suppose the volume-element OO' to contain the mass M of density ρ, so that $M=\rho dxdydz$. We then have from (e)

(g) $$\partial X/\partial x+\partial Y/\partial y+\partial Z/\partial z=-4\pi\rho,$$

or (h) $$\partial^2 V/\partial x^2+\partial^2 V/\partial y^2+\partial^2 V/\partial z^2+4\pi\rho=0.$$

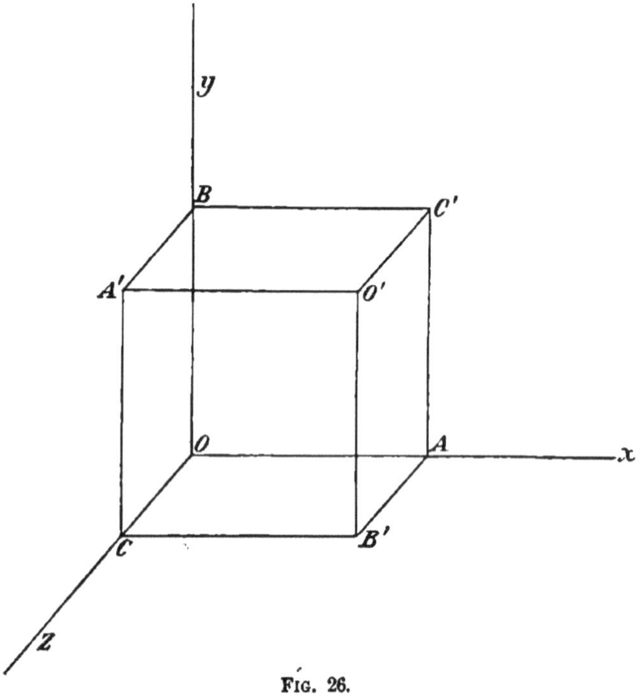

FIG. 26.

On account of the frequent use made of this equation in mathematical physics, we use for the sum of the first derivatives of a function f with respect to the three coordinates the symbol

$$\nabla f=\partial f/\partial x+\partial f/\partial y+\partial f/\partial z,$$

and for the sum of the second derivatives with respect to the same variables the symbol $\nabla^2 f=\partial^2 f/\partial x^2+\partial^2 f/\partial y^2+\partial^2 f/\partial z^2$. With this notation equation (h) may be written (i) $\nabla^2 V + 4\pi\rho = 0$. By the help of this equation, which was first used by Poisson, we can determine the density when the potential is known. If no matter is present in the region under consideration, that is, if $\rho = 0$, we have

(k) $$\partial^2 V/\partial x^2 + \partial^2 V/\partial y^2 + \partial^2 V/\partial z^2 = \nabla^2 V = 0.$$

This equation was first derived by Laplace. It may be obtained more simply in the following way. We start from

$$r^2 = (x-\xi)^2 + (y-\eta)^2 + (z-\zeta)^2,$$

where ξ, η, and ζ are constant, and obtain

$$\partial(1/r)/\partial x = -(x-\xi)/r^3, \quad \partial^2(1/r)/\partial x^2 = -1/r^3 + 3(x-\xi)^2/r^5.$$

Analogous expressions hold for $\partial^2(1/r)/\partial y^2$ and $\partial^2(1/r)\partial z^2$. Adding these equations, we have

$$\partial^2(1/r)/\partial x^2 + \partial^2(1/r)/\partial y^2 + \partial^2(1/r)/\partial z^2 = 0.$$

Since the potential $V = \Sigma m/r$ [(b) XII.], this is equivalent to $\nabla^2 V = 0$.

Poisson's equation may also be obtained in the following way. Let the density at the point P be ρ. Describe a sphere of infinitely small radius R so as to contain the point P, and suppose the density in the interior of the sphere to be constant. The potential V at the point P consists of two parts, V_i and V_a; V_a is due to the mass outside the sphere, and V_i to the mass within the sphere. The potential at P is $V = V_a + V_i$. If P is at the distance r from the centre of the sphere, we have from (d) XIII.

(k′) $$V_i = 2\pi\rho(R^2 - \tfrac{1}{3}r^2); \quad V = V_a + 2\pi\rho(R^2 - \tfrac{1}{3}r^2).$$

If ξ, η, ζ and x, y, z are the coordinates of the centre of the sphere and of P respectively, we will have $r^2 = (x-\xi)^2 + (y-\eta)^2 + (z-\zeta)^2$. By differentiation with respect to x, we obtain

$$\partial(r^2)/\partial x = 2(x-\xi) \text{ and } \partial^2(r^2)/\partial x^2 = 2,$$

therefore $\nabla^2 r^2 = 6$, and from equation (k′) $\nabla^2 V = \nabla^2 V_a - 4\pi\rho$. Now V_a is the potential due to the mass lying outside of the sphere, and therefore $\nabla^2 V_a = 0$ and $\nabla^2 V + 4\pi\rho = 0$. This is Poisson's equation.

In the parts of the region where ρ is infinitely great, Poisson's equation loses its meaning. In this case we return to the fundamental equation (e). For example, let a mass be distributed on a surface S with surface density σ. Draw the normals ν_i and ν_a to the element dS on both sides of the surface, and construct right cylinders on both sides of the surface on dS as base, and with the heights $d\nu_i$ and $d\nu_a$; the linear elements of these cylinders are lines of force. By applying equation (e) to the volume included in the cylinders, we obtain $\partial V_i/\partial \nu_i \,.\, dS + \partial V_a/\partial \nu_a \,.\, dS = -4\pi\sigma dS$, where V_i and V_a represent the values of the potential on both sides of the surface. Hence

(l) $$\partial V_i/\partial \nu_i + \partial V_a/\partial \nu_a + 4\pi\sigma = 0.$$

This equation finds an application in the theory of electricity.

Comparing formulas (e) and (h), and noticing that $M = \iiint \rho dxdydz$, we obtain the relation (m) $\iiint \nabla^2 V dxdydz = \iint \partial V/\partial n \,.\, dS$. The triple integral in (m) must be extended over the volume bounded by the surface S, and the double integral over the surface S. This theorem may also be proved by integration by parts.

Section XV. Examples of the Application of Laplace's and Poisson's Equations.

The potential V at the point x, y, z is a function of the three coordinates, and, from the previous discussion, has the form

(a) $$V = \iiint \rho d\xi d\eta d\zeta / \sqrt{(x-\xi)^2 + (y-\eta)^2 + (z-\zeta)^2},$$

where the density ρ at the point (ξ, η, ζ) is a function of the coordinates.

We may, however, use the differential equation (b) $\nabla^2 V + 4\pi\rho = 0$ as the starting point for the determination of the potential; we thus often obtain the desired result by a more convenient method. The density ρ must be given as a function of x, y, and z. The integral of (b) is always given by (a), but V may often be found more conveniently by direct integration of Poisson's equation.

In the solution of problems in potential, special attention must be paid to the boundary conditions which serve to determine the functions which are obtained by integration. We shall investigate the equations of condition to which the potential V_i within a closed surface S and the potential V_a outside that surface must conform, if the surface S encloses all the masses which are present in the field, and if no mass is present outside the surface. Applying Poisson's equation to the region enclosed by S, we have (c) $\nabla^2 V_i + 4\pi\rho = 0$. Outside the surface S we have (d) $\nabla^2 V_a = 0$.

If O is any point within S, P a point outside S, and if we set $OP = r$, we have, when r is very great, (e) $V_a = M/r$. M represents the whole mass enclosed by S. Hence, for $r = \infty$, the potential $V_a = 0$ and (f) $\lim(rV_a)_{r=\infty} = M$, that is, the product rV_a approaches the finite limit M if the point P moves off to infinity.

If P_1 and P_2 are two points which lie infinitely near each other on different sides of the surface S, the potentials at both points are equal, and we have for all points on the surface S, (g) $\bar{V}_i = \bar{V}_a$. The dash drawn over V is used to denote the value of V at the surface.

From (l) XIV. it follows further that for the points on the surface where $\sigma = 0$, we have (h) $\overline{\partial V_i}/\partial \nu = \overline{\partial V_a}/\partial \nu$, where the normal to S is designated by $\nu = -\nu_i = \nu_a$. The potential is therefore everywhere finite.

It is here assumed that ρ is everywhere finite. For the places where $\rho = \infty$ we obtain other equations of condition, which may readily be derived from those already given. For example, if σ is the surface density on a surface S, in which, therefore, ρ is infinitely great, and if $\rho = 0$ for all other points in the region, then, in our former notation, we have $\nabla^2 V_i = 0$ and $\nabla^2 V_a = 0$, but

(i) $$\bar{V}_i = \bar{V}_a, \quad \overline{\partial V_i}/\partial \nu_i + \overline{\partial V_a}/\partial \nu_a + 4\pi\sigma = 0$$

for all points on the surface S.

By these equations we may determine the potential if the density ρ within a sphere of radius R is constant. Outside the sphere ρ is supposed to be zero. The potential within the sphere is V_i, and outside of it V_a. We have then $\nabla^2 V_i + 4\pi\rho = 0$, $\nabla^2 V_a = 0$. Now we have

$$\partial V/\partial x = dV/dr \,.\, x/r,$$

$$\partial^2 V/\partial x^2 = d^2V/dr^2 \,.\, x^2/r^2 + dV/dr \,.\, 1/r - dV/dr \,.\, x^2/r^3.$$

Similar equations hold for the derivatives of V with respect to y and z. We thus obtain $\nabla^2 V = d^2V/dr^2 + 2/r \,.\, dV/dr$. Since, however,

$$d(rV)/dr = rdV/dr + V \text{ and } d^2(rV)/dr^2 = rd^2V/dr^2 + 2dV/dr,$$

we have (l) $$\nabla^2 V = 1/r \,.\, d^2(rV)/dr^2.$$

The differential equations which V_i and V_a must satisfy are therefore

(m) $$d^2(rV_i)/dr^2 + 4\pi\rho r = 0, \quad d^2(rV_a)/dr^2 = 0.$$

From these we obtain by integration

$$rV_i + \tfrac{2}{3}\pi\rho r^3 = Cr + C' \text{ and } V_a = C_1 + C_1'/r.$$

For $r = \infty$ it is assumed that $V_a = 0$, so that $V_a = C_1'/r$. Since V_i cannot become infinite for $r = 0$, we have $C' = 0$, and hence

$$V_i + \tfrac{2}{3}\pi\rho r^2 = C.$$

Since the force is a continuous function of the coordinates, and since, therefore, for all points on the surface, $\overline{dV_i}/dr = \overline{dV_a}/dr$, we will also have, when $r = R$, $\tfrac{4}{3}\pi\rho R = C_1'/R^2$. Hence $C_1' = \tfrac{4}{3}\pi\rho R^3$, and therefore (n) $V_a = \tfrac{4}{3}\pi R^3\rho/r$. Since $V_i = V_a$ when $r = R$, we have $C = 2\pi R^2\rho$, and therefore (o) $V_i = 2\pi\rho(R^2 - \tfrac{1}{3}r^2)$. These formulas are the same as (c) and (d) XIII.

If the potential depends on the distance of the point under consideration from a straight line, we choose this line as the z-axis of a system of rectangular coordinates. Let the distance from the z-axis of the point for which the potential is to be determined be r. We have then $r^2 = x^2 + y^2$, and further

$$\partial V/\partial x = dV/dr \,.\, x/r,$$

$$\partial^2 V/\partial x^2 = x^2/r^2 \,.\, d^2V/dr^2 + 1/r \,.\, dV/dr - x^2/r^3 \,.\, dV/dr, \text{ etc.}$$

Therefore

(p) $$\nabla^2 V = d^2V/dr^2 + 1/r \,.\, dV/dr = 1/r \,.\, d(rdV/dr)/dr.$$

If we are dealing with an infinitely long circular cylinder of radius R and surface-density σ, the axis of which is taken as the z-axis, we have (q) $\nabla^2 V_i = 0$ and $\nabla^2 V_a = 0$, while for $r = R$ we have

(r) $$V_i = V_a, \quad dV_a/dr - dV_i/dr = -4\pi\sigma.$$

It follows from equations (p) and (q) that

$$d(rdV_i/dr)/dr = 0 \text{ and } d(rdV_a/dr)/dr = 0.$$

Hence $$dV_i/dr = C_1/r \text{ and } dV_a/dr = C_2/r,$$

$$V_i = C_1 \log r + C_1' \text{ and } V_a = C_2 \log r + C_2'.$$

C_1 must be equal to zero, since no force acts at points in the axis of the cylinder. Further, for $r = R$, we have $C_1' = C_2 \log R + C_2'$. From equation (r) we have $C_2 = -4\pi R\sigma$, and therefore

$$V_i = C_2' - 4\pi R\sigma \log R; \quad V_a = C_2' - 4\pi R\sigma \log r.$$

These equations are the same as those given in (m) XIII.

SECTION XVI. ACTION AND REACTION. ON THE MOLECULAR AND ATOMIC STRUCTURE OF BODIES.

In our discussions up to this point we have considered the motion of a body under the action of given forces; but nothing has yet been said as to the origin of these forces. A body upon which no forces act moves forward, by the principle of inertia, in a straight line with a uniform velocity. A change in the motion can arise only from outside causes. We learn from experience that the motion of one body in the presence of another undergoes a change, and we are therefore led to assume that in the mutual action of these bodies is to be found the reason for the change of motion. We will first consider the mutual action of two bodies. We thus obtain the means of investigating the more general case in which three or more bodies act on one another. The mutual action may be of different kinds. If two bodies collide their motion changes. A similar change occurs when the bodies slide over each other. In both cases the bodies are at least momentarily in contact. Bodies also act on each other without contact; thus, for example, a magnet attracts a piece of iron, or a piece of amber when rubbed attracts a feather. The first serious effort to explain these mutual actions or so-called actions at a distance was made by Descartes. His explanation was based on the assumption that all space is filled with very small particles in motion, and that all observed motions of bodies are due to collisions between them and these invisible particles.

Hence the discovery of the laws of collision became one of the most important tasks in the study of physics. Descartes investigated this question, but without success. It was not until the close of the 17th century that Huygens, Wallis, and Wren contemporaneously succeeded in solving it. A sphere in motion can set in motion a sphere at rest; the moving sphere, therefore, possesses energy of itself. Let the collision be central, that is, let the direction of motion coincide with the line joining the centres of the spheres. An iron sphere produces a greater effect on the sphere at rest than a wooden sphere of equal size moving with the same velocity. Of two equally large spheres whose mass is the same, the one produces the greater effect which has the greater velocity. Hence the force which the moving sphere possesses increases with its mass and with its velocity jointly. The product of the mass and the velocity gives a measure for the force residing in the body, and is called its *momentum* or *quantity of motion.*

The principal result which Huygens, Wallis, and others obtained was the following: If two bodies collide, they undergo changes of momentum which are equally great and in opposite directions, or, they act on each other with equal but oppositely directed forces. The *action* and *reaction* are therefore equal and oppositely directed.

This is one of the most important laws of natural philosophy, and we will discuss the grounds upon which it is founded. It was first derived from observations on collision, without its thereby becoming apparent how far it holds for other interactions between bodies. Newton first recognized in this law a universal law of nature, which always applies when bodies act on one another. By careful investigation of the collisions of different bodies (steel, glass, wool, cork) he found that the action and reaction are equal, if allowance is made for the resistance of the air. In order to examine whether the same law holds for actions at a distance, he mounted a magnet and a piece of iron on corks and floated them on water. The iron and the magnet approached each other and remained at rest after they had come in contact, so that the forces by which the iron and the magnet were mutually attracted were oppositely directed, and equal. He showed further by the following argument that action and reaction are equal in the case of attraction or repulsion: If two bodies acting on each other are rigidly connected, they should both move in the direction of the greater force if action and reaction were not equal; this would contradict the principle of inertia.

Since Newton's time this law has been established in many ways, and many discoveries in physics have furnished proofs of its correctness. It has led in many cases to new discoveries, and there is no longer any doubt of its universal applicability.

The simplest conception of the structure of bodies is that, according to which bodies are composed of discrete particles, for whose mutual action the law of action and reaction holds. Starting from this view, Newton calculated the action of gravity. Gravity is a function of distance alone; its value is therefore the same so long as the distance is unchanged. This conception of the structure of bodies has led to important results in other branches of physics. There are, however, many cases in which it seems inadequate. Chemistry teaches that bodies are composed of molecules, which themselves may be groups of smaller particles or atoms. These molecules have certainly a very complex structure, and the mutual actions among them, especially if the distances between them are great in comparison with their size, must therefore be of a very complicated nature. As yet we have little knowledge on this subject. In what follows we will confine ourselves to the treatment of the motions of particles acting on each other with forces which are functions only of the distances between them.

SECTION XVII. THE CENTRE OF GRAVITY.

Gravity acts on all parts of a body; the forces thus arising may be considered parallel for all parts of the same body. The action of gravity on all the particles of a body may be combined in a resultant whose point of application is at *the centre of gravity.* If the centre of gravity is rigidly connected with the body and rests on a support, the body is in equilibrium in any position. Since gravity is proportional to the mass, the centre of gravity coincides with the *centre of mass.* The resultant applied at the centre of gravity is the *weight of the body.* The straight lever is in equilibrium if there is applied to its centre of gravity a force equal to its weight and acting in the opposite direction; the particles on the one side of the centre of gravity tend by their weight to produce rotation in one sense which is equal to that produced in the opposite sense by the particles on the other side.

Let the masses m_1 and m_2, whose velocities are represented by AA' and BB' (Fig. 27), be situated at the points A and B. Let the point C be so determined on the line joining A and B that $m_1 AC = m_2 BC$. The point C is then called the *centre of gravity* of the two masses m_1 and m_2. If the point C' is so determined on the line joining A' and B' that $m_1 A'C' = m_2 B'C'$, we may consider CC' the *velocity of*

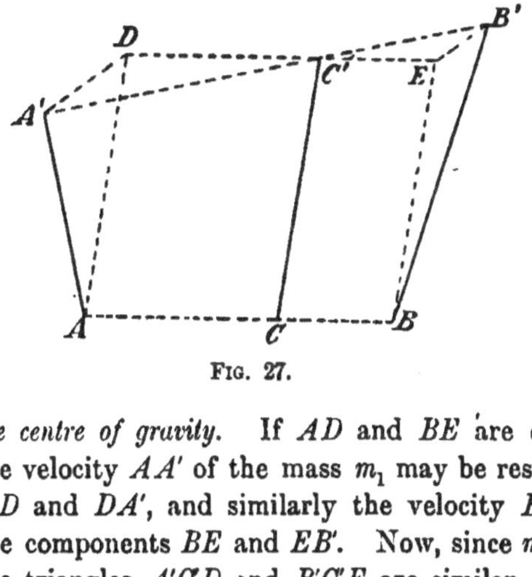

FIG. 27.

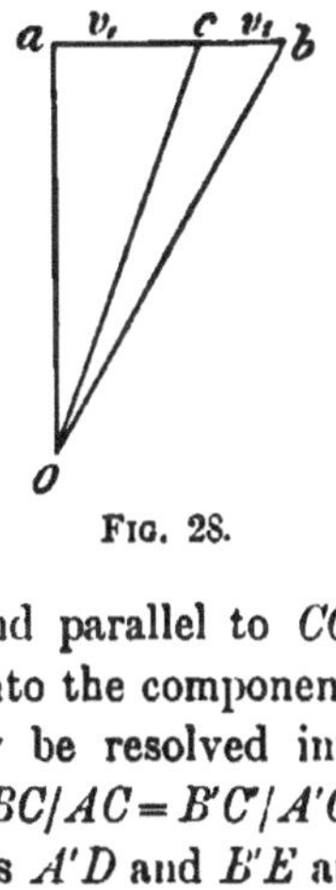

FIG. 28.

the centre of gravity. If AD and BE are equal and parallel to CC', the velocity AA' of the mass m_1 may be resolved into the components AD and DA', and similarly the velocity BB' may be resolved into the components BE and EB'. Now, since $m_1/m_2 = BC/AC = B'C'/A'C'$, the triangles $A'C'D$ and $B'C'E$ are similar, the sides $A'D$ and $B'E$ are parallel, and hence $m_1/m_2 = B'E/A'D$. The velocities of the masses may be considered as compounded of the velocity of the common centre of gravity and two velocities v_1 and v_2, which are parallel to each other and inversely proportional to the masses; so that $m_1 v_1 = m_2 v_2$.

If, therefore, oa and ob (Fig. 28) represent the velocities of the masses m_1 and m_2, and if ab is divided by the point c into the parts ac and bc, which are inversely proportional to the masses, then oc represents the velocity of the centre of gravity, and ca and cb

represent the velocities of the masses m_1 and m_2 relative to the centre of gravity. It is convenient to resolve the velocity in this way, because the velocity of the centre of gravity is changed by external forces only.

If momenta are resolved and compounded like forces, then from Fig. 28 the *momentum of the centre of gravity*, in which we may consider both masses united, equals the resultant of the momenta of the separate masses m_1 and m_2. The momentum $m_1 . \overline{oa}$ may be resolved into $m_1\overline{oc}+m_1\overline{ca}$, the momentum $m_2\overline{ob}$ into $m_2\overline{oc}+m_2\overline{cb}$. Now, $m_1 ca$ and $m_2 cb$ are equal but opposite in direction. Hence we have for the resultant momentum $m_1\overline{oc}+m_2\overline{oc}=(m_1+m_2)\overline{oc}$.

The velocity of the centre of gravity remains unchanged if the bodies m_1 and m_2 act on each other according to the law of action and reaction. In this case both bodies receive momenta which are equal but oppositely directed, and which annul each other. This result may be derived analytically in the following way. If x_1 and x_2 are the coordinates of the particles m_1 and m_2, the line joining which is taken for the x-axis, and if the x-components of the forces with which the masses act on each other are X_1 and X_2, the equations of motion are (a) $m_1\ddot{x}_1=X_1$ and $m_2\ddot{x}_2=X_2$. Adding these equations, we have (b) $d^2(m_1x_1+m_2x_2)/dt^2=X_1+X_2$. Since X_1 and X_2 arise from the mutual action of the masses on each other, they are equal but opposite in direction, and hence $X_1+X_2=0$. Setting

$$\text{(c)} \qquad m_1x_1+m_2x_2=(m_1+m_2)\xi,$$

we have $\ddot{\xi}=0$, $\dot{\xi}=\text{Const.}$ Hence the point determined by the ξ-coordinate moves with the constant velocity $\dot{\xi}$. The x-coordinate of the centre of gravity is ξ, because $m_1(x_1-\xi)=m_2(\xi-x_2)$. Differentiating equation (c) with respect to t, we have

$$m_1\dot{x}_1+m_2\dot{x}_2=(m_1+m_2)\dot{\xi}.$$

That is, *the momentum of the centre of gravity equals the sum of the momenta of the separate masses.*

By Newton's law of universal attraction two masses m_1 and m_2 act on each other with a force $-fm_1m_2/r^2$, where r is the distance between the two masses. Their motion may be determined in the following way. The velocity of the centre of gravity and the velocities of the masses relative to the centre of gravity are determined from the velocities of the masses. These act on each other with forces directed toward the centre of gravity, and we can therefore consider this as the attracting point. If r_1 is the distance of the mass m_1 from the centre of gravity, then $m_1r_1=m_2(r-r_1)$, and therefore $m_2 r = (m_1 + m_2) r_1$. By substitution of this value of r we obtain for the force the expression $fm_1 m_2^3/(m_1 + m_2)^2 r_1^2$. The mass m_1 therefore moves round the centre of gravity as if the force acting on it were due to a mass $M_1 = m_2^3/(m_1 + m_2)^2$ situated at that point.

SECTION XVIII. A MATERIAL SYSTEM.

We will now consider a system of separate masses in vacuo, which act on each other with forces which are functions of the distances of the masses from each other, and obey the law of action and reaction. The forces which act in such a way within the system are called *internal forces*. *External forces*, proceeding from bodies which do not belong to the system, may also act on it. The masses are designated by m_1, m_2, m_3, etc., and the positions of the masses are determined by the coordinates x, y, z, with appropriate indices. We may determine the position of the system by supposing each mass to be made up of different numbers of units of mass; the mean values ξ, η, ζ of the x-, y-, z-coordinates will then be

(a) $$\xi = (m_1 x_1 + m_2 x_2 + m_3 x_3 + \ldots)/(m_1 + m_2 + \ldots). \text{ etc.}$$

ξ, η, ζ are the coordinates of the *centre of gravity of the system of masses.* If the equation (a) is differentiated with respect to the time t, it appears that the velocity of the centre of gravity depends on the velocities of the particles. That is,

(b) $$\dot{\xi} = (m_1 \dot{x}_1 + m_2 \dot{x}_2 + m_3 \dot{x}_3 + \ldots)/(m_1 + m_2 + m_3 + \ldots), \text{ etc.}$$

The internal forces cannot change the motion of the centre of gravity, since by the law of action and reaction two masses impart to each other equal and opposite momenta, the sum of whose projections on any axis is equal to zero.

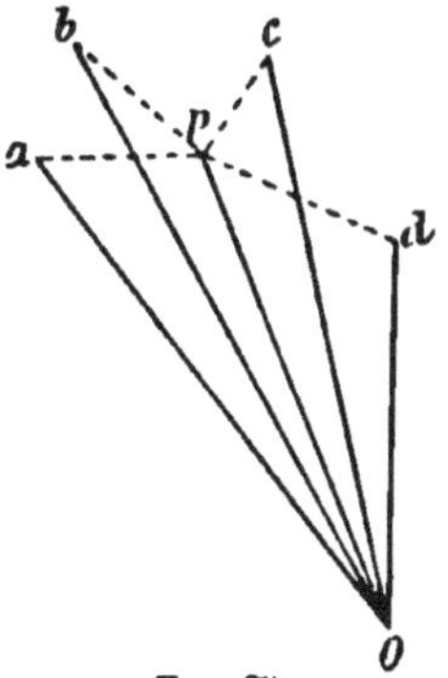

FIG. 29.

This result may be represented geometrically in the following way. From any point O (Fig. 29) draw the lines Oa, Ob, Oc, etc., which represent the velocities of the masses m_1, m_2, m_3, etc. Then if the masses m_1, m_2, m_3, etc., are placed at the points a, b, c, etc., respectively, and if p is the centre of gravity of the masses, Op is the velocity of the centre of gravity.

In order to determine the motion of the separate particles we must know the separate forces which act on them. If the components of the external forces acting on the mass m_a are designated by X_a, Y_a, Z_a, if F_{ab} is the force with which m_a is attracted by m_b, and if r_{ab} is the distance between m_a and m_b, the x-component of the forces acting on m_a is

$$X_a + F_{ab} . (x_a - x_b)/r_{ab} + F_{ac}(x_a - x_c)/r_{ac} + \dots .$$

We obtain similar expressions for m_b, m_c, etc. Hence we have

$$\text{(c)} \qquad m_a \ddot{x}_a = X_a + F_{ab} . (x_a - x_b)/r_{ab} + F_{ac} . (x_a - x_c)/r_{ac} + \dots, \text{ etc.}$$

Since by the law of action and reaction $F_{ab} = F_{ba}$, $F_{ac} = F_{ca}$, we have

$$\text{(d)} \qquad \begin{cases} d^2(m_a x_a + m_b x_b + \dots)/dt^2 = d^2 \Sigma mx/dt^2 \\ \qquad = X_a + X_b + \dots = \Sigma X. \end{cases}$$

If we now introduce the coordinates of the centre of gravity, we have, using equation (a),

$$(m_a + m_b + m_c + \dots)\ddot{\xi} = X_a + X_b + X_c + \dots \quad \text{or} \quad \ddot{\xi}\Sigma m = \Sigma X.$$

This equation contains the *law of the motion of the centre of gravity*, which may be thus stated: *The centre of gravity of a system of masses moves like a material point in which all the masses of the system are united, and at which all the forces are applied.*

The momentum of the whole system is compounded of the momenta of the separate masses. From a point O (Fig. 30) draw $OA = m_a v_a$ parallel to the direction of the velocity v_a. In the same way draw $AB = m_b v_b$, $BC = m_c v_c$, etc. Taking account of all particles we reach a point D. The line OD then represents the momentum of the system. The sums of the projections of the momenta on the coordinate axes are

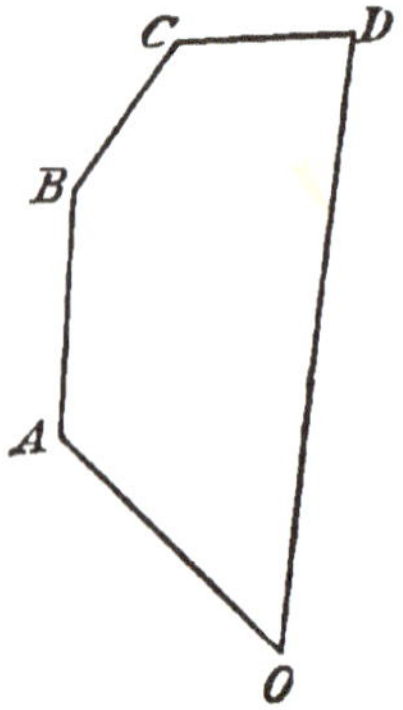

FIG. 30.

$$\Sigma m_\nu \dot{x}_\nu, \quad \Sigma m_\nu \dot{y}_\nu, \quad \Sigma m_\nu \dot{z}_\nu.$$

By equation (b) these sums equal the components of the momentum of the centre of gravity. In the time-element dt the motions of the separate masses are changed by the forces which act on them; nevertheless, the internal forces do not change the momentum, since the resultant of the momenta which these forces occasion is zero, by the law of action and reaction. On the other hand, changes of momentum are occasioned by the action of the external forces. A force K produces the momentum $K . dt$ in the time dt. If all the momenta which the external forces produce are determined in this way, and combined with those originally given, the actual momentum is obtained. This also appears from equation (d), which may be thus written:

$$(f)\qquad d(m_a\dot{x}_a + m_b\dot{x}_b + m_c\dot{x}_c + \ldots) = (X_a + X_b + X_c + \ldots)dt.$$

Sir William Rowan Hamilton introduced the word *vector* to represent magnitudes which have direction, and which may be compounded like motions, velocities, forces, etc. The sum of vectors is called their resultant. If we consider momentum and force as vectors, the increase of the momentum which a system receives in the time dt equals the product of the resultant of the external forces and the time dt. Since the momentum of the system equals the momentum of the centre of gravity, the law just stated holds also for this latter.

SECTION XIX. MOMENT OF MOMENTUM.

If the mass m at the point A (Fig. 31) moves in the direction AB with the velocity v, its momentum is mv. If O is an arbitrary fixed point, and $OC=p$ a line perpendicular to AB, the product mvp is called the moment of momentum with respect to O. The value of the moment depends on the position of the point O. If we erect a perpendicular on the plane determined by O and AB, and lay off on it from O a length proportional to mvp, the vector determined in this way is called the *moment of momentum*. This vector is to be so constructed that it points in the direction of the thumb, if the right hand points in the direction OC, and the palm is turned toward the direction of the force.

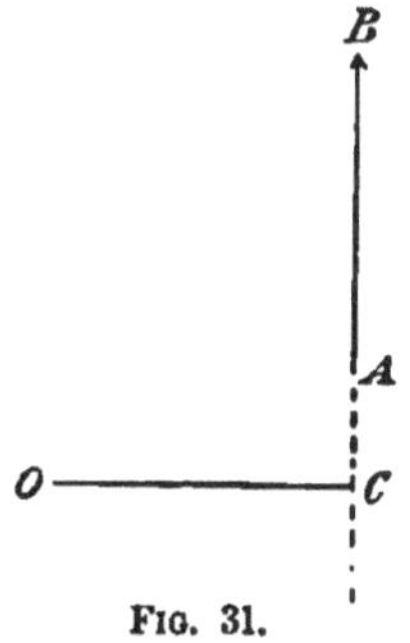

FIG. 31.

In the same way the vectors corresponding to all parts of the system can be determined, and compounded by the method given in Fig 30. If neither external nor internal forces act on the parts of the system, the moment of momentum of the entire system is invariable, since the separate moments remain invariable. The moment of momentum of the system is also not changed by the action of internal forces. If, for example, A and B (Fig. 32) are the points occupied by two masses m_1 and m_2, which repel each other with the force K, then A receives in the time dt the momentum $K.dt$ in the direction AA', and B receives the same momentum in the opposite direction. The moments of momentum of A and B

annul each other. On the other hand, the moment of momentum of the system will in general be changed by external forces, but it will remain constant in case the directions of all the external forces always pass through the fixed point O.

Hence, if the moments of momentum and the moments of the external forces are considered as vectors, the increment of the moment of momentum of the system in the time dt equals the resultant of the moments of the external forces multiplied by dt.

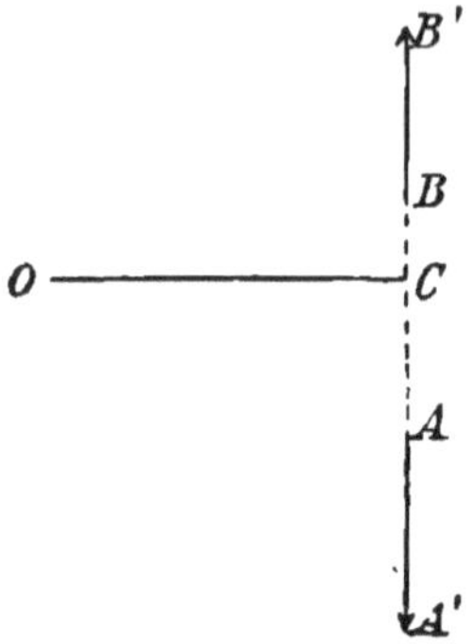

FIG. 32.

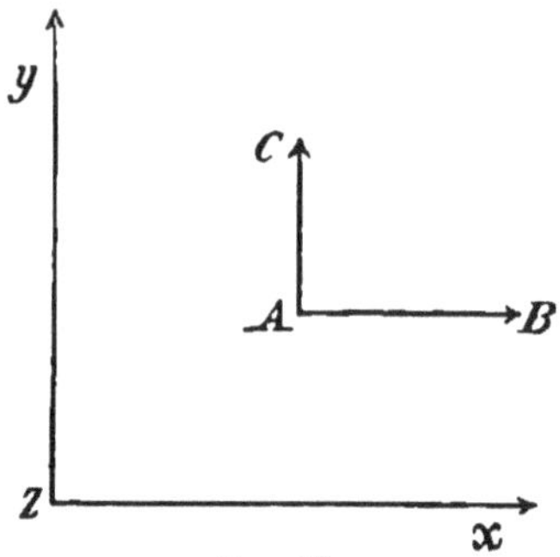

FIG. 33.

This may be represented analytically in the following way. Let $AB=\dot{x}$ and $AC=\dot{y}$ be the velocity components of the particle M situated at A (Fig. 33). The distance of the moving mass from the x-axis is y, and from the y-axis is x. Hence the moments of momentum with respect to the z-axis are $m\dot{x}y$ and $m\dot{y}x$. These being oppositely directed, their difference $m\dot{x}y - m\dot{y}x$ is the moment of momentum of m with respect to the z-axis. This moment receives in the time dt the increment $md(x\dot{y}-y\dot{x})=m(x\ddot{y}-y\ddot{x})dt$. Hence we have

(a) $\Sigma m(x\ddot{y}-y\ddot{x})=\Sigma(xY-yX)$, or (b) $d\Sigma m(x\dot{y}-y\dot{x})=dt\Sigma(xY-yX)$,

that is, *the increment which the moment of momentum about any axis receives in the time dt is equal to the product of the sum of the moments of the external forces about the same axis and the time-element dt.*

SECTION XX. THE ENERGY OF A SYSTEM OF MASSES.

If a particle m moves with a velocity $v=ds/dt$, its kinetic energy [V.] is $\frac{1}{2}mv^2=\frac{1}{2}m(ds/dt)^2=\frac{1}{2}m\dot{s}^2$. Since $ds^2=dx^2+dy^2+dz^2$, this may be written $\frac{1}{2}mv^2=\frac{1}{2}m(\dot{x}^2+\dot{y}^2+\dot{z}^2)$. The *kinetic energy* of the system is determined from the velocities of the separate particles of the system. It is expressed by $T = \frac{1}{2} . \Sigma m(\dot{x}^2 + \dot{y}^2 + \dot{z}^2)$. If x, y, z are the coordinates of a particle, and ξ, η, ζ the coordinates of the centre of gravity, the coordinates of the particle with respect to the centre of gravity are $x - \xi = x'$, $y - \eta = y'$, $z - \zeta = z'$. Using these new coordinates, we obtain

$$\Sigma m\dot{x}^2 = \Sigma m(\dot{\xi} + \dot{x}')^2 = \dot{\xi}^2 \Sigma m + \Sigma m\dot{x}'^2 + 2\dot{\xi}\Sigma m\dot{x}', \text{ etc.}$$

From XVIII. (a) we may set $\Sigma mx' = 0$, if the centre of gravity is chosen as the origin of coordinates. Then

$$T = \frac{1}{2} . (\dot{\xi}^2 + \dot{\eta}^2 + \dot{\zeta}^2) . \Sigma m + \frac{1}{2} . \Sigma m(\dot{x}'^2 + \dot{y}'^2 + \dot{z}'^2).$$

The kinetic energy of the system is equal to the sum of the kinetic energy of the masses due to the motion of the centre of gravity, and the kinetic energy of the masses due to their motion relative to the centre of gravity.

The increment of the kinetic energy of the system in the time-element dt equals the work done by the forces during that time. This is divisible into two parts, that of the external and that of the internal forces. If we designate the components of the motion of a particle parallel to the axes by dx, dy, dz, the work done by the external forces is $\Sigma(Xdx + Ydy + Zdz)$.

If r is the distance between two particles and the repulsive force F acts between them, the work done by the internal forces is ΣFdr. Hence we have $dT = \Sigma(Xdx + Ydy + Zdz) + \Sigma Fdr$. If the force with which the masses act on each other is a function of the distance r only, we can set $Fdr = d\psi$, where ψ is a function of r only. Now setting $\Sigma d\psi = dU$, we have finally

(d) $$dT = \Sigma(Xdx + Ydy + Zdz) + dU.$$

The function U depends only on the distance between the particles or on the configuration of the system. U is *the potential of the system on itself* or *the internal potential energy of the system.* Further, U is the work which would be done by the internal forces if the particles were to move from their positions at any instant into other positions in which their mutual actions are zero.

If, for example, the given masses act on each other according to Newton's law, we have $F = -fm_1m_2/r^2$, and therefore

$$Fdr = -fm_1m_2dr/r^2 = +fd(m_1m_2/r).$$

If several masses m_1, m_2, m_3, ... are present, whose distances from each other are r_{12}, r_{13}, r_{23}, ... respectively, we have

(e) $$dU = fd(m_1m_2/r_{12} + m_1m_3/r_{13} + m_2m_3/r_{23} + \dots).$$

If the system passes from one configuration to another, the work done by the internal forces is determined only by the initial and final positions of the particles, and does not depend on the paths traversed by them. If no external forces act on the system, we have

(f) $$dT = dU \text{ or } T - T_0 = U - U_0.$$

Now, from the discussion in VII., we may set $U_0 = 0$, and so obtain $T = U + T_0$, that is, the kinetic energy of the system equals the original kinetic energy T_0 increased by the work U done by the forces. In case of a change in the relative positions of the particles, supposing no external forces to act, a transformation of the one form of energy into the other occurs without causing a change in the total energy of the system, that is, *the sum of the kinetic and potential energies of such a system is constant.*

SECTION XXI. CONDITIONS OF EQUILIBRIUM. RIGID BODIES.

We will now consider the conditions of equilibrium of a system. If the positions of the separate masses at a definite instant, and also the internal and external forces are given, the system is in equilibrium, when the resultant of all the forces acting on each particle is zero. If the internal forces are in equilibrium, no change occurs in the motion of the system, so long as no external forces act on it. External forces will as a rule set the system in motion; but it is also possible that they will not change the equilibrium of the system as a whole even if its separate parts are set in motion. If the resultant of the external forces is zero, the motion of the centre of gravity remains unchanged [XVIII.]; so that, for example, if the centre of gravity is at rest, it remains at rest. But even when this is the case, the external forces may set the separate masses in motion; the relative positions of the particles may be changed, and changes of form or rotations may occur. The conditions for such changes are developed in the theory of elasticity and in hydrodynamics. At present we will consider only the behaviour of rigid bodies.

The particles of such bodies are so conditioned that the distances between them are constant or nearly so. If the positions of three particles of the body are given, the positions of all the other particles are also given, and the position of the body is determined. If the body (Fig. 34) is moved from its position so that the points A, B, C are brought to the points A', B', C' respectively, it can be brought back to its original position by a series of simple operations. The body may first be displaced parallel with itself through the distance AA', so that the point A' coincides with A, and the points B' and C' are brought to b and c. $C'c$, $B'b$, and $A'A$ are equal and parallel. The body may then be turned about an axis passing through A, perpendicular to the plane determined by BA and bA, through the angle BAb, so that b coincides with B and c is brought to c'. By a second rotation about the axis AB, c' may be made to coincide with C. The motion of the body is thus reduced to a translation and two rotations. A rigid body, therefore, cannot be moved, if it can neither be displaced nor rotated.

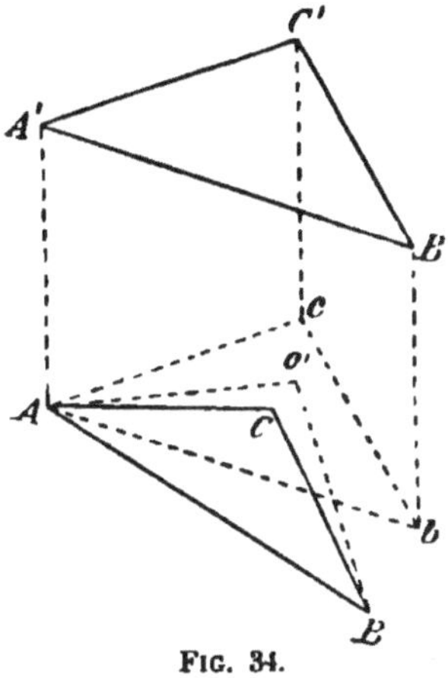

FIG. 34.

In order that a body acted on by external forces shall be in equilibrium, its centre of gravity must remain at rest. The necessary condition for this is that the resultant of the external forces is zero. It must also have no rotation about any axis. If such rotations exist, its particles receive a certain momentum, which has a moment with respect to the axis. Since each of the elements in this moment is positive, because the parts of the body all move in the same sense, and, therefore, the momenta of the separate particles have the same sign, the sum of the momenta can vanish only when each one of them is separately zero. Now the sum of the moments of momentum is [XIX.] equal to the product of the moment of force and the time during which the force acts. Hence it is required for equilibrium that the forces which act on the body have no moment with respect to the axis. This must hold for each axis about which the body can turn. Now, if the moment of the forces equals zero, the sum of the moments of momentum equals zero, therefore the moment of momentum of each particle equals zero; that is, each particle is in equilibrium. Furthermore, since moments can be compounded like forces, equilibrium will exist if the moments with respect to three arbitrary axes are zero.

Section XXII. Rotation of a Rigid Body. The Pendulum.

Let a solid body revolve around an invariable axis, which is chosen as the z-axis of a system of rectangular coordinates. Let the angular velocity of the body be ω. If r represents the distance of any particle m from the z-axis, the velocity of this particle is $r\omega$ and its kinetic energy $\frac{1}{2}mr^2\omega^2$. Since ω has the same value for all particles, the kinetic energy T equals $T=\frac{1}{2}\omega^2\Sigma mr^2$. The factor Σmr^2 is called the *moment of inertia* J of the body with respect to the z-axis; the moment of inertia is equal to the sum of the products of the particles into the squares of their respective distances from the z-axis. Hence we have $T=\frac{1}{2}\omega^2 J$, that is, *the kinetic energy of a rotating body is equal to its moment of inertia multiplied by half the square of its angular velocity.*

A length K may always be found, such that $\Sigma mr^2 = K^2\Sigma m$. This length is called the *radius of gyration* of the body. It is the distance from the axis at which a mass equal to the mass of the body would have the same moment of inertia with respect to the axis as that of the body.

If the only external forces which act on the body pass through the axis, the work done by them is zero, since the axis does not move. Since the internal forces also do no work, the kinetic energy and therefore also the angular velocity ω must remain constant.

Since [XVIII.] the centre of gravity moves as if the resultant of all the forces acted on the mass of the body concentrated at the centre of gravity, this resultant R can be determined. If we represent the distance OP (Fig. 35) of the centre of gravity from the z-axis by a, we have [IV. (b)] $R=\Sigma ma^2\omega^2/a=\Sigma ma\omega^2$. R is the resultant of the forces with which the body acts on the axis of rotation. In general, the forces applied to the body so act that they have no resultant; they tend only to produce rotation about the axis. In order to determine them, the theorem in XIX. concerning moments of momentum must be used.

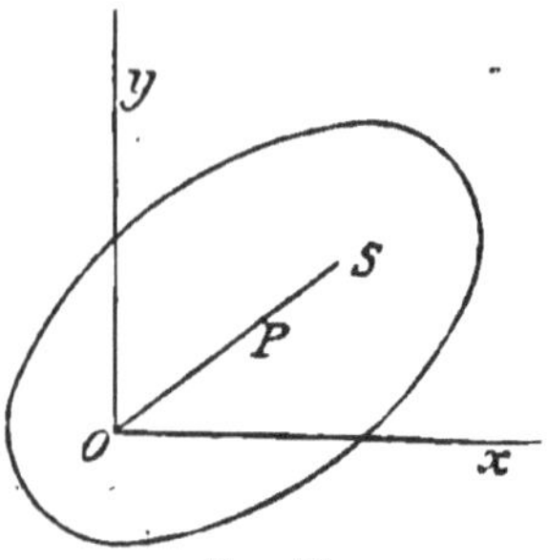

Fig. 35.

If external forces act on the body, its angular velocity changes. The amount of this change is determined from XIX. The momentum of a particle m is represented by $mr\omega$, and its moment of momentum by $mr\omega r$. Hence the moment for all particles of the body is $\omega\Sigma mr^2=\omega J$. If the moment of the forces with respect to the z-axis is represented by M, we have, from XIX., $d(\omega J) = Mdt$ or (c) $Jd\omega/dt = M$.

If, for example, the moment is constant, the angular velocity increases in direct ratio with the time.

If the moment is due to gravity, the body under certain conditions performs oscillations. We suppose the x-axis taken parallel to the direction of gravity and represent the force of gravity by g. The position of the centre of gravity P (Fig. 35) is determined by the angle $POX = \Theta$ and the angular velocity ω by $d\Theta = \omega dt$. The moment of force with respect to the z-axis is $-(m_1y_1 + m_2y_2 + \ldots)g = -\eta \, . \, g\Sigma m$, where η is the y-coordinate of the centre of gravity. Σm denotes the sum of all the masses. Since $\eta = a \sin \Theta$, we have from (c)

$$J \, . \, \ddot{\Theta} = -a \sin \Theta \, . \, g\Sigma m.$$

If Θ is very small, so that we may set $\sin \Theta = \Theta$, this becomes

$$\text{(d)} \qquad \ddot{\Theta} = -a\Theta g\Sigma m/J.$$

Comparing this equation with that given in VIII. (g), we find that they are identical when we set $1/l = a\Sigma m/J$. The period of oscillation of the physical pendulum is therefore (e) $t = \pi\sqrt{l/g} = \pi\sqrt{J/ga\Sigma m}$. Since $J = \Sigma m(x^2 + y^2 + z^2)$, we get by transferring the origin of the system of coordinates to the centre of gravity (ξ, η, ζ), if x', y', z' are the coordinates with respect to the new origin,

$$\text{(f)} \qquad J = \Sigma m\{(x' + \xi)^2 + (y' + \eta)^2 + (z' + \zeta)^2\} = \Sigma m a^2 + \Sigma m(x'^2 + y'^2 + z'^2),$$

since the terms $\xi\Sigma mx'$, $\eta\Sigma my'$ and $\zeta\Sigma mz'$ vanish.

Now, if we set $J = a^2\Sigma m + k^2\Sigma m$, where k is the radius of gyration, we obtain from (e) (g) $t = \pi\sqrt{(a^2 + k^2)/ga}$. We call $l = \dfrac{a^2 + k^2}{a}$ the *reduced length of the pendulum* or *the length of the equivalent simple pendulum*. The point S which is at the extremity of the line $OS = l$ (Fig. 35), drawn through O and P, is called the centre of oscillation. If an axis is passed through S parallel to the z-axis, and the body oscillates about it, the reduced length of the pendulum l' is

$$l' = \big((l - a)^2 + k^2\big)/(l - a).$$

Since, however, $l - a = k^2/a$, we have $l' = (a^2 + k^2)/a = l$. The reduced length of the pendulum and therefore the time of oscillation are the same for this new axis as for the former one.

CHAPTER II.

THE THEORY OF ELASTICITY.

SECTION XXIII. INTERNAL FORCES.

IF all parts of a body are in equilibrium and if no tensions or pressures act on them, yet internal forces must be present acting between the separate parts of the body. Every action produces changes of form in the body, and thus develops forces in its interior, which act in a sense opposite to the external forces. These internal forces condition the nature of the body, determining, for example, the difference between solids and fluids. No sharp distinction can be drawn, however, between these two classes of bodies. Viscous fluids and jelly-like solids are bodies which seem to be transition forms between true solids and fluids.

If a pressure acts on the surface of a fluid, it must be equally great on equal areas of the surface at all points, and it must be perpendicular to the surface, if the fluid is to be in equilibrium. This pressure is exerted throughout the whole mass; all equal surface-elements at a point are subjected to equal pressures, which are always perpendicular to the surface-elements. We call such a pressure *hydrostatic pressure*. A similar pressure may also be present in solids. If a solid, a piece of glass, for example, which fills the volume enclosed by its external surface, is immersed in a fluid on which a pressure is exerted, the same pressure exists at every point in the surface of the glass as in the fluid. The pressure is everywhere the same, and perpendicular to the surface-elements. We may therefore speak of hydrostatic pressure in solids also.

Yet, in general, internal forces in solids are very different from those in fluids. Let a cylindrical rod be fastened at one end, and let the force V be applied at the other end so as to lengthen the rod. In a cross section perpendicular to the axis of the cylinder the internal forces are everywhere equal. Let the area of the cross section be A (Fig. 36), then the force V/A acts on unit of area in A.

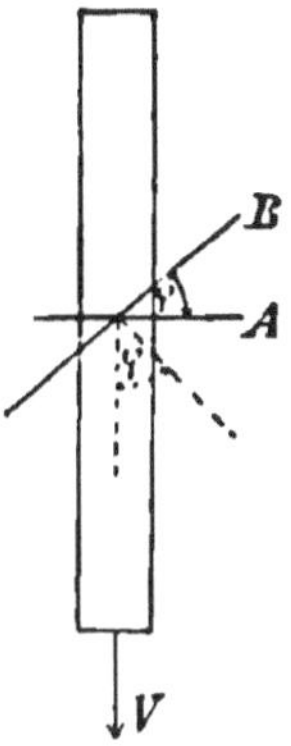

FIG. 36.

This quotient represents the stress S in the rod. If another plane cross section B is taken in the rod, which makes the angle ϕ with A, the force S' acts on each unit of area of B, so that

$$S'.B = S.A = S.B\cos\phi,$$

and hence (a) $S' = S\cos\phi$. The stress S' is no longer perpendicular to the surface B on which it acts; its magnitude decreases with $\cos\phi$ and vanishes for $\phi = \frac{1}{2}\pi$. A surface-element within the cylinder and parallel to its axis is therefore subjected neither to pressure nor to tension; this conclusion holds for an element of the surface of the cylinder. We may resolve S' into two components, one of which, T, is tangent, and the other, N, normal to B, and have

(b) $$N = S\cos^2\phi, \quad T = S\cos\phi\sin\phi.$$

If internal forces of this type exist within a body, we call the stresses *axial*. In the direction of the axis the stress is S; a unit of surface whose normal makes the angle ϕ with the axis is acted on by a force $S\cos\phi$ in the direction of the axis.

We will consider a rectangular parallelepiped (Fig. 37), of which the lines OA, OB, and OC are adjacent edges. The stresses which act on each unit of area of the faces which are perpendicular to OA, OB, and OC are S_a, S_b, S_c respectively. If the normal to an arbitrarily situated unit of surface f makes the angles α, β, γ, with the edges OA, OB, OC respectively, the force acting on f is the resultant of the forces $S_a\cos\alpha$, $S_b\cos\beta$, $S_c\cos\gamma$, which are parallel to OA, OB, OC respectively. If the stresses S_a, S_b, S_c have the same value S, this resultant is $S\sqrt{\cos^2\alpha + \cos^2\beta + \cos^2\gamma} = S$. Hence three equal stresses which are perpendicular to each other cause a hydrostatic stress, since their resultant has the same value whatever may be the position of the surface. Since the components of this stress are $S\cos\alpha$, $S\cos\beta$, and $S\cos\gamma$, it is perpendicular to the unit area f.

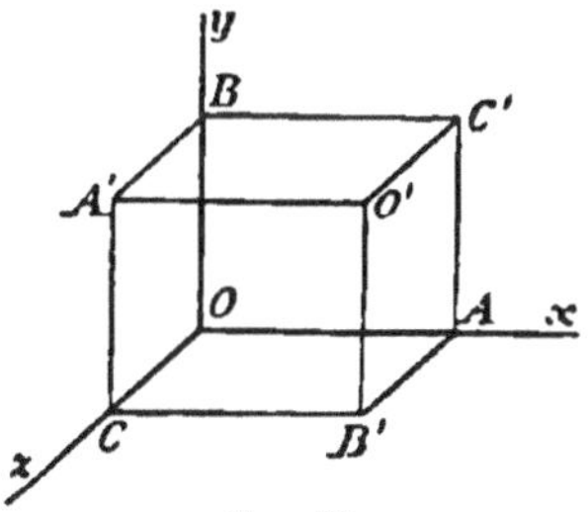

FIG. 37.

On the other hand, if $S_c = 0$ and $S_a = S_b = S$, that is, if two stresses act at right angles to each other, while the stress perpendicular to them both is zero, the components in the directions OA, OB, OC respectively are $S \cos \alpha$, $S \cos \beta$, 0. Hence the force acting on f is

$$S\sqrt{\cos^2\alpha + \cos^2\beta} = S\sqrt{1 - \cos^2\gamma} = S \sin \gamma,$$

and is perpendicular to OC. Such a state of stress in a body may be called *equatorial*. The plane which contains OA and OB, or rather, every plane parallel to both these lines, may be called an equatorial plane. The same stress S acts on each unit area perpendicular to the equatorial plane. If the normal to the surface f makes the angle ϕ with the equatorial plane, the stress on it is proportional to $\cos \phi$.

Section XXIV. Components of Stress.

Let the surface F (Fig. 38) divide a body into two parts, A and B. If the portion of A which touches the element dF of the surface F is removed, a force must act on dF to keep B in equilibrium. This force SdF is not, as a rule, perpendicular to the element dF. The forces acting at the various points of F are, in general, different. If the force tends to move the element dF into the space occupied by B, it is called a *pressure* on the surface dF; if it tends to move the element dF into the space occupied by A, it is called a *tension*. In all cases we call the force S a *stress*; if this acts as a tension, it is a *positive stress*, if as a pressure, it is a *negative stress*. If the part of B which touches dF is removed, then to maintain equilibrium in A a force SdF must act on dF, since action and reaction are equal. Hence both forces which act on an element of surface within a body are equal, but oppositely directed. It is characteristic of a stress that it may be looked on as made up of two equal and opposite forces.

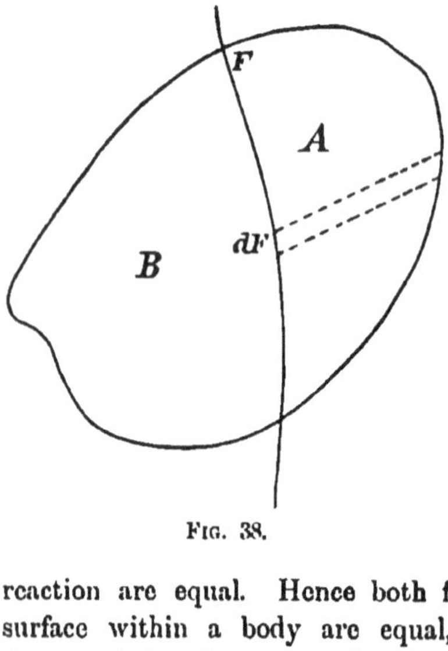

Fig. 38.

If the surface-element dF remains in its original place in the body, but is turned about one of its points, a particular value of the stress corresponds to every one of its positions; for special positions the stress may be zero. When the body in which the surface is drawn is a fluid, the stress is independent of the position of the surface. We assume in the body a system of rectangular coordinates. The stresses in the surface-elements, which are perpendicular to the directions of the axes, are determined by their components.

Let the surface-element dF be perpendicular to the x-axis, and let $dF = dy \,.\, dz$. If that part of the body is removed which lies on the positive side of the surface-element $dydz$, the positive side being determined by the positive direction of the x-axis, then, to maintain equilibrium, a force $Sdydz$ must act on the surface $dydz$. The force S is resolved into the components X_x, Y_x, Z_x, which are respectively parallel to the coordinate axes. The index indicates that the forces act on an element which is perpendicular to the x-axis. X_x is perpendicular to the surface-element; it is therefore called the *normal force*; Y_x and Z_x are *tangential forces*. Now, let the element dF remain in the same place, but be turned so that it is perpendicular to the y-axis. We may then set $dF = dzdx$. As before, there are three components of force X_y, Y_y, Z_y acting on the surface-element $dzdx$, of which Y_y is the normal force, X_y and Z_y are the tangential forces. If the surface-element dF is turned so as to be perpendicular to the z-axis, we have as components X_z, Y_z, Z_z, of which Z_z is the normal force and X_z and Y_z are the tangential forces. There are therefore, in all, nine components,

$$X_x,\ Y_x,\ Z_x;\quad X_y,\ Y_y,\ Z_y;\quad X_z,\ Y_z,\ Z_z.$$

By these components the stress on any surface is determined. Let OA, OB, OC (Fig. 39) represent line-elements, parallel respectively to the x-, y-, z-axes. Let a plane be passed through A, B, and C, so as to form the tetrahedron $OABC$. Let P, Q, and R be the components of the stress in the directions of the coordinate axes at a point in the base ABC of the tetrahedron. We now form the equation of condition, which must hold that the tetrahedron shall not move in the direction of the x-axis. The forces which tend to move the tetrahedron in that direction are $P \,.\, ABC$ acting on its base, and $-X_x \,.\, OBC$, $-X_y \,.\, OAC$, $-X_z \,.\, OAB$ acting on its faces. Hence the force which urges the tetrahedron in

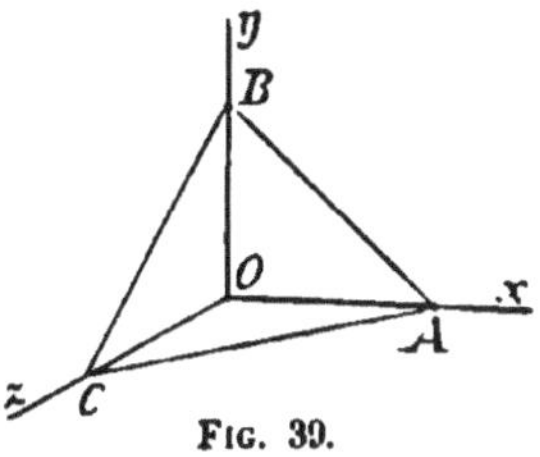

FIG. 39.

the direction of the x-axis is $P.ABC - X_x.OBC - X_y.OAC - X_z.OAB$. Designate by α, β, γ the angles made with the axes by the normal to the surface ABC drawn outward from the tetrahedron; then the expression for the force in the direction of the x-axis becomes

$$(P - X_x \cos\alpha - X_y \cos\beta - X_z \cos\gamma) . ABC.$$

Now, if no external attractions or repulsions act on any part of the body, the conditions of equilibrium, obtained by setting this, and the two similar expressions which hold for the other axes, equal to zero, are

$$\text{(a)} \qquad \begin{cases} P = X_x \cos\alpha + X_y \cos\beta + X_z \cos\gamma, \\ Q = Y_x \cos\alpha + Y_y \cos\beta + Y_z \cos\gamma, \\ R = Z_x \cos\alpha + Z_y \cos\beta + Z_z \cos\gamma. \end{cases}$$

If other forces besides the stresses act on the parts of the body, these must be taken into account in equations (a). If the force X acts on the unit of mass in the direction of the x-axis, the force acting in that direction on the tetrahedron is $X\rho dv$, if dv represents its volume, and ρ its density. The condition of equilibrium in the direction of the x-axis then becomes

$$(P - X_x \cos\alpha - X_y \cos\beta - X_z \cos\gamma) . ABC + X\rho dv = 0.$$

Now, since $dv = \frac{1}{3}h . ABC$, where h is the height of the tetrahedron, this equation is equivalent to

$$P - X_x \cos\alpha - X_y \cos\beta - X_z \cos\gamma + \tfrac{1}{3}h\rho X = 0.$$

Since the height h of the tetrahedron is infinitely small, we may neglect the term containing it, and again obtain the first of equations (a), which hold generally.

In order to exhibit the meaning of equations (a), we will consider the following case. Suppose a tension S to act in the direction of the x-axis, and a pressure of the same value to act in the direction of the y-axis. Then $X_x = S$, $Y_y = -S$, and all other components of stress are equal to zero. Hence $P = S\cos\alpha$, $Q = -S\cos\beta$, $R = 0$.

The resultant A of these components is $A = S\sin\gamma$. If λ, μ, ν are the angles between A and the axes, we have $\cos\lambda = \dfrac{\cos\alpha}{\sin\gamma}$, $\cos\mu = -\dfrac{\cos\beta}{\sin\gamma}$, $\cos\nu = 0$. The angle ϵ between A and the normal to the surface-element considered is determined by $\cos\epsilon = \dfrac{\cos^2\alpha - \cos^2\beta}{\sin\gamma}$.

If the surface-element is parallel to the z-axis, $\gamma = \dfrac{\pi}{2}$, $A = S$, $\cos\epsilon = \cos 2\alpha$, $\epsilon = \pm 2\alpha$.

If $a=\frac{\pi}{4}$, then $\epsilon=\frac{\pi}{2}$, the resultant is a tangential force. Thus the surface of a prism whose axis is parallel to the z-axis, and whose

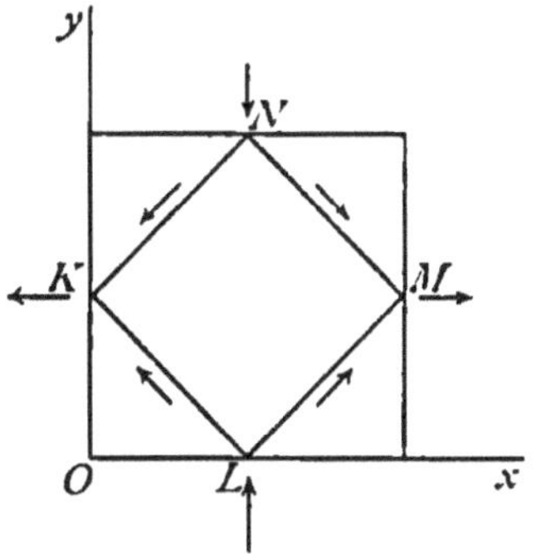

FIG. 39 *a*.

sides make angles of 45° with the xz- and yz-planes, is acted on only by tangential forces, each equal to S.

SECTION XXV. RELATIONS AMONG THE COMPONENTS OF STRESS.

The force which acts on the volume-element $dxdydz$ (Fig. 40) is determined from the components of stress. Let the components acting at the point O be given, and let the force which acts on OA' in the direction of the x-axis be equal to $-X_x dydz$. By development by Maclaurin's theorem we obtain for the force acting on AO' the expression

$$(X_x+\partial X_x/\partial x \,.\, dx)dydz.$$

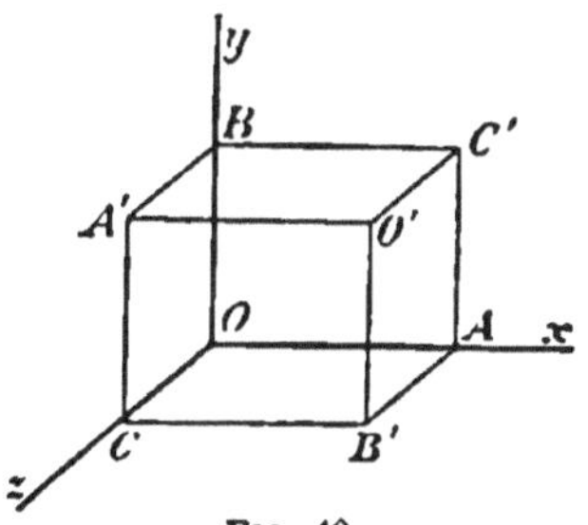

FIG. 40.

The resultant of these two forces is $\partial X_x/\partial x \,.\, dxdydz$. The forces $-X_y dxdz$ and $(X_y+\partial X_y/\partial y \,.\, dy)dxdz$, whose resultant is $\partial X_y/\partial y \,.\, dxdydz$, act on the surfaces OB' and $O'B$ respectively in the direction of the x-axis. The resultant of the forces acting in the same direction on the surfaces $O'C$ and OC' is $\partial X_z/\partial z \,.\, dxdydz$. Hence the total force acting on the parallelepiped $dxdydz$ in the direction of the x-axis is

$$(\partial X_x/\partial x+\partial X_y/\partial y+\partial X_z/\partial z)dxdydz.$$

If (X), (Y), and (Z) represent the components of the force with which the stresses act on unit of volume, we have

$$\text{(b)} \qquad \begin{cases} (X) = \partial X_x/\partial x + \partial X_y/\partial y + \partial X_z/\partial z, \\ (Y) = \partial Y_x/\partial x + \partial Y_y/\partial y + \partial Y_z/\partial z, \\ (Z) = \partial Z_x/\partial x + \partial Z_y/\partial y + \partial Z_z/\partial z. \end{cases}$$

If the body is acted on only by stresses, equilibrium will exist if the three components (X), (Y), (Z) are each equal to zero. The equations (b) in this case are three differential equations which the components of stress must satisfy. If a force whose components are X, Y, Z acts on each unit of mass, and if the density of the body is ρ, we obtain the conditions of equilibrium,

$$\text{(c)} \qquad \begin{cases} \partial X_x/\partial x + \partial X_y/\partial y + \partial X_z/\partial z + \rho X = 0, \\ \partial Y_x/\partial x + \partial Y_y/\partial y + \partial Y_z/\partial z + \rho Y = 0, \\ \partial Z_x/\partial x + \partial Z_y/\partial y + \partial Z_z/\partial z + \rho Z = 0. \end{cases}$$

Internal forces produce both *translations* and *rotations* in the body. The tangential components tend to rotate the parallelepiped OO' about the z-axis. The tangential force X_y acts on the surface OB' in the negative direction, while the tangential force $X_y + \partial X_y/\partial y \,.\, dy$ acts on the opposite surface $O'B$ in the positive direction. These two forces form a couple acting on the parallelepiped with a moment $X_y \,.\, dxdz \,.\, dy$, if terms of an order higher than the third are neglected. This moment tends to turn the parallelepiped about the z-axis in the negative direction. The tangential forces acting on the surfaces OA' and $O'A$ have the moment $Y_x \,.\, dydz \,.\, dx$, which tends to turn the parallelepiped in the positive direction. The total moment which tends to rotate the parallelepiped about the z-axis is $(Y_x - X_y)dxdydz$. If the body is in equilibrium under the action of the stresses considered, this moment must be zero, that is, (d) $Y_x = X_y$, and similarly $Z_y = Y_z$, $X_z = Z_x$. The last two equations are derived in the same way as the first. If attractive forces, such as gravity, or in general, if any forces acting at a distance act on the body, equations (d) will still be applicable. The point of application of such forces, in infinitely small bodies, coincides with the centre of gravity; such forces, therefore, cannot produce rotations, and, therefore, cannot make equilibrium with the forces which tend to rotate the body.

It appears from equations (d) that six quantities are sufficient to determine the stress at a point in a body, namely, X_x, Y_y, Z_z; $Z_y = Y_z$, $X_z = Z_x$, $Y_x = X_y$. The first three are *normal forces*, the other three *tangential forces*. It is possible to express these forces by a simpler notation, but we will retain the above, which has the advantage that it exhibits more clearly than any other the true significance of the quantities involved. It must be borne in mind that the value of a component of stress remains unchanged if the direction of the force and the direction of the normal to the surface-element, on which the stress acts, are interchanged.

SECTION XXVI. THE PRINCIPAL STRESSES.

In order to obtain a better understanding of the nature of internal forces, we will examine if it is possible to pass a surface through a given point in a body in such a position that no tangential force acts on it. We may anticipate our conclusion by the statement that three such surfaces may be drawn through any point and that they are perpendicular to each other. To show this, we proceed from the equations [XXIV. (a)]

$$\text{(a)} \qquad \begin{cases} P = X_x \cos\alpha + X_y \cos\beta + X_z \cos\gamma, \\ Q = Y_x \cos\alpha + Y_y \cos\beta + Y_z \cos\gamma, \\ R = Z_x \cos\alpha + Z_y \cos\beta + Z_z \cos\gamma, \end{cases}$$

in which α, β, γ are the angles between the normal to the surface and the axes, and determine the position of the surface on which the components of stress P, Q, R act. It is to be shown that this surface may have such a position in the body that the stress acts perpendicularly to it; we will call the stress in this case the *principal stress* S. The angles which the direction of S makes with the axes are as before, α, β, γ, and

$$\text{(b)} \qquad P = S\cos\alpha, \quad Q = S\cos\beta, \quad R = S\cos\gamma.$$

Introducing these values in (a), we have

$$\text{(c)} \qquad \begin{cases} (X_x - S)\cos\alpha + X_y \cos\beta + X_z \cos\gamma = 0, \\ Y_x \cos\alpha + (Y_y - S)\cos\beta + Y_z \cos\gamma = 0, \\ Z_x \cos\alpha + Z_y \cos\beta + (Z_z - S)\cos\gamma = 0, \end{cases}$$

If $\cos\alpha$, $\cos\beta$, $\cos\gamma$ are eliminated from these equations, we obtain

$$\text{(d)} \qquad \begin{cases} S^3 - (X_x + Y_y + Z_z)S^2 + (X_xY_y + Y_yZ_z + Z_zX_x - Z_y^2 - X_z^2 - Y_x^2)S \\ \quad - (X_xY_yZ_z + 2Z_yX_zY_x - X_xZ_y^2 - Y_yX_z^2 - Z_zY_x^2) = 0. \end{cases}$$

This equation has always *one* real root A, and we can find the corresponding values of α, β, γ from equations (c) and the relation

$\cos^2\alpha+\cos^2\beta+\cos^2\gamma=1$. Therefore, through any point in the body there may be passed at least one plane having the property that no tangential forces act on it. We call such a plane a *principal plane.*

Let the system of coordinates be so rotated that this principal plane is parallel to the *yz*-plane. On this supposition, we have $X_x=A$, $Y_x=0$, $Z_x=0$. The equations (c) then become

$$(A-S)\cos\alpha=0;\quad (Y_y-S)\cos\beta+Y_z\cos\gamma=0;$$

$$Z_y\cos\beta+(Z_z-S)\cos\gamma=0.$$

These equations are satisfied when we set

$$S=A,\quad \cos\alpha=1,\quad \cos\beta=\cos\gamma=0.$$

We thus return to the principal plane already found, with its appropriate normal stress A. The same equations are also satisfied if we set $\cos\alpha=0$; $\cos\beta/\cos\gamma=-Y_z/(Y_y-S)=-(Z_z-S)/Z_y$.

Since $\cos\alpha=0$, and $\alpha=\frac{1}{2}\pi$, the new principal planes are perpendicular to the first one. We have further,

$$S=\tfrac{1}{2}(Y_y+Z_z\pm\sqrt{(Y_y-Z_z)^2+4Z_y^2})$$

and $$\cos\beta/\cos\gamma=\tfrac{1}{2}(Y_y-Z_z\pm\sqrt{(Y_y-Z_z)^2+4Z_y^2})/Z_y.$$

These equations present two values of S and two values each of β and γ. If we represent the values of β and γ by β' and β'', γ' and γ'' respectively, we have

$$\cos\beta'\cos\beta''\,/\,\cos\gamma'\cos\gamma''=-1,$$

and hence $$\cos\beta'\cos\beta''+\cos\gamma'\cos\gamma''=0.$$

Since the corresponding values of α are equal to $\frac{1}{2}\pi$, it follows that the two new principal planes are perpendicular to each other.

It is thus proved that, in general, through any point in a body, there may be drawn three surface-elements, and only three, on which only normal forces act, and that they are perpendicular to one another. The normal stresses corresponding to the three planes may be designated by A, B, and C. From (d) the following relations hold among these normal stresses and the components of stress,

$$\text{(e)}\quad\left\{\begin{aligned} A+B+C&=X_x+Y_y+Z_z,\\ BC+AC+AB&=Z_zY_y+X_xZ_z+Y_yX_x-Z_y^2-X_z^2-Y_x^2,\\ ABC&=X_xY_yZ_z+2Z_yX_zY_x-X_xZ_y^2-Y_yX_z^2-Z_zY_x^2.\end{aligned}\right.$$

The first of these equations should be especially noticed; it shows, *that the sum of the normal forces for three planes perpendicular to each other is constant.*

If the axes of the system of coordinates are parallel to the directions of the principal stresses A, B, and C, equations (a) become

$$P = A\cos\alpha,\quad Q = B\cos\beta,\quad R = C\cos\gamma.$$

If $A > B > C$, and we set $A = B + S_1$, $C = B - S_2$, the principal stresses can be replaced by a hydrostatic stress B and two axial stresses S_1 and S_2, the first of which is a tension, the second a pressure.

This investigation shows that through any point in a body three planes can always be passed which are acted on only by normal stresses, equal to the principal stresses A, B, and C. A, B, and C are the three roots of equation (d); their directions may be determined by the help of equations (c). A makes the angles α, β, γ, with the coordinate axes. We write $\cos\alpha_1 = l_1$, $\cos\beta_1 = m_1$, and $\cos\gamma_1 = n_1$. The corresponding notation for B and C is exhibited in the following table:

(g)

	x	y	z
A	l_1	m_1	n_1
B	l_2	m_2	n_2
C	l_3	m_3	n_3

From equations (c) the following relations hold among these quantities:

$$\text{(h)}\quad \begin{cases} Al_1 = X_x l_1 + X_y m_1 + X_z n_1; & Bl_2 = X_x l_2 + X_y m_2 + X_z n_2; \\ Am_1 = Y_x l_1 + Y_y m_1 + Y_z n_1; & Bm_2 = Y_x l_2 + Y_y m_2 + Y_z n_2; \\ An_1 = Z_x l_1 + Z_y m_1 + Z_z n_1; & Bn_2 = Z_x l_2 + Z_y m_2 + Z_z n_2; \\ \quad Cl_3 = X_x l_3 + X_y m_3 + X_z n_3, & \\ \quad Cm_3 = Y_x l_3 + Y_y m_3 + Y_z n_3, & \\ \quad Cn_3 = Z_x l_3 + Z_y m_3 + Z_z n_3. & \end{cases}$$

These equations can be solved for the components of stress X_x, Y_y, etc. These quantities may, however, be determined more easily in the following way. Through a point P draw the lines PA', PB', and PC' parallel to the directions of the principal stresses A, B, and C. These three lines, together with a plane F parallel to the yz-plane, determine a tetrahedron. The plane F is so placed that the tetrahedron is infinitely small; its base is dF. The areas of the faces which meet at P are l_1dF, l_2dF, and l_3dF. The force acting on unit area in l_1dF in the direction of the x-axis is Al_1; the forces acting on unit area in the two other faces are Bl_2, Cl_3, respectively, and the force acting on unit area in dF is X_x. That the tetrahedron shall not move in the direction of the x-axis we must have

$$l_1A \, . \, l_1dF + l_2B \, . \, l_2dF + l_3C \, . \, l_3dF = X_xdF \quad \text{or} \quad X_x = Al_1^2 + Bl_2^2 + Cl_3^2.$$

By a similar process we obtain for the other components the following equations:

$$\text{(i)} \qquad \begin{cases} X_x = Al_1^2 + Bl_2^2 + Cl_3^2; & Z_y = Am_1n_1 + Bm_2n_2 + Cm_3n_3, \\ Y_y = Am_1^2 + Bm_2^2 + Cm_3^2; & X_z = Al_1n_1 + Bl_2n_2 + Cl_3n_3, \\ Z_z = An_1^2 + Bn_2^2 + Cn_3^2; & Y_x = Al_1m_1 + Bl_2m_2 + Cl_3m_3. \end{cases}$$

It may easily be seen that these values of the components of stress satisfy equations (h), if the known relations among the quantities given in (g) are taken into account.

SECTION XXVII. FARADAY'S VIEWS ON THE NATURE OF FORCES ACTING AT A DISTANCE.

Newton considered the action between two masses as an action at a distance which is not propagated from particle to particle of the medium surrounding the masses. Faraday, on the other hand, in discussing electrical action, held that the intervening medium is the seat of the action between two charged bodies, and that the action is transferred from particle to particle. In each of these particles electricity is displaced in the direction of a line of force, one end of which becomes positively and the other negatively electrified.

In a body thus polarized the particles are so arranged that poles of opposite name are contiguous. Hence the lines of force tend to contract, and a state of stress arises in the medium. This stress is similar to the elastic stress, and was called by Maxwell *electrical elasticity*. In Chapter V. of his *Treatise on Electricity*, Maxwell, using Faraday's hypothesis, developed a theory which we will now proceed to discuss. Since electrical and magnetic forces conform to the same law as that of universal attraction, the discussion may be made perfectly general, and applicable to all forces between bodies which are inversely proportional to the squares of the distances separating the bodies.

Let the potential ψ be given for all points of the region. The density ρ is determined from the potential by Poisson's equation

$$\text{(a)} \qquad \partial^2\psi/\partial x^2 + \partial^2\psi/\partial y^2 + \partial^2\psi/\partial z^2 + 4\pi\rho = 0.$$

The mass ρdv contained in the volume-element dv is acted on by a force whose components are

$$\mp \rho dv \partial\psi/\partial x; \quad \mp \rho dv \partial\psi/\partial y; \quad \mp \rho dv \partial\psi/\partial z.$$

The upper sign holds for magnetic or electrical attractions, the lower for mass attractions. Introducing the value of ρ given in (a) the component acting in the direction of the x-axis becomes

$$\pm \partial\psi/\partial x . (\partial^2\psi/\partial x^2 + \partial^2\psi/\partial y^2 + \partial^2\psi/\partial z^2) . dv/4\pi.$$

This quantity must be capable of representation as the sum of three differential coefficients with respect to x, y, and z. We have

$$\partial\psi/\partial x . \partial^2\psi/\partial x^2 = \tfrac{1}{2}\partial(\partial\psi/\partial x)^2/\partial x,$$

$$\partial\psi/\partial x . \partial^2\psi/\partial y^2 = \partial(\partial\psi/\partial x . \partial\psi/\partial y)/\partial y - \partial\psi/\partial y . \partial^2\psi/\partial x\partial y$$
$$= \partial(\partial\psi/\partial x . \partial\psi/\partial y)/\partial y - \tfrac{1}{2}\partial(\partial\psi/\partial y)^2/\partial x,$$

$$\partial\psi/\partial x . \partial^2\psi/\partial z^2 = \partial(\partial\psi/\partial x . \partial\psi/\partial z)/\partial z - \partial\psi/\partial z . \partial^2\psi/\partial x\partial z$$
$$= \partial(\partial\psi/\partial x . \partial\psi/\partial z)/\partial z - \tfrac{1}{2}\partial(\partial\psi/\partial z)^2/\partial x.$$

Hence the force which acts in the direction of the x-axis on the volume-element dv is

$$\pm \partial\{(\partial\psi/\partial x)^2 - (\partial\psi/\partial y)^2 - (\partial\psi/\partial z)^2\}/\partial x . dv/8\pi$$
$$\pm \partial(\partial\psi/\partial x . \partial\psi/\partial y)/\partial y . dv/4\pi \pm \partial(\partial\psi/\partial x . \partial\psi/\partial z)/\partial z . dv/4\pi.$$

If we designate the components of force which act on the unit of volume by (X), (Y), and (Z) [XXV.], and if, for brevity, we set

$$X = -\partial\psi/\partial x, \quad Y = -\partial\psi/\partial y, \quad Z = -\partial\psi/\partial z,$$

we obtain

$$\text{(b)} \quad \begin{cases} (X) = \pm 1/8\pi . [\partial(X^2 - Y^2 - Z^2)/\partial x + 2\partial(XY)/\partial y + 2\partial(XZ)/\partial z], \\ (Y) = \pm 1/8\pi . [2\partial(XY)/\partial x + \partial(Y^2 - X^2 - Z^2)/\partial y + 2\partial(YZ)/\partial z], \\ (Z) = \pm 1/8\pi . [2\partial(XZ)/\partial x + 2\partial(YZ)/\partial y + \partial(Z^2 - X^2 - Y^2)/\partial z]. \end{cases}$$

Since these equations are perfectly analogous to those which determine the force with which stresses act on the unit of volume, we may consider forces acting at a distance as arising from stresses in the medium. If we are dealing with universal mass attraction, the ether may be assumed to be the intervening medium; if we are discussing electrical actions, the dependence of the stress in the ether on the matter which fills the region, air, water, etc., must be taken into account. It is not necessary to enter upon this question in our treatment of the subject.

A comparison of equation (b) with equation XXV. (b) shows that

$$\text{(c)} \quad \begin{cases} X_x = \pm(X^2 - Y^2 - Z^2)/8\pi, & Y_z = Z_y = \pm(YZ)/4\pi, \\ Y_y = \pm(Y^2 - X^2 - Z^2)/8\pi, & Z_x = X_z = \pm(XZ)/4\pi, \\ Z_z = \pm(Z^2 - X^2 - Y^2)/8\pi, & X_y = Y_x = \pm(XY)/4\pi. \end{cases}$$

To determine the principal stresses in the medium, we use equations XXVI. (e), which give

$$A+B+C=\mp(X^2+Y^2+Z^2)/8\pi,$$
$$BC+AC+AB=-\left((X^2+Y^2+Z^2)/8\pi\right)^2,$$
$$ABC=\pm\left((X^2+Y^2+Z^2)/8\pi\right)^3,$$

If we set (d) $(X^2+Y^2+Z^2)/8\pi=S$, A, B, and C are the roots of the equation $D^3\pm SD^2-S^2D\mp S^3=0$ or $(D\mp S)(D\pm S)^2=0$. We have therefore either (e) $A=+S$, $B=C=-S$ or $A=-S$, $B=C=+S$. Hence two principal stresses are always equal. In order to determine their directions, α, β, and γ must be calculated from XXVI. (c). It is easiest to determine the directions of the equal stresses B and C. If the values of $\pm S$, given in (d), are substituted for S in the equations referred to [XXVI. (c)], using the negative value of S in combination with the positive value of X_x, etc., and *vice versa*, we obtain (f) $X\cos\alpha+Y\cos\beta+Z\cos\gamma=0$. Hence both of the equal principal stresses are perpendicular to the direction of the force; the third principal stress is in the direction of the force, and is equal to the square of the force divided by 8π.

It has thus been shown that all forces acting at a distance may be explained by a state of stress in an intervening medium. From this point of view universal mass attraction is replaced by a negative stress, that is, a pressure, in the direction of the lines of force, and a positive stress, that is, a tension in all directions perpendicular to the force. A surface-element which lies perpendicular to the direction of the force is acted on by a tension which is equal to the force. In the case of magnetic and electrical attractions the opposite holds true. There is no independent evidence for the existence of such stresses in the case of gravity; but several phenomena in electricity indicate that the medium between two electrified bodies is in a state of stress, and no facts are known that are inconsistent with the assumption that this stress is the cause of the forces acting on the bodies.

SECTION XXVIII. DEFORMATION.

If a body changes its shape or its position in space, one of its points, whose coordinates are originally x, y, z, may be so displaced that its coordinates become $x+\xi$, $y+\eta$, $z+\zeta$. ξ, η, ζ are the projections of the path which P has traversed or the components of the displacement. If ξ, η, ζ are given as functions of the time, the position of the point P at any instant is determined. The motions of the separate points of the body are in general different, that is, ξ, η, ζ are functions of x, y, z. We will first consider some simple motions of the body.

If ξ, η, ζ are equal for all points of the body, the points all move through equal distances and in the same direction; the motion is a *translation*. In this motion all parts of the body remain at fixed distances from each other, and there are no internal forces developed. This holds also in the case of a *rotation* of the body about an axis. Let the axis of rotation be parallel to the x-axis, and pass through the point P (Fig. 41), whose coordinates are x, y, z. Let a point Q,

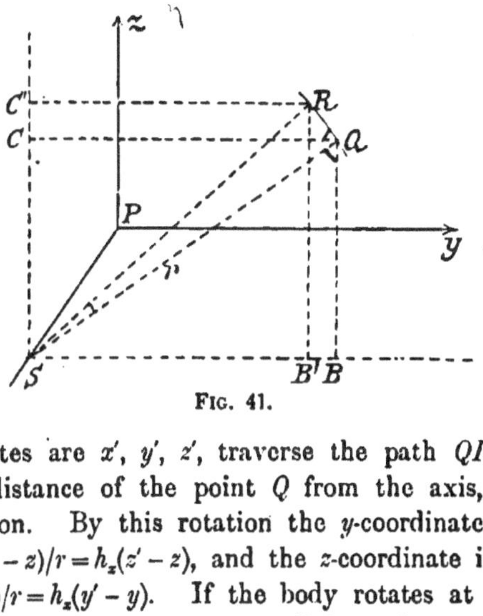

FIG. 41.

whose coordinates are x', y', z', traverse the path $QR = h_x . r$, where $r = QS$ is the distance of the point Q from the axis, and h_x is the angle of rotation. By this rotation the y-coordinate is diminished by $BB' = QR(z' - z)/r = h_x(z' - z)$, and the z-coordinate is increased by $CC' = QR(y' - y)/r = h_x(y' - y)$. If the body rotates at the same time about two other axes, which are parallel to the y- and z-axes, and if the angles of rotation are designated by h_y and h_z respectively, the coordinates of Q are increased by ξ, η, ζ, which have the following values:

$$\text{(a)} \qquad \begin{cases} \xi = (z' - z)h_y - (y' - y)h_z, \\ \eta = (x' - x)h_z - (z' - z)h_x, \\ \zeta = (y' - y)h_x - (x' - x)h_y. \end{cases}$$

We may now proceed to the discussion of the general case, in which the points of the body change their relative positions. Let the point P, whose coordinates are x, y, z, pass during this motion to the point P', whose coordinates are $x + \xi$, $y + \eta$, $z + \zeta$; let another point Q, whose coordinates are originally x', y', z', pass to the point Q, whose coordinates are $x'+\xi'$, $y'+\eta'$, $z'+\zeta'$. If ξ is a known function of x, y, z, we will have

$$\xi' = \xi + (x'-x)\partial\xi/\partial x + (y'-y)\partial\xi/\partial y + (z'-z)\partial\xi/\partial z + \ldots.$$

We may assume that P and Q are infinitely near, so that

$$x'-x = dx, \quad y'-y = dy, \quad z'-z = dz.$$

Neglecting terms of the second order we obtain the following relations,

$$\begin{aligned} \xi' &= \xi + \partial\xi/\partial x \,.\, dx + \partial\xi/\partial y \,.\, dy + \partial\xi/\partial z \,.\, dz, \\ \eta' &= \eta + \partial\eta/\partial x \,.\, dx + \partial\eta/\partial y \,.\, dy + \partial\eta/\partial z \,.\, dz, \\ \zeta' &= \zeta + \partial\zeta/\partial x \,.\, dx + \partial\zeta/\partial y \,.\, dy + \partial\zeta/\partial z \,.\, dz. \end{aligned}$$

By introducing the following notation,

$$\text{(b)} \quad \begin{cases} x_x = \partial\xi/\partial x; & z_y = y_z = \tfrac{1}{2}(\partial\zeta/\partial y + \partial\eta/\partial z); & h_x = \tfrac{1}{2}(\partial\zeta/\partial y - \partial\eta/\partial z); \\ y_y = \partial\eta/\partial y; & x_z = z_x = \tfrac{1}{2}(\partial\xi/\partial z + \partial\zeta/\partial x); & h_y = \tfrac{1}{2}(\partial\xi/\partial z - \partial\zeta/\partial x); \\ z_z = \partial\zeta/\partial z; & y_x = x_y = \tfrac{1}{2}(\partial\eta/\partial x + \partial\xi/\partial y); & h_z = \tfrac{1}{2}(\partial\eta/\partial x - \partial\xi/\partial y); \end{cases}$$

we obtain

$$\text{(c)} \quad \begin{cases} \xi' = \xi + x_x dx + x_y dy + x_z dz - h_z dy + h_y dz, \\ \eta' = \eta + y_x dx + y_y dy + y_z dz - h_x dz + h_z dx, \\ \zeta' = \zeta + z_x dx + z_y dy + z_z dz - h_y dx + h_x dy. \end{cases}$$

These equations determine the motion of a point in the neighbourhood of P. This motion is compounded of a translation, whose components are ξ, η, ζ, a rotation, whose components are h_x, h_y, h_z, and two motions, determined by x_x, y_y, z_z and z_y, x_z, y_x. If we confine our attention to the way in which the form of the body changes, we need only consider the motion whose components $d\xi$, $d\eta$, $d\zeta$ are determined by the following equations:

$$\text{(d)} \quad \begin{cases} d\xi = x_x dx + x_y dy + x_z dz, \\ d\eta = y_y dy + y_x dx + y_z dz, \\ d\zeta = z_z dz + z_x dx + z_y dy. \end{cases}$$

To interpret the coefficients x_x, y_y, z_z and z_y, x_z, y_x, we assume that all except x_x are equal to zero. Then $d\xi = x_x \,.\, dx$ and $d\eta = d\zeta = 0$. The change of form corresponding to this is a *dilatation* of the body in the direction of the x-axis, by which dx increases by $d\xi$. The coefficient x_x therefore represents the dilatation of a unit of length parallel to the x-axis, or is the dilatation in the direction of the x-axis. Hence y_y and z_z are the dilatations in the directions of the y- and z-axis respectively.

If, on the other hand, all the coefficients vanish with the exception of z_y, we have $d\xi = 0$, $d\eta = z_y . dz$, $d\zeta = z_y . dy$. The particles are displaced in a plane parallel to the yz-plane, and their distances from the yz-plane remain unchanged. Let the original coordinates of the point P (Fig. 42) be x, y, z; let $ABCD$ be a square, the length of whose sides is $2a$. The point A, whose original coordinates were x, $y+a$, $z+a$, referred to the axes PY and PZ, is displaced to A', whose coordinates are $a + z_y a$, $a + z_y a$. A' therefore lies on PA produced. The points B and D are displaced to B' and D', which lie on BD; C is displaced along AC produced to C'. The square $ABCD$ becomes the rhombus $A'B'C'D'$. This change of form is called a *shear*; the quantities z_y, x_z, y_x are called *components of shear*.

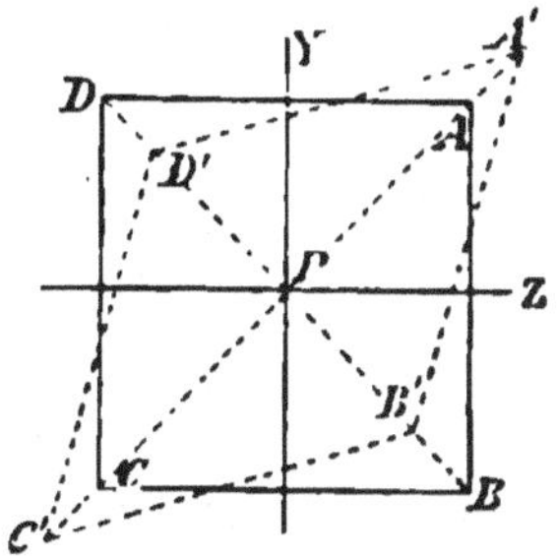

FIG. 42.

In the theory of elasticity we consider only very small deformations of the body; the components x_z, y_y, etc., are consequently small quantities, whose second and higher powers may be neglected. The volume of the body is not changed by a shear; the square whose area is $4a^2$ will become a rhombus $A'B'C'D'$ whose area is

$$2PA' . PB' = 2(a + z_y a)\sqrt{2}(a - z_y a)\sqrt{2} = 4a^2(1 - z_y^2).$$

If we neglect z_y^2, the area of the square is equal to that of the rhombus; hence the volume will not be changed by the shear.

From Fig. 42 it is evident that the infinitely small angle between AB and $A'B'$ is equal to $az_y/a = z_y$; hence the right angle DAB is diminished by the shear by $2z_y$, so that

$$\angle D'A'B' = \angle DAB - 2z_y.$$

As the result of a dilatation determined by x_x, y_y, z_z, the volume of the parallelepiped $dxdydz$ becomes $dxdydz(1+x_x)(1+y_y)(1+z_z)$. If the components of dilatation are supposed infinitely small, we may neglect their second and higher powers. Hence the increase in unit volume is $\Theta = x_x + y_y + z_z$. Θ is called the *volume dilatation*. Substituting the values of x_x, y_y, z_z, we have also

(e) $$\Theta = \partial\xi/\partial x + \partial\eta/\partial y + \partial\zeta/\partial z.$$

Let dr be an element of a straight line which makes the angles α, β, γ, with the coordinate axes; then

$$dx = dr\cos\alpha, \quad dy = dr\cos\beta, \quad dz = dr\cos\gamma.$$

By the deformation dr becomes dr', and makes the angles α', β', γ' with the axis, so that

$$dx + d\xi = dr' \cos\alpha'; \quad dy + d\eta = dr' \cos\beta'; \quad dz + d\zeta = dr' \cos\gamma',$$

from which $d\xi$, $d\eta$, $d\zeta$ may be determined by equations (d). If the direction of the line dr remains unchanged, we have $\alpha = \alpha'$, $\beta = \beta'$, and $\gamma = \gamma'$, and hence $d\xi = d\rho \cos\alpha$, $d\eta = d\rho \cos\beta$, $d\zeta = d\rho \cos\gamma$, where $d\rho = d(r' - r)$. The length $d\rho$ is the elongation of dr, and $d\rho/dr$ is the dilatation s in the direction of the line dr. Hence we have

$$s = d\rho/dr.$$

Equations (d) then assume the following form:

$$\text{(f)} \qquad \begin{cases} (x_x - s)\cos\alpha + x_y \cos\beta + x_z \cos\gamma = 0, \\ y_x \cos\alpha + (y_y - s)\cos\beta + y_z \cos\gamma = 0, \\ z_x \cos\alpha + z_y \cos\beta + (z_z - s)\cos\gamma = 0. \end{cases}$$

A comparison of these relations with those of XXVI. (c) shows that they both may be interpreted in a similar way.

There are therefore three directions perpendicular to each other, called the *principal axes of dilatation*, in which only dilatations occur; every line-element which is parallel to one of these three directions contains after deformation the particles which were in it before the deformation. This conclusion holds only on the supposition that the body does not rotate, a supposition which has been made in deducing equations (d). If the principal dilatations thus determined are called a, b, c, we have, as in XXVI. (e),

$$\text{(g)} \qquad \begin{cases} a + b + c = x_x + y_y + z_z, \\ bc + ac + ab = z_z y_y + x_x z_z + y_y x_x - z_y^2 - x_z^2 - y_x^2, \\ abc = x_x y_y z_z + 2 z_y x_z y_x - x_x z_y^2 - y_y x_z^2 - z_z y_x^2. \end{cases}$$

The first of these equations shows that *the volume dilatation does not depend on the position of the system of coordinates.*

In the same way as that in which the components of stress are expressed in terms of the principal stresses [XXVI. (i)] x_x, x_y,... may be expressed in terms of the principal dilatations a, b, and c. Denoting the cosines of the angles which the direction of a makes with the axes by l_1, m_1, n_1, and the cosines of the angles which b and c make with the axes by l_2, m_2, n_2; l_3, m_3, n_3, we obtain

$$\text{(h)} \qquad \begin{cases} x_x = al_1^2 + bl_2^2 + cl_3^2; & z_y = am_1 n_1 + bm_2 n_2 + cm_3 n_3, \\ y_y = am_1^2 + bm_2^2 + cm_3^2; & x_z = al_1 n_1 + bl_2 n_2 + cl_3 n_3, \\ z_z = an_1^2 + bn_2^2 + cn_3^2; & y_x = al_1 m_1 + bl_2 m_2 + cl_3 m_3. \end{cases}$$

SECTION XXIX. RELATIONS BETWEEN STRESSES AND DEFORMATIONS.

The study of the deformations of an elastic body has shown that a parallelepiped which is stretched by forces applied to its ends, is increased in length and diminished in cross section. If we only consider forces which are so small that the limits of elasticity are not exceeded, the elongation s per unit of length is $s = S/E$, where E is the *coefficient of elasticity* and S the force acting on the unit of surface. The contraction s' per unit of length parallel to the end surfaces, is given by $s' = k \cdot S/E$, where k is a constant. It is assumed that the body is isotropic, that is, equally elastic in all directions and at all points.

We will first consider a rectangular parallelepiped, whose edges are parallel with the coordinate axes. The normal forces are denoted by X_x, Y_y, Z_z and a unit of length which is parallel to the x-axis increases by x_x; the units of length which are parallel to the y- and z-axes respectively increase by y_y and z_z. We then have

$$x_x = X_x/E - k(Y_y + Z_z)/E,$$
$$y_y = Y_y/E - k(X_x + Z_z)/E,$$
$$z_z = Z_z/E - k(X_x + Y_y)/E.$$

From XXVIII. (c) the volume dilatation Θ is

$$\Theta = x_x + y_y + z_z = (1 - 2k)(X_x + Y_y + Z_z)/E.$$

From these equations we deduce

$$X_x = kE\Theta/(1+k)(1-2k) + Ex_x/(1+k).$$

Setting $\lambda = kE/(1+k)(1-2k), \quad \mu = \frac{1}{2}E/(1+k),$

we obtain (a) $X_x = \lambda\Theta + 2\mu x_x;\ \ Y_y = \lambda\Theta + 2\mu y_y;\ \ Z_z = \lambda\Theta + 2\mu z_z,$

and by addition

(b) $$X_x + Y_y + Z_z = (3\lambda + 2\mu)\Theta.$$

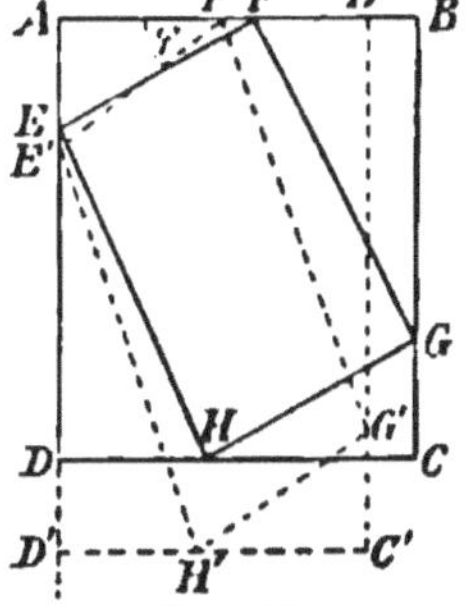

FIG. 43.

To investigate the relation between the shears and the tangential forces we may use the following method, due to V. v. Lang.* If the prism $ABCD$ (Fig. 43) is stretched by the tension S applied to each unit of surface of its ends AB and CD, it takes the form $AB'C'D'$. Four plane sections EF, FG, GH, and HE are passed through the prism, which mark out the rectangle $EFGH$ on a plane parallel to the axis; the rectangle $EFGH$ becomes by deformation

* V. v. Lang, *Theoretische Physik*, § 411.

the parallelogram $E'F'G'H'$. The angle AFE is represented by ϕ. The tangential stress T, which acts on the surface EF in the direction EF, is given [XXIII. (b)] by $T = S \sin\phi \cos\phi$.

Since $\angle BFG = \frac{1}{2}\pi - \phi$, the same tangential stress T acts on GF in the direction GF. On deformation the angle AFE becomes $AF'E' = \phi + d\phi$, and we have

$$\operatorname{tg}(\phi + d\phi) = AE\,.\,(1+s)/AF\,.\,(1-s') = \operatorname{tg}\phi(1+s)/(1-s').$$

Now since $s = S/E$ and $s' = kS/E$ are infinitely small, we have

$$(1+s)/(1-s') = 1 + s + s' = 1 + (1+k)S/E.$$

Further, we have

$$\operatorname{tg}(\phi + d\phi) = \big(1 + (1+k)S/E\big)\operatorname{tg}\phi$$

and

$$\operatorname{tg}(\phi + d\phi) = \operatorname{tg}\phi + d\phi/\cos^2\phi,$$

so that

$$d\phi = (1+k)S \sin\phi \cos\phi/E = (1+k)T/E.$$

Hence the change of the angle ϕ is proportional to the tangential stress T. Since the same tangential stress acts on GF as on EF, the angle BFG increases by $d\phi$, the angle EFG diminishes by $2d\phi$, and the angle FGH increases by $2d\phi$. The shear is thus equal to $2d\phi$, and we have $2d\phi = 2(1+k)T/E$. But $2d\phi$ is the quantity $2z_y$ previously introduced, when the rectangle $EFGH$ is parallel to the yz-plane; and hence $T = Z_y$ and $z_y = (1+k)Z_y/E$. If we set

$$\mu = \tfrac{1}{2}E/(1+k),$$

we have (c) $\quad Z_y = 2\mu z_y, \quad X_z = 2\mu x_z, \quad Y_x = 2\mu y_x.$

The equations (a) and (c) are the solution of the problem, to find the components of stress, when the deformations are given, and conversely. They contain only two constants, λ and μ, which involve the deformations caused by simple dilatation in the following way:

(d) $$\begin{cases} \lambda = kE/(1+k)(1-2k)\,; & \mu = \tfrac{1}{2}E/(1+k), \\ E = (3\lambda\mu + 2\mu^2)/(\lambda+\mu)\,; & k = \tfrac{1}{2}\lambda/(\lambda+\mu). \end{cases}$$

Since λ and μ are positive, k must be less than $\frac{1}{2}$.

The relations between the elastic forces and the deformations may also be derived by another method. Let the principal stresses A, B, and C, at the point P, be known in magnitude and direction [cf. XXVI. (g)]. An infinitely small parallelepiped, whose edges are parallel to the directions of the stresses A, B, and C, is extended in those three directions. The increments a, b, and c of the unit of length are parallel to A, B, and C, and as in (a), we have

(e) $\quad A = \lambda\Theta + 2\mu a, \quad B = \lambda\Theta + 2\mu b, \quad C = \lambda\Theta + 2\mu c,$

when $\Theta = a + b + c$, or [XXVIII. (g)], $\Theta = x_x + y_y + z_z$.

By applying the formula XXVI. (i), we obtain the equation

$$X_x = \lambda\Theta + 2\mu(al_1^2 + bl_2^2 + cl_3^2),$$

which, from XXVIII. (h), becomes $X_x = \lambda\Theta + 2\mu x_x$. The expressions for Y_y and Z_z are obtained in a similar way.

From XXVI. (i), we have

$$Z_y = 2\mu(am_1n_1 + bm_2n_2 + cm_3n_3),$$

and hence [XXVIII. (h)] $Z_y = 2\mu z_y$. We obtain the expressions for X_z and Y_x in a similar way.

The coefficients E and k depend on the nature of the body. It was at one time believed that k had the same value for all bodies. This opinion was first expressed by Navier. He assumed that bodies are made up of material points which repel one another, and on this assumption concluded that $k = \frac{1}{4}$. Poisson also had the same opinion.

While k is a mere number, the coefficient of elasticity E is determined by $E = S/s$; the fraction $1/E$ is called the *modulus of elasticity*. S is the force which acts on the unit of surface, and [III.] its dimensions are $LT^{-2}M/L^2 = L^{-1}T^{-2}M$. Since s is the ratio between the elongation and the original length, it is also a mere number. Hence the dimensions of E are $L^{-1}T^{-2}M$.

In practical units E denotes the number of kilograms which would produce an elongation in a rod of one square millimetre cross section, such that its length is doubled. In order to transform it into absolute measure, we notice that the weight of one gram is about equal to 981 dynes, and that therefore the weight of one kilogram is equal to 981 000 dynes. The cross section must be taken equal to 1 sq. cm., and the number must therefore be multiplied by 100, so that the factor of transformation becomes 98,100,000. According to Wertheim, E equals 17278 in practical units for English steel; therefore, in absolute units it equals $17278 \,.\, 981 \,.\, 10^5 = 1{,}695 \,.\, 10^{12}$.

In the case of fluids, the discussion is simplified by the condition that a fluid always yields to tangential forces, so that, when it is in equilibrium, there are no tangential forces acting in it. This condition, from (c), enables us to set $\mu = 0$. If the fluid is subjected to the pressure p, and if its volume v is thereby diminished by dv, we have from (b)

$$\text{(f)} \qquad \begin{cases} -3p = -(3\lambda + 2\mu)dv/v, \\ \text{or, since } \mu = 0, \\ dv = pv/\lambda. \end{cases}$$

If, for example, the unit volume of water is diminished by 0,000 046, when the pressure is increased by 1 atmosphere, we have

$$\lambda = pv/dv = 76 \,.\, 13{,}596 \,.\, 981/0{,}000{,}046 = 2{,}204 \,.\, 10^{10}.$$

In the case of gases, if we represent the original pressure by P, and its increase by p, Mariotte's law gives the equation

$$Pv = (P+p)(v-dv).$$

Assuming that p is very small in comparison with P, we obtain

$$dv = pv/P,$$

and therefore, for gases, we have from (f), (g) $P = \lambda$.

Section XXX. Conditions of Equilibrium of an Elastic Body.

If a force whose components are X, Y, Z acts on the unit of mass of a body, we have from XXV. (c)

(a) $$\partial X_x/\partial x + \partial X_y/\partial y + \partial X_z/\partial z + \rho X = 0, \text{ etc.}$$

Further [XXIX. (a) and (c)] we have

(b) $$X_x = \lambda\Theta + 2\mu \,.\, \partial\xi/\partial x, \quad Z_y = Y_z = \mu(\partial\zeta/\partial y + \partial\eta/\partial z), \text{ etc.}$$

If the values for X_x, etc., are substituted in (a), it follows that

(c) $$\begin{cases} (\lambda+\mu) \,.\, \partial\Theta/\partial x + \mu\nabla^2\xi + \rho X = 0, \\ (\lambda+\mu) \,.\, \partial\Theta/\partial y + \mu\nabla^2\eta + \rho Y = 0, \\ (\lambda+\mu) \,.\, \partial\Theta/\partial z + \mu\nabla^2\zeta + \rho Z = 0. \end{cases}$$

By the use of our former symbols for the components of rotation, viz.,

(d) $$\begin{cases} 2h_x = \partial\zeta/\partial y - \partial\eta/\partial z, \quad 2h_y = \partial\xi/\partial z - \partial\zeta/\partial x, \\ 2h_z = \partial\eta/\partial x - \partial\xi/\partial y, \end{cases}$$

equations (c) become

(e) $$\begin{cases} (\lambda+2\mu) \,.\, \partial\Theta/\partial x + 2\mu(\partial h_y/\partial z - \partial h_z/\partial y) + \rho X = 0, \\ (\lambda+2\mu) \,.\, \partial\Theta/\partial y + 2\mu(\partial h_z/\partial x - \partial h_x/\partial z) + \rho Y = 0, \\ (\lambda+2\mu) \,.\, \partial\Theta/\partial z + 2\mu(\partial h_x/\partial y - \partial h_y/\partial x) + \rho Z = 0. \end{cases}$$

If the first equation is differentiated with respect to x, the second with respect to y, and the third with respect to z, we have by addition when ρ is constant,

$$(\lambda+2\mu)\nabla^2\Theta + \rho(\partial X/\partial x + \partial Y/\partial y + \partial Z/\partial z) = 0.$$

If X, Y, and Z are the derivatives of a potential ψ, and if $\nabla^2\psi = 0$ everywhere within the body, then (f) $\nabla^2\Theta = 0$.

This result must be supplemented by the conditions of equilibrium of the surface of the body. The force acting on the surface-element dS, whose components are $P'Q'R'$, is in equilibrium with the elastic forces which act on the parts of the body contiguous to dS. If X_x', Y_y', etc., are the components of the elastic forces, we have

(g) $$P' = X_x'\cos\alpha + X_y'\cos\beta + X_z'\cos\gamma.$$

There are similar values for Q' and R'. The symbols α, β, γ represent the angles which the normal to the surface directed outward makes with the coordinate axes.

We assume (h) $\xi = ax$, $\eta = by$, $\zeta = cz$, where a, b, and c are constants. ξ, η, and ζ are therefore linear functions of x, y, and z, and ξ depends only on x, η only on y, and ζ only on z. On this supposition [XXVIII. (b)] the deformation of the body is made up of dilatations only. The volume dilatation is $\Theta = a + b + c$; hence the values assumed for ξ, η, ζ satisfy equations (c), if we neglect the action of external forces. We have further

$$X_x = \lambda(a+b+c) + 2\mu a; \quad Y_y = \lambda(a+b+c) + 2\mu b; \quad Z_z = \lambda(a+b+c) + 2\mu c.$$

$$Y_z = 0, \quad Z_x = 0, \quad X_y = 0.$$

If we set $$X_x = S, \quad Y_y = 0, \quad Z_z = 0,$$

we have $b = c$ and

$$S = \lambda(a+2b) + 2\mu a; \quad 0 = \lambda(a+2b) + 2\mu b.$$

The last equation gives

$$b/a = -\tfrac{1}{2}\lambda/(\lambda+\mu) = -k,$$

and the first $$S = a(3\lambda\mu + 2\mu^2)/(\lambda+\mu) = Ea.$$

The equations thus obtained give the law of the expansion of an elastic prism.

Section XXXI. Stresses in a Spherical Shell.

Suppose a spherical shell, bounded by two concentric spheres, whose radii are r_1 and r_2. Of these we assume $r_2 > r_1$. Suppose a constant hydrostatic pressure p_1 applied to the inner surface, and a similar pressure p_2 applied to the outer surface. The pressures p_1 and p_2 are perpendicular to the surfaces. Let the centre O of the sphere be the origin of coordinates, and let the distance from O of any point in the shell be r. On the hypothesis that has been made with respect to the pressures, all points lying in the same spherical surface having the centre O, receive equal displacements from the centre. Let the displacement of the point considered be ϵr, where ϵ is a very small quantity. We then have

(a) $$\xi = \epsilon x, \quad \eta = \epsilon y, \quad \zeta = \epsilon z.$$

Since ϵ is a function of r only, we may set

$$\xi = \epsilon r \,.\, x/r = d\phi/dr \,.\, \partial r/\partial x = \partial\phi/\partial x,$$

where ϕ is a new function of r. We may represent η and ζ in a similar way, so that

(b) $$\xi = \partial\phi/\partial x, \quad \eta = \partial\phi/\partial y, \quad \zeta = \partial\phi/\partial z.$$

Hence we have (c) $$\Theta = \nabla^2\phi.$$

The equations XXX. (c), if the action of gravity is neglected, become

$$(\lambda + 2\mu) \,.\, \nabla^2\partial\phi/\partial x = 0, \quad (\lambda + 2\mu) \,.\, \nabla^2\partial\phi/\partial y = 0, \quad (\lambda + 2\mu) \,.\, \nabla^2\partial\phi/\partial z = 0,$$

so that (d) $\Theta = \nabla^2\phi = a$, where a is a constant.

From XXX. (b) the components of stress are

(e) $$\begin{cases} X_x = \lambda a + 2\mu \,.\, \partial^2\phi/\partial x^2 \,; & Z_y = 2\mu \,.\, \partial^2\phi/\partial y\partial z \,; \\ Y_y = \lambda a + 2\mu \,.\, \partial^2\phi/\partial y^2 \,; & X_z = 2\mu \,.\, \partial^2\phi/\partial x\partial z \,; \\ Z_z = \lambda a + 2\mu \,.\, \partial^2\phi/\partial z^2 \,; & Y_x = 2\mu \,.\, \partial^2\phi/\partial x\partial y. \end{cases}$$

The stress in a surface-element perpendicular to r is given by XXIV. (a), if we set

$$\cos\alpha = x/r, \quad \cos\beta = y/r, \quad \cos\gamma = z/r.$$

If the components of stress are P, Q, and R, we have

$$P = \lambda a \,.\, x/r + 2\mu(x/r \,.\, \partial^2\phi/\partial x^2 + y/r \,.\, \partial^2\phi/\partial x\partial y + z/r \,.\, \partial^2\phi/\partial x\partial z).$$

Using the equations

$$\partial^2\phi/\partial x^2 = x^2/r^2 \,.\, d^2\phi/dr^2 - x^2/r^3 \,.\, d\phi/dr + 1/r \,.\, d\phi/dr,$$

$$\partial^2\phi/\partial x\partial y = xy/r^2 \,.\, d^2\phi/dr^2 - xy/r^3 \,.\, d\phi/dr,$$

$$\partial^2\phi/\partial x\partial z = xz/r^2 \,.\, d^2\phi/dr^2 - xz/r^3 \,.\, d\phi/dr,$$

we have $$P = (\lambda a + 2\mu \,.\, d^2\phi/dr^2) \,.\, x/r.$$

Similar expressions may be obtained for Q and R. Hence a principal stress (f) $A = \lambda a + 2\mu \,.\, d^2\phi/dr^2$ acts on the surface-element considered.

For a surface-element which contains r, the components are obtained in the same way. If α, β, γ are the angles which the normal to the surface-element makes with the axes, we have

$$P = \lambda a \cos\alpha + 2\mu(\partial^2\phi/\partial x^2 \,.\, \cos\alpha + \partial^2\phi/\partial x\partial y \,.\, \cos\beta + \partial^2\phi/\partial x\partial z \,.\, \cos\gamma).$$

If we notice that in this case

$$\cos\alpha \,.\, x/r + \cos\beta \,.\, y/r + \cos\gamma \,.\, z/r = 0,$$

and use the expressions given above for the differential coefficients, we have $$P = (\lambda a + 2\mu/r \,.\, d\phi/dr)\cos\alpha.$$

We may obtain Q and R by replacing α by β and γ respectively.

Hence the principal stress B acting on the element is

(g) $$B = \lambda a + 2\mu/r \,.\, d\phi/dr.$$

From (d) and XV. (1) we have

$$\nabla^2\phi = 1/r \,.\, d^2(r\phi)/dr^2 = a,$$

and therefore

(h) $$d\phi/dr = \tfrac{1}{3}ar + b/r^2; \quad d^2\phi/dr^2 = \tfrac{1}{3}a - 2b/r^3.$$

From (f) and (g) it follows that

$$A = (\lambda + \tfrac{2}{3}\mu)a - 4\mu b/r^3; \quad B = (\lambda + \tfrac{2}{3}\mu)a + 2\mu b/r^3.$$

For $r = r_1$, $A = -p_1$, and for $r = r_2$, $A = -p_2$, therefore

$$a = 3/(3\lambda + 2\mu) \,.\, (p_1 r_1^3 - p_2 r_2^3)/(r_2^3 - r_1^3),$$

$$b = 1/4\mu \,.\, (p_1 - p_2) r_1^3 \,.\, r_2^3/(r_2^3 - r_1^3),$$

and $$A = (p_1 r_1^3 - p_2 r_2^3)/(r_2^3 - r_1^3) - (p_1 - p_2) r_1^3 r_2^3/(r_2^3 - r_1^3) \,.\, 1/r^3,$$

$$B = (p_1 r_1^3 - p_2 r_2^3)/(r_2^3 - r_1^3) + \tfrac{1}{2}(p_1 - p_2) r_1^3 r_2^3/(r_2^3 - r_1^3) \,.\, 1/r^3.$$

SECTION XXXII. TORSION.

Let us consider a circular cylinder whose axis coincides with the z-axis; and let the circle in which the xy-plane cuts the cylinder be the end of the cylinder and be fixed in position. If torsion is applied to the cylinder, a point at the distance r from the axis describes an arc $r\phi$, parallel to the xy-plane, whose centre lies on the z-axis. This angle, in the case of pure torsion, is proportional to the distance of the point from the xy-plane, so that $\phi = kz$, where k is a constant. The displacement of this point is krz, and its components ξ, η, ζ are

(a) $$\xi = -kyz, \quad \eta = kxz, \quad \zeta = 0.$$

Using these values, we find that the volume-dilatation Θ is zero, that is, *pure torsion does not cause a change of volume.* We have further [XXX. (b)], $$X_x = 0, \quad Y_y = 0, \quad Z_z = 0,$$

and hence no normal forces act on the surfaces which are parallel to the coordinate planes. On the other hand, we have

$$Z_y = \mu kx, \quad X_z = -\mu ky, \quad Y_x = 0.$$

A surface-element perpendicular to the z-axis is acted on by the tangential forces $Y_z = +\mu kx$ and $X_z = -\mu ky$, whose resultant μkr is perpendicular to the radius r and to the z-axis.

By XXIV. (a) we reach the same result. That is, we get

(b) $$\begin{cases} P = -\mu ky \cos\gamma, \quad Q = \mu kx \cos\gamma, \\ R = -\mu ky \cos a + \mu kx \cos\beta. \end{cases}$$

For the stress on the surface of the cylinder we must set

$$\cos\alpha = x/r, \quad \cos\beta = y/r, \quad \cos\gamma = 0.$$

We will then have $\qquad P=0, \quad Q=0, \quad R=0.$

Hence a surface-element perpendicular to the radius, or which is part of the surface of a circular cylinder whose axis is the z-axis, is not acted on by a force.

To find the surface-elements on which the only forces which act are normal forces, we use equation XXVI. (d), which, in the case before us, becomes

$$S^3 - \mu^2 k^2 r^2 S = 0.$$

If A, B, and C are the roots of this equation, we can set

$$A = 0, \quad B = \mu kr, \quad C = -\mu kr.$$

If the angles between the axes and the normal to one of these surface-elements are represented by α, β, γ,

$$S\cos\alpha = -\mu ky\cos\gamma, \quad S\cos\beta = \mu kx\cos\gamma, \quad S\cos\gamma = -\mu ky\cos\alpha + \mu kx\cos\beta.$$

If we substitute in these equations the particular values of S given by A, B, and C, the values of α, β, γ thus obtained show that the stress $A=0$ acts on a surface-element perpendicular to r; and that B and C act in directions perpendicular to the radius r, and making angles of 45 degrees with the z-axis. B acts in the same direction as the torsion, C in the opposite direction.

For example, considering a point which lies in the surface of the cylinder and in the z-plane, and setting therefore $Y=0$ and $X=r$, we have $\quad S\cos\alpha = 0, \quad S\cos\beta = \mu kr\cos\gamma, \quad S\cos\gamma = \mu kr\cos\beta.$
When $S=0$ we have $\gamma = \beta = \frac{1}{2}\pi$; when $S = \pm\mu kr$ we have $\alpha = \frac{1}{2}\pi$, $\cos\beta = \cos\gamma$. Since $\cos^2\alpha + \cos^2\beta + \cos^2\gamma = 1$, we have $\cos\beta = \pm\sqrt{\frac{1}{2}}$.

The moment of force M to which the torsion of the cylinder is due is

$$M = \int_0^R \mu kr \,.\, 2\pi r dr \,.\, r.$$

The upper limit R of the integral is the radius of the cylinder. Integrating, we have $M = \frac{1}{2}\pi\mu k R^4 = \pi\phi\mu R^4/2l$, where l is the length of the cylinder and ϕ the *angle of torsion.*

The factor $\tau = \pi\mu R^4/2l$ is called the *moment of torsion* of the cylinder. It depends only on the dimensions of the cylinder and the constant of elasticity μ. For this reason μ is called the *coefficient of torsion.*

SECTION XXXIII. FLEXURE.

It is not possible to give a rigorous discussion of the flexure of a prism. We will, therefore, confine ourselves to an approximate calculation in one very simple case*.

Let *ABCD* (Fig. 44) be the prism considered. Its length is supposed horizontal and coincident with the axis Ox. The axis Oz is directed perpendicularly upward, and the axis Oy is therefore horizontal. After flexure, the cross section *AB* is displaced to *A'B'*, which may lie in the same plane as *AB*. Another plane cross section *FG*, also perpendicular to the axis, is displaced by the flexure to *F'G'*; we assume that the section *F'G'* is also plane, and that the plane *F'G'* cuts the plane *A'B'* in a horizontal line passing through *P*. This line of intersection is supposed to be common to the planes of all sections perpendicular to the axis. The parts of the prism which originally lay in *OQ* lie after the flexure in *OQ'*, which we will consider as the arc of a circle whose centre is *P*. Such a flexure is called circular. All the lines in the prism which were originally parallel to the x-axis become circles, whose centres lie on the straight line passing through *P*.

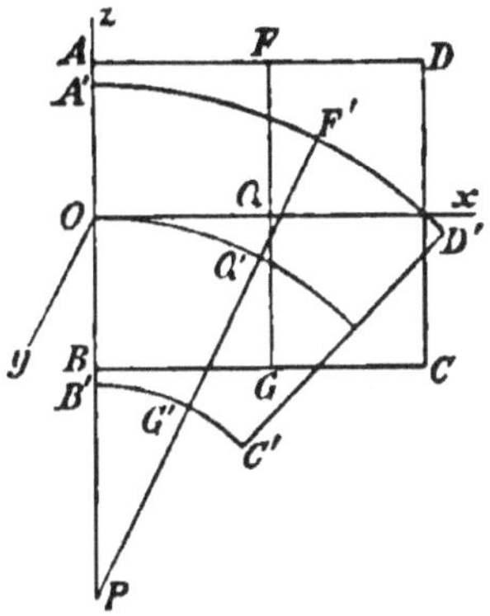

FIG. 44.

Represent the original coordinates of a point M in the section *AB* by $0, y, z$, and its coordinates after flexure by $0, y+\eta_0, z+\zeta_0$. The same changes occur in the other cross sections, for example in *FG*. If the coordinates of a point M' in *FG* are originally x, y, z, they will become by flexure $x+\xi$, $y+\eta$, $z+\zeta$. We set $\angle OPQ' = \phi$, $OP = \rho$ and $OQ = OQ'$. This last assumption is admissible, since there is always one line whose length does not change by flexure, and since we have as yet made no assumption as to the position of the x-axis. We therefore obtain

$$x+\xi = (\rho + z + \zeta_0)\sin\phi,$$
$$y+\eta = y+\eta_0,$$
$$z+\zeta = z+\zeta_0 - (\rho+z+\zeta_0)(1-\cos\phi).$$

If ρ is very great in comparison with x, z, and ζ_0, we may set

$$\sin\phi = x/\rho; \quad 1-\cos\phi = x^2/2\rho^2$$

and obtain (a) $\quad \xi = xz/\rho, \quad \eta = \eta_0, \quad \zeta = \zeta_0 - x^2/2\rho.$

* Barré de Saint-Venant, Mem. prés. par div. Savants. T. 14. Paris, 1856.

We may so determine η_0 and ζ_0 that all the components of stress except X_x vanish; hence we may write

(1) $X_x = \lambda\Theta + 2\mu z/\rho = S,$ (4) $Z_y = \mu(\partial\zeta_0/\partial y + \partial\eta_0/\partial z) = 0,$

(2) $Y_y = \lambda\Theta + 2\mu\partial\eta_0/\partial y = 0,$ (5) $X_z = \mu\partial\zeta_0/\partial x = 0,$

(3) $Z_z = \lambda\Theta + 2\mu\partial\zeta_0/\partial z = 0,$ (6) $Y_x = \mu\partial\eta_0/\partial x = 0.$

Further, we have $\Theta = z/\rho + \partial\eta_0/\partial y + \partial\zeta_0/\partial z.$

From (2) and (3) it follows that (b) $\partial\eta_0/\partial y = \partial\zeta_0/\partial z = -\frac{1}{2}\lambda/(\lambda+\mu) \,.\, z/\rho.$ Comparing (b) with XXIX. (d) it appears that the contraction of the cross section is to the increase in length in the ratio of $\frac{1}{2}\lambda$ to $\lambda+\mu$, or of k to 1. Further, since η_0 and ξ_0 do not involve x, we have from (b) $\eta_0 = -kyz/\rho + f(z), \quad \zeta_0 = -kz^2/2\rho + g(y),$

when f and g designate two unknown functions. From (4) we have $-ky/\rho + f'(z) + g'(y) = 0$, and hence $f'(z) = c$, where c is an unknown constant. It follows that

$$f(z) = cz + c', \quad g(y) = ky^2/2\rho - cy + c''$$

and $\eta_0 = -kyz/\rho + cz + c', \quad \zeta_0 = k(y^2 - z^2)/2\rho - cy + c''.$

At the point O, where $y = z = 0$, we have $\eta_0 = 0$, $\zeta_0 = 0$, and hence $c' = 0$ and $c'' = 0$. Since the prism does not turn about the x-axis during flexure, it follows that for $y = 0$, $\eta_0 = 0$ also, and consequently $c = 0$. We obtain therefore

(c) $$\eta_0 = -kyz/\rho, \quad \zeta_0 = k(y^2 - z^2)/2\rho,$$

and further, from (a),

(d) $$\xi = xz/\rho, \quad \eta = -kyz/\rho, \quad \zeta = k(y^2 - z^2)/2\rho - x^2/2\rho.$$

These values for ξ, η, and ζ satisfy the equations XXX. (c), since by hypothesis $X = Y = Z = 0$. The equations 1–6 show that the conditions of equilibrium are fulfilled. From (1) and (b) we get $X_x = S = (3\lambda\mu + 2\mu^2)/(\lambda+\mu) \,.\, z/\rho$. If we introduce the general coefficient of elasticity E [XXIX. (d)], we have (e) $S = Ez/\rho$.

The resultant R of the forces S is (f) $R = E/\rho \,.\, \int z \,.\, dydz$, and is equal to zero, if the x-axis passes through the centre of gravity of the prism. If we assume this and then determine the moment M of the forces S with respect to a horizontal line passing through the centre of gravity, we will have $M = \int Szdydz = E/\rho \,.\, \int z^2 dydz = EJ/\rho$, where J is the moment of inertia of the cross section. In order to bend the prism so that an axis passing through the centre of gravity of the prism becomes a circle of radius ρ, a rotating force of moment M must act on each end surface; the axes of the rotating forces are perpendicular to the plane of the circle, and are oppositely directed.

The cross section of the prism is noticeably altered by the flexure. Since the parts on the convex side of the prism are extended, and the parts on the concave side compressed, the former tend to contract in the directions of the y and z-axes, the latter to expand. If, for example, the cross section is a rectangle $ABCD$, as in Fig. 45, $ABCD$ takes the form $A'B'C'D'$. The two plane surfaces whose projections are represented in the figure by AB and CD are transformed into surfaces of double curvature. We may consider $A'B'$ and $C'D'$ as arcs with the centre E, while $A'D'$ and $B'C'$ are straight lines which intersect at E. The lines $A'D'$ and $B'C'$ are not changed in length, AB is shortened and CD lengthened. If $z=\frac{1}{2}BC$, it follows from the definition of k [cf. XXIX.] that

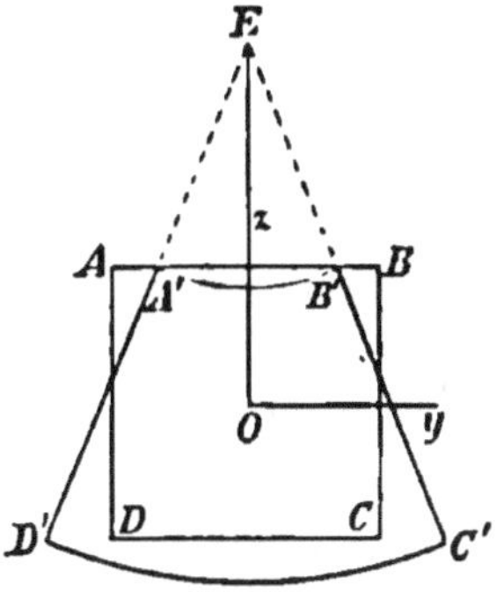

FIG. 45.

$$A'B' = AB(1-kz/\rho), \quad C'D' = CD(1+kz/\rho).$$

If $OE=\rho'$, then

$$A'B'/C'D' = (\rho'-z)/(\rho'+z) = (1-kz/\rho)/(1+kz/\rho),$$

from which it follows that $\rho = k\rho'$. This relation has been applied to the determination of k for glass prisms.

SECTION XXXIV. EQUATIONS OF MOTION OF AN ELASTIC BODY.

The resultant with which the elastic forces act on an infinitely small volume-element dv of an elastic body in the direction of the x-axis is [XXV.], $(\partial X_x/\partial x+\partial X_y/\partial y+\partial X_z/\partial z)dv$. If the body is acted on besides by attractions or repulsions, whose component in the direction of the x-axis is X, the element dv is also acted on by the component of force $X.dv.\rho$, where ρ is the density of the body. Hence the x-component of the acting forces is

$$(\partial X_x/\partial x+\partial X_y/\partial y+\partial X_z/\partial z+\rho X)dv.$$

If this resultant is not equal to zero, motion occurs in the direction of the x-axis, and the momentum imparted to the part of the body under consideration in unit time is $\rho dv d^2(x+\xi)/dt^2 = \rho dv d^2\xi/dt^2$, where t denotes the time. Hence we have

$$\rho d^2\xi/dt^2 = \partial X_x/\partial x+\partial X_y/\partial y+\partial X_z/dz+\rho X.$$

If the components of stress are expressed by ξ, η, and ζ, as in XXX. (b), we obtain the equation

(a) $$\rho\ddot{\xi} = (\lambda + \mu) \,.\, \partial\Theta/\partial x + \mu\nabla^2\xi + \rho X.$$

The equations for $\ddot{\eta}$ and $\ddot{\zeta}$ are similar.

As in XXX. (e) the equations (a) take the form

(b) $$\rho\ddot{\xi} = (\lambda + 2\mu) \,.\, \partial\Theta/\partial x + 2\mu(\partial h_y/\partial z - \partial h_z/\partial y) + \rho X.$$

If the force whose components are X, Y, and Z has a potential, and if therefore $X = -\partial\Psi/\partial x, \quad Y = -\partial\Psi/\partial y, \quad Z = -\partial\Psi/\partial z,$ by differentiation of equation (b) with respect to x, y, z respectively, and by addition, we have

(c) $$\rho\ddot{\Theta} = (\lambda + 2\mu)\nabla^2\Theta - \rho\nabla^2\Psi.$$

In what follows we assume that no external forces act, so that the components X, Y, Z are zero. Therefore $\nabla^2\Psi$ drops out of equation (c).

SECTION XXXV. PLANE WAVES IN AN INFINITELY EXTENDED BODY.

Lamé* treated this form of motion in the following way. Suppose a plane wave propagated in a direction which makes the angles α, β, γ with the axes; let the velocity of propagation be V, and let the direction of vibration make with the axes the angles a, b, c. If u represents the distance of a point from its position of equilibrium, U the *amplitude*, and T the *period of vibration*, the vibration at the origin may be expressed by $u = U\cos(2\pi t/T)$. At any other point, whose coordinates are x, y, z, we have

(a) $$u = U\cos\left\{2\pi/T \,.\, \left(t - \frac{(x\cos\alpha + y\cos\beta + z\cos\gamma)}{V}\right)\right\}.$$

We have further

(b) $$\xi = u\cos a, \quad \eta = u\cos b, \quad \zeta = u\cos c.$$

If the angle between the direction of propagation and the direction of vibration is represented by ϕ, we have

$$\cos\phi = \cos a\cos\alpha + \cos b\cos\beta + \cos c\cos\gamma.$$

For brevity we set

$$s = U\sin\left\{2\pi/T \,.\, \left(t - \frac{(x\cos\alpha + y\cos\beta + z\cos\gamma)}{V}\right)\right\},$$

* Lamé, Théorie de l'élasticité, p. 138. Paris, 1866.

and obtain

$$\Theta = 2\pi s/TV \,.\, \cos\phi, \quad \partial\Theta/\partial r = -4\pi^2 u/T^2V^2 \,.\, \cos u \,.\, \cos\phi,$$

$$\nabla^2\xi = -4\pi^2 u/T^2V^2 \,.\, \cos a, \quad \ddot{\xi} = -4\pi^2 u/T^2 \,.\, \cos a.$$

By the help of these relations and corresponding ones for η and ζ, we obtain from XXXIV. (a)

(c)
$$\begin{cases} (\lambda+\mu)\cos a \cos\phi + (\mu - \rho V^2)\cos a = 0, \\ (\lambda+\mu)\cos\beta \cos\phi + (\mu - \rho V^2)\cos b = 0, \\ (\lambda+\mu)\cos\gamma \cos\phi + (\mu - \rho V^2)\cos c = 0. \end{cases}$$

If these equations are multiplied by $\cos a$, $\cos\beta$, $\cos\gamma$ respectively and then added, we have $(\lambda + 2\mu - \rho V^2)\cos\phi = 0$. We therefore have either (d) (e) $\rho V^2 = \lambda + 2\mu$ or $\cos\phi = 0$. In the first case, equations (c) become

$$\cos a = \cos a \cos\phi, \quad \cos b = \cos\beta\cos\phi, \quad \cos c = \cos\gamma\cos\phi.$$

If the right and left sides of these equations are squared and added, we obtain (f) $\cos^2\phi = 1$, so that either $\phi = 0$ or $\phi = \pi$. The vibrations therefore occur in the direction of propagation; they are called *longitudinal vibrations.* In the second case $\phi = \frac{1}{2}\pi$, that is, the vibrations are perpendicular to the direction of propagation; they are called *transverse vibrations.*

Longitudinal Vibrations.—The velocity of propagation Ω of these vibrations is determined by (d), (g) $\Omega = \sqrt{(\lambda + 2\mu)/\rho}$. Hence condensations and rarefactions occur, since

$$\Theta = 2\pi/T\Omega \,.\, U\sin\left\{2\pi/T \,.\, \left(t - \frac{(x\cos a + y\cos\beta + z\cos\gamma)}{\Omega}\right)\right\}.$$

To determine the stresses we assume that the waves are propagated in the direction of one of the coordinate axes, say the z-axis. In this case we have

$$\xi = 0, \quad \eta = 0, \quad \zeta = U\cos\{2\pi/T \,.\, (t - z/\Omega)\}.$$

From XXX. (b) the tangential forces are zero; the normal forces are

(h) $$X_x = Y_y = 2\pi\lambda/T\Omega \,.\, U\sin\{2\pi/T \,.\, (t - z/\Omega)\}.$$

(i) $$Z_z = 2\pi(\lambda + 2\mu)/T\Omega \,.\, U\sin\{2\pi/T \,.\, (t - z/\Omega)\}.$$

Transverse Vibrations.—The velocity of propagation ω, from (c), equals (k) $\omega = \sqrt{\mu/\rho}$. Since, for these vibrations, we have $\cos\phi = 0$, we also have $\Theta = 0$, that is, neither condensations nor rarefactions occur. If the wave is propagated in the direction of the z-axis, and if the vibrations are parallel to the x-axis,

$$\xi = U\cos\{2\pi/T \,.\, (t - z/\omega)\}, \quad \eta = 0, \quad \zeta = 0.$$

All components of stress vanish with the exception of the tangential force Z_x, (l) $Z_x = 2\pi\mu/T\omega \,.\, U\sin\{2\pi/T \,.\, (t - z/\omega)\}$.

In a solid, therefore, two different wave motions may exist, which are propagated with different velocities Ω and ω. From formulas (g) and (k) the velocity Ω of the longitudinal vibrations is always greater than the velocity ω of the transverse vibrations. In liquids and gases the only vibrations which can occur are longitudinal, since for these bodies $\mu = 0$.

For gases we have $\lambda = P$ [XXIX. (g)], and hence the velocity of sound in air is (m) $\Omega = \sqrt{P/\rho}$. P must here be expressed in absolute units. According to Regnault the density of atmospheric air at Paris equals 0,0012 932 under a pressure of 76 cm. of mercury, and at a temperature of 0° C. Since the acceleration of gravity at Paris is 980,94, the pressure of the air on a square centimetre equals 76 . 13,596 . 980,94 in absolute units. Hence the density ρ of the air under a pressure P in absolute units is

(n) $$\rho = 0{,}001\,293\, P/76 \,.\, 13{,}596 \,.\, 980{,}94 = P \,.\, 1{,}2759 \,.\, 10^{-9}.$$

Using this value of P/ρ we obtain $\Omega = 27\,996$ cm, or approximately 280 metres per second at 0° C. Since the density of the air at t° C is $\rho = P \,.\, 1{,}2759 \,.\, 10^{-9}/(1 + at)$ the velocity of sound at t° is

$$\Omega = 27\,996/\sqrt{1 + at},$$

where a is the coefficient of expansion of air 0,00 366. The result obtained from this form of the theory does not agree with that found by observation. Observation shows that Ω is about 330. The reason why theory and observation are not in accord will be discussed later in the theory of heat.

The velocity of sound in water is obtained in a similiar manner. For water at 15° C. we have $\lambda = 2{,}22 \,.\, 10^{10}$. At the same temperature we have $\rho = 0{,}999\,173$, whence $\Omega = 149\,060$ cm.

In a research carried out on the Lake of Geneva, Colladon and Sturm found that the velocity of sound in water at 8,1° C. is $\Omega = 143\,500$ centimetres; the difference between the observed and calculated values is explained by the difference in temperature, since λ increases rapidly for water as the temperature rises.*

No observations have been made on wave motions in large masses of metal; but the velocity of sound has been determined in a metallic wire. In such a body, however, sound is propagated with a different velocity from that which it would have in an extended body. If the wire is parallel to the z-axis, and if we consider only the motion of the particles in the direction of this axis, the stress Z_z at the distance z from the xy-plane is, in our usual notation, $Z_z = E\partial\zeta/\partial z$.

* Fogliani und Vicentini, Wied. Beibl. Bd. 8. S. 794.

At the distance $(z+dz)$ the stress is

$$Z_z + \partial Z_z/\partial z \,.\, dz = E(\partial\zeta/\partial z + \partial^2\zeta/\partial z^2 \,.\, dz).$$

Hence a portion of the wire whose length is dz and whose cross section is A, is acted on by a force given by $AE\partial^2\zeta/\partial z^2 \,.\, dz$. The equation of motion is $\rho A \,.\, dz \,.\, \ddot{\zeta} = AE\partial^2\zeta/\partial z^2 \,.\, dz$ or (o) $\ddot{\zeta} = V^2\partial^2\zeta/\partial z^2$, where $V = \sqrt{E/\rho}$. The integral of the differential equation (o) is (p) $\zeta = \cos\{2\pi/T \,.\, (t - z/V)\}$; the velocity of propagation V is obtained from equation (p).

According to the researches of Wertheim the velocity calculated from (p) agrees fairly well with the results of observation.

SECTION XXXVI. OTHER WAVE MOTIONS.

Spherical Waves.—We will investigate the circumstances of the propagation of spherical waves in an infinitely extended elastic body, when the direction of vibration of every particle passes through the same point. We take this point as the origin of coordinates. As in XXXI. (b) we set

(a) $$\xi = \partial\phi/\partial x, \quad \eta = \partial\phi/\partial y, \quad \zeta = \partial\phi/\partial z,$$

where ϕ is an unknown function of t and of the distance r from the origin. The equations of motion [XXXIV. (b)] give (b) $\ddot{\phi} = \Omega^2\nabla^2\phi$.

In this case [XV. (l)] we may set $\nabla^2\phi = 1/r \,.\, \partial^2(r\phi)/\partial r^2$, and hence (c) $\partial^2(r\phi)/\partial t^2 = \Omega^2\partial^2(r\phi)/\partial r^2$. This equation is satisfied by

(d) $$\phi = a/r \,.\, \cos\{2\pi/T \,.\, (t - r/\Omega)\},$$

when a is a constant and T the period of vibration.

The distance u of a point from its position of equilibrium is

$$u = \partial\phi/\partial r = -a/r^2 \,.\, \cos\{2\pi/T \,.\, (t - r/\Omega)\} + 2\pi a/Br \,.\, \sin\{2\pi/T \,.\, (t - r/\Omega)\},$$

where $B = \Omega T$. If r is very much greater than the wave length we can neglect the first term on the right, and have

$$u = A/r \,.\, \sin\{2\pi/T \,.\, (t - r/\Omega)\}.$$

The wave motion is therefore one in which the wave surfaces are spheres propagated with the velocity Ω.

Since the expressions (a) satisfy the equations of motion, if ϕ has the value given in (d), these equations are also satisfied if ϕ is replaced by $\partial\phi/\partial x$, or by another differential coefficient taken with respect to one or more coordinates.

Vibrations Due to Torsion.—Let the axis of a circular cylinder coincide with the z-axis, and its separate parts oscillate in arcs about the same axis. The components of displacement of a particle from its position of equilibrium may be expressed [XXXII. (a)] by (e) $\xi = -\phi y$, $\eta = \phi x$, $\zeta = 0$, where ϕ is a function of z.

From XXVIII. (e) we have $\Theta = 0$; therefore condensation and rarefaction do not occur. The equations of motion are [XXXIV. (a) and XXXV. (k)],

$$\ddot{\xi} = \omega^2 \nabla^2 \xi, \quad \ddot{\eta} = \omega^2 \nabla^2 \eta,$$

whence we again obtain (f) $\ddot{\phi} = \omega^2 \partial^2 \phi / \partial z^2$. This equation is satisfied by (g) $\phi = a \sin \{2\pi/T \,.\, (t - z/\omega)\}$. Hence $\omega = \sqrt{\mu/\rho}$ is the velocity with which a wave motion is propagated in the direction of the axis of the cylinder. From XXX. (b) the components of stress are

$$Z_y = -A\mu x, \quad X_z = +A\mu y \text{ where } A = 2\pi a/T\omega \,.\, \cos\{2\pi/T \,.\, (t - z/\omega)\}.$$

The other components of stress are zero.

If the cylinder is of finite length, stationary waves can exist in it, that is, waves such that certain definite points of the cylinder called nodal points are at rest, while on both sides of a nodal point the vibrations are in opposite phase. The amplitude of the vibration is greatest half way between two nodal points, at the ventral segments. Stationary waves are formed when waves which have passed over a certain point return to that point again in the opposite direction. To find the period T of these vibrations, we notice that equation (f) will be satisfied not only by (g), but also by $\phi = b \sin\{2\pi/T \,.\, (t + z/\omega)\}$, and in general by

$$\phi = B \sin 2\pi t/T \,.\, \cos(2\pi z/T\omega) + C \cos(2\pi t/T) \,.\, \sin(2\pi z/T\omega),$$

where B, C, and T are constants. If the points for which $z = 0$ are fixed, the constant B will be zero and (i) $\phi = C \cos(2\pi t/T) \,.\, \sin(2\pi z/T\omega)$. If l represents the length of the cylinder, and if the points for which $z = l$ are also fixed, we will have $\phi = 0$ when $z = l$, and therefore $2\pi l/T\omega = p\pi$, where p is a whole number. Hence

$$T = 2l/p\omega = 2l/p \,.\, \sqrt{\rho/\mu}.$$

If, on the other hand, one end of the rod is free, $Y_z = X_z = 0$ when $z = l$. Since

$$X_z = -\mu y \,.\, \partial\phi/\partial z, \quad Y_z = +\mu x \,.\, \partial\phi/\partial z,$$

we have $\partial\phi/\partial z = 0$ when $z = l$.

In this case we obtain from equation (i) $2\pi l/T\omega = \frac{1}{2}(2p+1) \,.\, \pi$, where p is a whole number; and hence $T = 4l/(2p+1) \,.\, \sqrt{\rho/\mu}$. If both ends of the rod are free, $T = 2l/p \,.\, \sqrt{\rho/\mu}$.

SECTION XXXVII. VIBRATING STRINGS.

Although the problem of the motion of vibrating strings is only slightly connected with the theory of elasticity, a simple example of this form of motion will be considered here. We suppose a perfectly flexible string stretched between two fixed points A and B. If P is the stress in the string, l_0 the length of the string before the application of the stress, and l its length while the stress is applied, p the cross section of the string, E the coefficient of elasticity, we have $l - l_0 = Pl_0/FE$. Let the string be slightly moved from its position of equilibrium, that is, the straight line which joins A and B, and let the new form of the string be designated by $ACDB$. By this deformation the length of the string is increased by $dl = dP \,.\, l_0/FE$. It is here assumed that dP is infinitely small in comparison with P, so that we may set the stress in the string everywhere equal to P.

For the sake of simplicity we suppose that the motion of the string is always in one plane, say the xy-plane. Let A be the origin of coordinates, and let B lie on the x-axis at the distance l from A. The distance of any point C of the string from A may be represented by s, and that of the infinitely near point D by $s + ds$. The components of stress at C in the directions of the x- and y-axis respectively are $P\partial x/\partial s$ and $P\partial y/\partial s$. For the point D the similar components are

$$P(\partial x/\partial s + \partial^2 x/\partial s^2 \,.\, ds) \text{ and } P(\partial y/\partial s + \partial^2 y/\partial s^2 \,.\, ds).$$

The infinitely short portion CD of the string is therefore acted on by the force $P\partial^2 x/\partial s^2 \,.\, ds$ in the direction of the x-axis, and by the force $P\partial^2 y/\partial s^2 \,.\, ds$ in the direction of the y-axis. If the string is displaced only very slightly from its position of equilibrium, we can set $s = x$; the x-component then vanishes and the particles of the string oscillate perpendicularly to the x-axis. If m represents the mass of unit length of the string, the equation of motion is

$$m \,.\, ds \,.\, \ddot{y} = P \,.\, \partial^2 y/\partial x^2 \,.\, ds,$$

or, if we set $ma^2 = P$, (a) $\ddot{y} = a^2 \partial^2 y/\partial x^2$. The integral of this differential equation is (b) $y = A_n \cos(n\pi at/l) \,.\, \sin(n\pi x/l)$, where n is a whole number. When $x = 0$ and $x = l$ we have $y = 0$, and when $t = 0$,

$$y = A_n \sin(n\pi x/l);$$

this is the equation of a sinusoid.

If, in the general case, the form of the string when $t = 0$ is given by the equation $y = f(x)$, then

$$f(x) = A_1 \sin(\pi x/l) + A_2 \sin(2\pi x/l) + A_3 \sin(3\pi x/l) + \ldots .$$

The coefficients A_1, A_2, A_3 ... are determined in the following way. Let the general term of the series be $A_n \sin n\phi$, where $\phi = \pi x/l$. If both sides of the last equation are multiplied by $\sin n\phi$, we will have

$f(l\phi/\pi)\sin n\phi = A_1 \sin\phi \sin n\phi + A_2 \sin 2\phi \sin n\phi + \ldots + A_n \sin^2 n\phi + \ldots .$

If this equation is multiplied by $d\phi$, and integrated between the limits 0 and π, we will have

$$\int_0^\pi f(l\phi/\pi)\sin n\phi \, . \, d\phi = A_n \int_0^\pi \sin^2 n\phi \, d\phi.$$

For if m and n are different numbers, we have

$$\int_0^\pi \sin m\phi \sin n\phi \, d\phi = \tfrac{1}{2}\int_0^\pi \big(\cos(m-n)\phi - \cos(m+n)\phi\big) d\phi = 0.$$

But when they are equal, $\int_0^\pi \sin^2 n\phi \, d\phi = \frac{1}{2}\pi$. Hence

$$\text{(c)} \qquad A_n = 2/\pi \, . \int_0^\pi f(l\phi/\pi)\sin n\phi \, d\phi = 2/l \, . \int_0^l f(x)\sin(n\pi x/l) \, . \, dx.$$

If, for example, the string is so displaced from its position of equilibrium that a point in it at the distance p from the end A is moved through the distance h in the direction of the y-axis, we have $f(x) = hx/p$ for $0 < x < p$, but $f(x) = h(l-x)/(l-p)$ for $p < x < l$. Hence

$$A_n = 2/l \, . \int_0^p \frac{hx}{p} \, . \sin\frac{n\pi x}{l} \, . \, dx + 2/l \, . \int_p^l \frac{h(l-x)}{l-p} \, . \sin\frac{n\pi x}{l} \, . \, dx,$$

and therefore

$$A_n = 2hl^2/p(l-p) \, . \left(\sin\frac{n\pi p}{l}\right) \Big/ n^2\pi^2.$$

We obtain for y,

$$y = 2a^2h/(a-1)\pi^2 \, . \left[1/1^2 \, . \sin\frac{\pi}{a} \, . \sin\frac{\pi x}{l} \, . \cos\frac{\pi at}{l} \right.$$

$$\left. + 1/2^2 \, . \sin\frac{2\pi}{a} \, . \sin\frac{2\pi x}{l} \, . \cos\frac{2\pi at}{l} + \ldots \right],$$

where $a = l/p$. If the string is struck in the middle, we have $a = 2$ and

$$y = 8h/\pi^2 \left\{ \sin\frac{\pi x}{l} \, . \cos\frac{\pi at}{l} - 1/3^2 \, . \sin\frac{3\pi x}{l} \, . \cos\frac{3\pi at}{l} + \ldots \right\}.$$

SECTION XXXVIII. POTENTIAL ENERGY OF AN ELASTIC BODY.

When an elastic body changes its form work will be done. This is stored up in the body as potential energy if the body is perfectly elastic, which we will assume to be the case. The work necessary to bring about a particular change is equal to the potential energy gained by the body, and may be determined in the following way.

Let A', B' and C' represent the principal stresses at a point in the body; about this point construct the infinitely small parallelepiped, whose edges u, v and w are parallel to the principal stresses. When the stresses are applied, the edges of the parallelepiped are extended, u becoming $u(1+a')$, v becoming $v(1+b')$, and w becoming $w(1+c')$. From XXIX. (e) we have

$$A' = \lambda\Theta' + 2\mu a', \quad B' = \lambda\Theta' + 2\mu b', \quad C' = \lambda\Theta' + 2\mu c'.$$

If a', b' and c' change by the increments da', db' and dc' respectively, the edges of the parallelepiped are increased by uda', vdb' and wdc', and the parallelepiped undergoes an infinitely small change of form. The forces which act in the directions of the edges are vwA', uwB' and uvC'. Hence the work done by the stresses during the change of form is

$$(A'da' + B'db' + C'dc')uvw = \big(\lambda\Theta' d\Theta' + 2\mu(a'da' + b'db' + c'dc')\big)uvw,$$

since

$$\Theta' = a' + b' + c'.$$

To change the form of the parallelepiped by an amount which is determined by the elongations a, b, c, the work

$$\tfrac{1}{2}\big(\lambda\Theta^2 + 2\mu(a^2 + b^2 + c^2)\big)uvw$$

must be done. If we set dv for the volume uvw of the parallelepiped, the potential energy E_p of the whole body is given by

(a) $$E_p = \tfrac{1}{2}\int\big(\lambda\Theta^2 + 2\mu(a^2 + b^2 + c^2)\big)dv.$$

If we introduce the principal stresses A, B, C [XXIX. (e)], we have

(b) $$E_p = \tfrac{1}{2}\int\{(A + B + C)^2/E - (AB + BC + CA)/\mu\}dv.$$

By this equation (b) the potential energy is determined, the components of stress and of elongation are known. We confine ourselves to the statement of the following relation,

(c) $$E_p = \tfrac{1}{2}\int(X_x x_x + Y_y y_y + Z_z z_z + 2Z_y z_y + 2X_z x_z + 2Y_x y_x)dv,$$

from which the others can easily be deduced.

Galileo was the first to study the properties of elastic bodies; he failed, however, to reach correct results. The physical basis for the theory of elasticity was given by Robert Hooke, who in 1678 published a treatise, *De potentia restitutiva*, in which he showed by experiment that the changes of form of an elastic body are proportional to the forces applied to it. Among earlier investigations those of Mariotte and Coulomb deserve especial mention. More recently the theory of elasticity has been developed principally by the French mathematicians, Cauchy, Poisson, Lamé, Barré de Saint-Vénant, and others.

We owe to Cauchy the theory of the components of stress in the form here given. For more extended accounts of the theory of elasticity we may mention Lamé, *Théorie Mathématique de l'Élasticité des Corps Solides.* Paris, 1866. Clebsch, *Theorie der Elasticität fester Körper.* Leipzig, 1862. Among the more important recent treatises on the theory of elasticity we mention: *Boussinesq, Application des Potentiel à l'Étude de l'Équilibre et du mouvement des Solides Élastiques.* Paris, 1885. Barré de Saint-Vénant, *Mémoire sur la Torsion des Prismes. Mém. d. sav. étr.* T. XIV. Paris, 1856; *Mémoire sur la Flexion des Prismes.* Liouville I., 1856. William Thomson, *Elements of a Mathem. Theory of Elasticity.* Phil. Tr. London, 1856; *Dynamical Problems on Elastic Spheroids.* Phil. Tr. London, 1864.

Further researches on the theory of elasticity have been carried out in recent years by W. Voigt.

CHAPTER III.

EQUILIBRIUM OF FLUIDS.

SECTION XXXIX. CONDITIONS OF EQUILIBRIUM.

THE principal difference between solids on the one hand and liquids and gases on the other consists in the fact that the latter do not, like the former, offer a great resistance to change of form. A force is always needed to change the form of a fluid mass, but the resistance offered by the fluid is determined by the rate at which the change of form proceeds, and will be infinitely small if it proceeds very slowly. We assume that the motion by which the condition of equilibrium is attained proceeds very slowly, and we may therefore assume, in *hydrostatics*, that a fluid offers no resistance to change of form, so long as this does not involve change of volume.

Each infinitely small change of form of an infinitely small part of the body may [XXVIII.] be treated as if it were produced by the dilatations a, b, c in three directions perpendicular to each other. The lengths u, v, w drawn in these three directions become $u(1+a)$, $v(1+b)$, $w(1+c)$. If A, B, C are the corresponding normal forces per unit of surface which act on the surfaces vw, uw, wv respectively, the work done by the normal forces in this change of form is

$$Avwua + Buwvb + Cuvwc \text{ or } (Aa + Bb + Cc)u \,.\, v \,.\, w.$$

The change of form considered will, in general, involve an increase of volume, given by $uvw(1+a)(1+b)(1+c) - uvw$. Since a, b, c are infinitely small, the increment of the volume equals $(a+b+c)u \,.\, v \,.\, w$, if we neglect infinitely small quantities of a higher order.

If we start from the assumption that the work done by the forces equals zero if the volume is not changed, we have at the same time $Aa + Bb + Cc = 0$ and $a + b + c = 0$. These equations can both be true only if $A = B = C$.

The equations for the components of stress [XXVI. (i)] give

$$X_x = Y_y = Z_z \text{ and } Z_y = 0,\ X_z = 0,\ Y_x = 0.$$

There are, therefore, no tangential forces in a fluid in equilibrium.

If we start from the condition that the only forces which act on fluids in equilibrium are perpendicular to their surfaces, we reach the same result, namely, that the normal stresses are all equal. To show this we set $Z_y = 0,\quad X_z = 0,\quad Y_x = 0,$
and have, from XXVI. (a),

$$P = X_x \cos a,\quad Q = Y_y \cos\beta,\quad R = Z_z \cos\gamma.$$

P, Q and R are the components of stress for a surface whose normal makes the angles a, β, γ with the axes. The stress acting on this surface is $\sqrt{P^2+Q^2+R^2}$, the normal force N is determined by

$$N = P\cos a + Q\cos\beta + R\cos\gamma.$$

The tangential force T is

$$T^2 = (P^2+Q^2+R^2) - (P\cos a + Q\cos\beta + R\cos\gamma)^2.$$

Introducing the values of P, Q and R, we have

$$(X_x - Y_y)^2\cos^2 a\cos^2\beta + (Y_y - Z_z)^2\cos^2\beta\cos^2\gamma + (Z_z - X_x)^2\cos^2 a\cos^2\gamma = 0,$$

and hence $X_x = Y_y = Z_z$.

From the expression given above for N, it follows that $N = X_x$, that is, *the normal force acting on a surface-element in the interior of the fluid is independent of the position of the element.*

If we neglect the force of cohesion of the fluid, which will be treated later, the normal force will be a pressure; if this is designated by p, we have

(a) $$X_x = Y_y = Z_z = -p;\quad Z_y = 0,\quad X_z = 0,\quad Y_x = 0.$$

If ρ is the density of the fluid, we obtain from XXV. (c) the conditions of equilibrium

(b) $$\partial p/\partial x = \rho X,\quad \partial p/\partial y = \rho Y,\quad \partial p/\partial z = \rho Z.$$

The components of the force acting on the unit of mass are X, Y, Z; we may consider ρ as constant in liquids; in gases ρ is a function of the pressure.

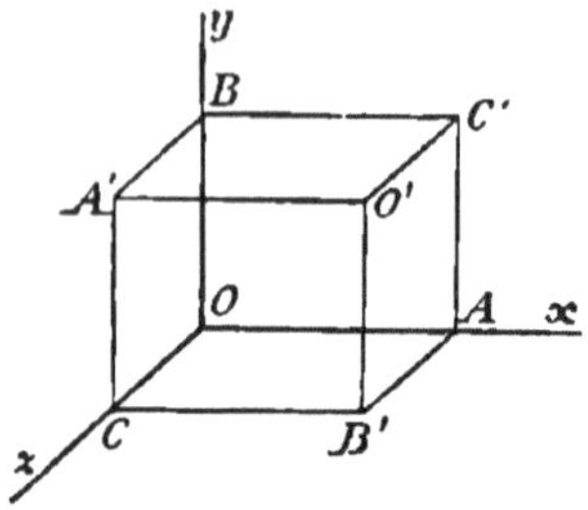

FIG. 46.

Equations (b) may also be developed in the following way. Represent the sides of the parallelepiped OO' (Fig. 46) by

$$OA = dx,\quad OB = dy,\quad OC = dz.$$

The pressure on OA' is $pdydz$, the pressure on $O'A$ is $(p + \partial p/\partial x \,.\, dx)dydz$. The resultant of the pressures is the pressure $-\partial p/\partial x \,.\, dx \,.\, dy \,.\, dz$ in the direction of the x-axis. The force $\rho X dxdydz$ also acts on the parallelepiped in the same direction. The condition of equilibrium is therefore $(-\partial p/\partial x + \rho X)dxdydz = 0$, from which we obtain the first of equations (b).

If equilibrium obtains, p must satisfy equations (b); the conditions for this are

$$\text{(c)} \qquad \begin{cases} \partial(\rho X)/\partial y = \partial(\rho Y)/\partial x, \quad \partial(\rho Y)/\partial z = \partial(\rho Z)/\partial y, \\ \partial(\rho Z)/\partial x = \partial(\rho X)/\partial z. \end{cases}$$

These equations will hold if a function Φ exists, such that

$$\text{(d)} \qquad \partial\Phi/\partial x = \rho X, \quad \partial\Phi/\partial y = \rho Y, \quad \partial\Phi/\partial z = \rho Z.$$

Equations (c) are the essential conditions of equilibrium; if they are satisfied, p may be determined from the equation

$$dp = \rho(Xdx + Ydy + Zdz).$$

If the forces have a potential ψ, so that

$$X = -\partial\psi/\partial x, \quad Y = -\partial\psi/\partial y, \quad Z = -\partial\psi/\partial z,$$

we will have (e) $dp = -\rho d\psi$.

In gases ρ is a function of p; in liquids ρ may be considered constant. In the latter case we obtain (f) $p = c - \rho\psi$, where c is constant.

Section XL. Examples of the Equilibrium of Fluids.

The conditions of equilibrium of a liquid mass contained in a vessel, and acted on by gravity only, may be determined in the following way: Suppose the position of the particles of the liquid referred to a system of rectangular coordinates, whose z-axis is directed perpendicularly upward; we then have

$$X = 0, \quad Y = 0, \quad Z = -g,$$

and therefore $\psi = gz$. Since the density ρ is considered constant, equilibrium can obtain under the action of gravity. From XXXIX. (f) we have $p = c - g\rho z$. Hence the pressure at the same level is everywhere the same.

We now determine the pressure in a liquid contained in a vessel, which rotates about a perpendicular axis A with constant angular velocity ω. The fluid will turn, like a solid, about the axis A with the same angular velocity as that of the vessel.

A particle at the distance r from the axis A is acted on both by gravity and by a centrifugal force whose acceleration is $\omega^2 r$. We refer it to a system of rectangular coordinates whose z-axis is directed perpendicularly upward, and coincides with the axis of rotation. We then have

$$X=\omega^2 x, \quad Y=\omega^2 y, \quad Z=-g,$$

and the potential ψ is $\psi=-\frac{1}{2}\omega^2 r^2+gz$. From XXXIX. (f) the pressure is $p=c+\rho(\frac{1}{2}\omega^2 r^2-gz)$. The surfaces of constant pressure are paraboloids of revolution with the common axis A.

We will make a third application of the conditions of equilibrium to the *determination of the pressure in the atmosphere.* We suppose gravity directed toward the centre of the earth, its acceleration γ may then be expressed by $\gamma=-ga^2/r^2$, where g is the acceleration at the surface of the earth, a the earth's radius, and r the distance of the point considered from the centre of the earth. We then have $\psi=-ga^2/r$. If the temperature is constant $\rho=k\,.\,p$, where k is constant. We have then [XXXIX. (e)]

$$dp=-k\,.\,pd\psi \text{ or } \log p=c-k\psi.$$

If, at the earth's surface, the pressure is p_0, and the potential ψ_0, we have $\log p_0=c-k\psi_0$ and $\log(p_0/p)=k(\psi-\psi_0)=kg(ar-a^2)/r$. If the difference $r-a$ is very small in comparison with a, we can set $\log(p_0/p)=kgh$, where $h=r-a$ is the height of the point considered above the earth's surface, and k is equal to $1,2759\,.\,10^{-9}$ for dry air at 0°C.

CHAPTER IV.

MOTION OF FLUIDS.

SECTION XLI. EULER'S EQUATIONS OF MOTION.

IN the study of the motion of fluids very serious difficulties are encountered, and thus far only a few problems have been completely solved. In the following chapter we will consider only the so-called ideal fluids, and therefore neglect the friction between their moving particles and the forces of adhesion and cohesion, which will be treated later. We further assume that the fluids considered are incompressible, and thus limit our discussion to liquids, which are only slightly compressible. On these assumptions several characteristics of the motion of liquids may be derived. As the study of the motion of gases is extremely difficult, and as little success has so far been obtained in it, we will not enter upon it here.

For a complete determination of the motion of a fluid the path of each separate particle, as well as the position of the particle in the path at any instant, may be given. The coordinates x, y, z of the particle M may be given as functions of the time t.

An easier method is one in which the motion is expressed in terms of the components of velocity, which according to circumstances shall be designated by U, V, W, or u, v, w. Suppose U, V, W to be the components of velocity of a particular particle of the fluid; if the particle at the time t is situated at P and at the time $t+dt$ at P', U, V, W will be the components of velocity at the point P, and

$$U+\frac{dU}{dt}dt, \quad V+\frac{dV}{dt}dt, \quad W+\frac{dW}{dt}dt$$

the components of velocity at P'. The quantities dU, dV, dW represent the increments which U, V, W respectively receive during the time dt, if our attention is confined to the motion of a particular particle.

The other symbols, u, v, w, represent the components of velocity at a definite point in space, where one particle replaces another in the course of the motion. If x, y, z are the coordinates of the point considered, u, v, w are the components of velocity of a particle situated at that point at the time t. After the lapse of the time dt the same point is occupied by another particle, whose components of velocity are

$$u + \partial u/\partial t \,.\, dt, \quad v + \partial v/\partial t \,.\, dt, \quad w + \partial w/\partial t \,.\, dt.$$

A particle situated, at the time t, at a point whose coordinates are $x + dx$, $y + dy$, $z + dz$, has a velocity whose projection on the x-axis is

$$u + \partial u/\partial x \,.\, dx + \partial u/\partial y \,.\, dy + \partial u/\partial z \,.\, dz.$$

The velocities u, v, w are everywhere functions of x, y, z and t. If u, v, w are the components of velocity, at the time t, at the point P, whose coordinates are x, y, z, the components at the time $t + dt$ at another point P', whose coordinates are $x + dx$, $y + dy$, $z + dz$, will be

$$u + \partial u/\partial t \,.\, dt + \partial u/\partial x \,.\, dx + \partial u/\partial y \,.\, dy + \partial u/\partial z \,.\, dz, \text{ etc.}$$

If the fluid particle is situated at the time t at P and at the time $t + dt$ at P', then we have $U = u$ and

$$U + dU/dt \,.\, dt = u + \partial u/\partial t \,.\, dt + \partial u/\partial x \,.\, dx + \partial u/\partial y \,.\, dy + \partial u/\partial z \,.\, dz,$$

or (a) $$dU/dt = \partial u/\partial t + \partial u/\partial x \,.\, dx/dt + \partial u/\partial y \,.\, dy/dt + \partial u/\partial z \,.\, dz/dt.$$

The particle considered traverses the distance PP' in the time dt, hence its velocity is PP'/dt, the projections of which on the coordinate axis are evidently $dx/dt = u$, $dy/dt = v$, $dz/dt = w$. Thus we obtain

$$dU/dt = \partial u/\partial t + u \,.\, \partial u/\partial x + v \,.\, \partial u/\partial y + w \,.\, \partial u/\partial z.$$

The equations for dV/dt and dW/dt are similar.

To find the equations of motion of a fluid let us cut from it a parallelepiped $d\omega$, whose edges are dx, dy, dz, and on which a force acts whose components are X, Y, Z. In the time dt the parallelepiped receives an increase of momentum, whose components are

$$\rho X d\omega dt, \quad \rho Y d\omega dt, \quad \rho Z d\omega dt,$$

when ρ denotes the density of the fluid.

The pressure p acting at the point x, y, z, as has been shown in a former chapter, imparts to $d\omega$ the components of momentum

$$-\partial p/\partial x \,.\, d\omega dt, \quad -\partial p/\partial y \,.\, d\omega dt, \quad -\partial p/\partial z \,.\, d\omega dt.$$

Under the action of these forces the body receives, in unit time, an increment of velocity whose components are dU/dt, dV/dt, dW/dt, and hence we have

(b) $$\rho dU/dt = \rho X - \partial p/\partial x\,; \quad \rho dV/dt = \rho Y - \partial p/\partial y\,; \quad \rho dW/dt = \rho Z - \partial p/\partial z.$$

By the help of equations (a) we then obtain

$$
\text{(c)} \quad \begin{cases} \partial u/\partial t + u\partial u/\partial x + v\partial u/\partial y + w\partial u/\partial z = X - 1/\rho \,.\, \partial p/\partial x, \\ \partial v/\partial t + u\partial v/\partial x + v\partial v/\partial y + w\partial v/\partial z = Y - 1/\rho \,.\, \partial p/\partial y, \\ \partial w/\partial t + u\partial w/\partial x + v\partial w/\partial y + w\partial w/\partial z = Z - 1/\rho \,.\, \partial p/\partial z. \end{cases}
$$

These equations are due to Euler, and are known as *Euler's Equations of Motion.* To them must be added the so-called *equation of continuity*, which is found in the following way: The parallelepiped receives in the time dt, through the face $dydz$, the quantity of fluid $\rho udydzdt$; and loses, through the opposite face, the quantity $(\rho u + \partial(\rho u)/\partial x \,.\, dx)dydzdt$. The difference between the quantities flowing through the two surfaces, which indicates a loss of fluid, if $\partial(\rho u)/\partial x \,.\, dx$ is positive, will be

$$\partial(\rho u)/\partial x \,.\, d\omega \,.\, dt.$$

By a similar argument applied to the two other pairs of faces it appears that the total difference between the quantities of fluid which leave and enter the parallelepiped is

$$(\partial(\rho u)/\partial x + \partial(\rho v)/\partial y + \partial(\rho w)/\partial z)d\omega \,.\, dt.$$

The parallelepiped at first contained the quantity $\rho \,.\, d\omega$; after the lapse of the time dt it contains the quantity $(\rho + \partial\rho/\partial t \,.\, dt)d\omega$; the difference of these two quantities is $-\partial\rho/\partial t \,.\, d\omega \,.\, dt.$ By equating these two expressions for the same quantity we get the equation of continuity (d) $\partial\rho/\partial t + \partial(\rho u)/\partial x + \partial(\rho v)/\partial y + \partial(\rho w)/\partial z = 0.$ If the density ρ of the fluid is constant, the equation of continuity becomes

$$\text{(e)} \qquad \partial u/\partial x + \partial v/\partial y + \partial w/\partial z = 0.$$

Euler's equations are specially suited to investigations of the motion in fluid masses with fixed boundaries. If the surface of the fluid changes there will be points which will lie sometimes within and sometimes without the fluid; the velocity at such a point cannot be determined by the method here given. Lagrange's method is the one then employed. To this we will return later.

In equations (c) and (e) there are contained four unknown quantities u, v, w and p, for whose determination we have four equations given. To determine the constants of integration the conditions of the motion of the fluid must be given at a definite time. If the fluid is bounded by a fixed surface, the components of velocity in the direction of the normals to the bounding surface are zero. If $\bar{u}$, $\bar{v}$, $\bar{w}$ are the components of velocity of a particle at the boundary of the fluid, and if the normal to the bounding surface makes the angles a, β, γ with the axes, we have

$$\text{(f)} \qquad \bar{u}\cos a + \bar{v}\cos\beta + \bar{w}\cos\gamma = 0.$$

SECTION XLII. TRANSFORMATION OF EULER'S EQUATIONS.

In a fluid in motion an elementary parallelepiped, whose edges are originally dx, dy, dz, not only changes its position in space but may also rotate and change its form at the same time. Its motion at any instant is determined by the components of velocity u, v, w; the rotations and changes of form may be determined in the following way: In the theory of elasticity the component of rotation h_x of such an element is expressed by $h_x = \frac{1}{2}(\partial\zeta'/\partial y - \partial\eta'/\partial z)$, if ζ' and η' are the infinitely small changes of the coordinates z and y introduced by the motion. We may set $\zeta' = w\,.\,dt$, $\eta' = v\,.\,dt$, and obtain

$$h_x = \tfrac{1}{2}(\partial w/\partial y - \partial v/\partial z)\,.\,dt.$$

If ξ is the corresponding angular velocity, we will have $h_x = \xi\,.\,dt$ and hence

(a) $\xi = \frac{1}{2}(\partial w/\partial y - \partial v/\partial z)$; $\eta = \frac{1}{2}(\partial u/\partial z - \partial w/\partial x)$; $\zeta = \frac{1}{2}(\partial v/\partial x - \partial u/\partial y)$.

The equations for η and ζ may be derived in the same way as the first; ξ, η, ζ are the components of angular velocity in a rotation about the three coordinate axes.

If no rotation exists in the fluid, we have $\xi = \eta = \zeta = 0$ or

$$\partial w/\partial y = \partial v/\partial z, \quad \partial u/\partial z = \partial w/\partial x, \quad \partial v/\partial x = \partial u/dy.$$

These equations are the conditions for the existence of a function ϕ of x, y, z and t which has the property that

$$u = -\partial\phi/\partial x, \quad v = -\partial\phi/\partial y, \quad w = -\partial\phi/\partial z.$$

This function ϕ is called by v. Helmholtz the *velocity potential*, since the components of velocity u, v, w, are related to each other in the same way as the components of a force if it has a potential.

The equation of continuity [XLI. (e)], on the assumption that a velocity potential exists and that the fluid is *incompressible*, becomes $\partial^2\phi/\partial x^2 + \partial^2\phi/\partial y^2 + \partial^2\phi/\partial z^2 = \nabla^2\phi = 0$. The velocity h of a particle is

$$h^2 = u^2 + v^2 + w^2 = (\partial\phi/\partial x)^2 + (\partial\phi/\partial y)^2 + (\partial\phi/\partial z)^2.$$

From equations a it follows that $\partial u/\partial y = \partial v/\partial x - 2\zeta$, $\partial u/\partial z = \partial w/\partial x + 2\eta$. The first of equations XLI. (c) becomes

$$\partial u/\partial t + 2(w\eta - v\zeta) + u\,.\,\partial u/\partial x + v\,.\,\partial v/\partial x + w\,.\,\partial w/\partial x = X - 1/\rho\,.\,\partial p/\partial x.$$

This equation may be written

(b)
$$\begin{cases} \partial u/\partial t + 2(w\eta - v\zeta) = X - 1/\rho\,.\,\partial p/\partial x - \frac{1}{2}\partial h^2/\partial x. \\ \text{We have similarly} \\ \partial v/\partial t + 2(u\zeta - w\xi) = Y - 1/\rho\,.\,\partial p/\partial y - \frac{1}{2}\partial h^2/\partial y, \\ \partial w/\partial t + 2(v\xi - u\eta) = Z - 1/\rho\,.\,\partial p/\partial z - \frac{1}{2}\partial h^2/\partial z, \end{cases}$$

where h is the velocity of a particle. We may eliminate p from equations (b) by differentiating the second of those equations with respect to z, and the third with respect to y and subtracting. We thus obtain

$$\partial\xi/\partial t + \partial(v\xi - u\eta)/\partial y - \partial(u\zeta - w\xi)/\partial z = \tfrac{1}{2}(\partial Z/\partial y - \partial Y/\partial z).$$

If we use the equation of continuity $\partial u/\partial x + \partial v/\partial y + \partial w/\partial z = 0$, and the relation following from (a) $\partial\xi/\partial x + \partial\eta/\partial y + \partial\zeta/\partial z = 0$, we obtain

$$\partial\xi/\partial t + u\,.\,\partial\xi/\partial x + v\,.\,\partial\xi/\partial y + w\,.\,\partial\xi/\partial z - \xi\,.\,\partial u/\partial x - \eta\,.\,\partial u/\partial y - \zeta\,.\,\partial u/\partial z$$
$$= \tfrac{1}{2}(\partial Z/\partial y - \partial Y/\partial z).$$

If ξ, η, ζ represent the components of rotation at a point in the region containing the fluid at the time t, we may use Ξ, H, Z to represent the components of rotation of a particle at the time $t + dt$, whose components at the time t were ξ, η, ζ.

The connection between the components ξ, η, ζ and Ξ, H, Z may be established in the same way as that previously used to find the relation between the velocity at a point in space and the velocity of a particle of the fluid. We have

$$\xi = \Xi, \quad d\Xi/dt = \partial\xi/\partial t + u\,.\,\partial\xi/\partial x + v\,.\,\partial\xi/\partial y + w\,.\,\partial\xi/\partial z.$$

Using this equation we obtain

(c) $$d\Xi/dt = \xi\,.\,\partial u/\partial x + \eta\,.\,\partial u/\partial y + \zeta\,.\,\partial u/\partial z + \tfrac{1}{2}(\partial Z/\partial y - \partial Y/\partial z).$$

If at any time no rotation exists in the fluid, and if therefore $\xi = \eta = \zeta = 0$ at any point in the fluid, a rotation may still be set up if Z and Y have no potential. If, on the other hand, Z and Y have a potential so that $Z = -\partial\Psi/\partial z$ and $Y = -\partial\Psi/\partial y$, we will have $d\Xi/dt = 0$. If besides $X = -\partial\Psi/\partial x$, we have $d\mathrm{H}/dt = 0$ and $d\mathrm{Z}/dt = 0$. *Hence no rotation can be set up in an ideal fluid if the forces have a potential. In this case the particles which rotate already continue to rotate, but the particles which do not rotate from the beginning will never rotate. This theorem was first given by v. Helmholtz.*

SECTION XLIII. VORTEX MOTIONS AND CURRENTS IN A FLUID.

In researches on the motion of fluids it is important to observe whether the particles rotate or not. If there is rotation it is called *vortex motion.* We then have

(a) $\xi = \tfrac{1}{2}(\partial w/\partial y - \partial v/\partial z)$, $\eta = \tfrac{1}{2}(\partial u/\partial z - \partial w/\partial x)$, $\zeta = \tfrac{1}{2}(\partial v/\partial x - \partial u/\partial y)$.

From this it follows at once that (b) $\partial\xi/\partial x + \partial\eta/\partial y + \partial\zeta/\partial z = 0$. The equation of continuity is (c) $\partial u/\partial x + \partial v/\partial y + \partial w/\partial z = 0$.

If the forces have a potential it follows from equations XLII. (c) that

(d) $$\begin{cases} d\Xi/\partial t = \xi \,.\, \partial u/\partial x + \eta \,.\, \partial u/\partial y + \zeta \,.\, \partial u/\partial z, \\ d\mathrm{H}/\partial t = \xi \,.\, \partial v/\partial x + \eta \,.\, \partial v/\partial y + \zeta \,.\, \partial v/\partial z, \\ d\mathrm{Z}/\partial t = \xi \,.\, \partial w/\partial x + \eta \,.\, \partial w/\partial y + \zeta \,.\, \partial w/\partial z. \end{cases}$$

In these equations ξ, η, ζ are the components of rotation at the point x, y, z; Ξ, H, Z are the same components for a particle which at the time t is situated at the point x, y, z, but which at the time $t+dt$ is situated at the point $x+dx$, $y+dy$, $z+dz$.

On the other hand, if the components ξ, η, ζ are zero at every point in the fluid at a definite instant they are equal to zero at any time, from equations (d). In this case we call the motion a *flow*. It is characterized by the equations

(e) $$\partial w/\partial y = \partial v/\partial z, \quad \partial u/\partial z = \partial w/\partial x, \quad \partial v/\partial x = \partial u/\partial y.$$

From XLII. u, v and w then have a *velocity potential* ϕ, which depends in general on x, y, z and t. The equation of continuity is

(g) $$\partial^2\phi/\partial x^2 + \partial^2\phi/\partial y^2 + \partial^2\phi/\partial z^2 = \nabla^2\phi = 0.$$

Euler's equations XLI. (c) take the form

(h) $$\begin{cases} X = -\partial^2\phi/\partial t\partial x + \tfrac{1}{2}\partial h^2/\partial x + 1/\rho \,.\, \partial p/\partial x; \\ Y = -\partial^2\phi/\partial t\partial y + \tfrac{1}{2}\partial h^2/\partial y + 1/\rho \,.\, \partial p/\partial y; \\ Z = -\partial^2\phi/\partial t\partial z + \tfrac{1}{2}\partial h^2/\partial z + 1/\rho \,.\, \partial p/\partial z. \end{cases}$$

Hence such a motion can exist only when the forces have a potential Ψ. In case this condition holds, we obtain from (h) by integration
(i) $\Psi + T = \partial\phi/\partial t - \tfrac{1}{2}h^2 - p/\rho$, where T is a function of the time only.

In order to have a simple example of the two classes of motion just described, we consider the case of an infinite fluid mass, all particles of which move in circles parallel to the xy-plane whose centres lie on the z-axis. All particles at the same distance from the z-axis move with the same velocity and in the same sense. We have then from XXXVI. (e) $u = -\omega y$, $v = +\omega x$, $w = 0$. ω depends only on the distance r of the particle from the z-axis. Since we have $\partial u/\partial x = -xy/r \,.\, d\omega/dr$ and $\partial v/\partial y = +xy/r \,.\, d\omega/dr$. the equation of continuity is satisfied, because $\partial u/\partial x + \partial v/\partial y = 0$..

In general there is rotation of the separate particles, since

$$\partial u/\partial y = -\omega - y^2/r \,.\, d\omega/dr; \quad \partial v/\partial x = \omega + x^2/r \,.\, d\omega/dr,$$

and therefore $\zeta = \omega + \tfrac{1}{2}r \,.\, d\omega/dr$. Since $\xi = 0$, $\eta = 0$ and u, v, w are independent of z, the equations of motion (d) are satisfied.

We assume $\zeta = \zeta_0$ for $r < r_0$ and $\zeta = 0$ for $r > r_0$, where ζ_0 is a constant. In the first case, we have $\omega = \zeta_0 + C/r^2$, where C is a new constant.

C must vanish, because otherwise the particles at the axis would have an infinitely great velocity. Hence $\omega = \zeta_0$ for the part of the fluid lying within a circular cylinder whose radius is r_0, and whose axis coincides with the z-axis. These fluid particles therefore rotate about the z-axis, just as if they formed a solid body. If, on the other hand, $r > r_0$, and hence $\zeta = 0$, the angular velocity ω' will be $\omega' = C'/r^2$. The linear velocity is $r\omega'$ or C'/r, and therefore inversely proportional to the distance of the particle from the axis. On the condition that there is no discontinuity in the motion of the fluid, we have for $r = r_0$, $\zeta_0 = C'/r_0^2$. Hence for $r > r_0$, we have $r\omega' = r_0^2\zeta_0/r$. If r_0 is infinitely small and ζ_0 infinitely great, we obtain a so-called *vortex filament*.

The action of the vortex filament on the surrounding fluid depends on its cross-section and its angular velocity. If we set $m = \pi r_0^2 \zeta_0$, the velocity h of a fluid particle which does not belong to the vortex is $h = r\omega' = m/\pi r$.

Vortex filaments may have other forms; they were first investigated by v. Helmholtz,* and afterwards by William Thomson, and several others. We see from this example, that the separate parts of the fluid do not need to turn about themselves as their centres of gravity describe circles; although the fluid surrounding the vortex filament revolves about the z-axis, the separate drops, into which the mass may be divided, do not rotate about themselves.

SECTION XLIV. STEADY MOTION WITH VELOCITY-POTENTIAL.

If the components of velocity are independent of the time, or if the condition of motion at any definite point in the fluid does not change, the motion is called *steady*. If a velocity-potential ϕ exists, we have (a) $u = -\partial\phi/\partial x$, $v = -\partial\phi/\partial y$, $w = -\partial\phi/\partial z$, where ϕ is a function of x, y, z only. The same holds for the potential Ψ of the forces; hence the function T in XLIII. (i) must be constant. If we set $T = -C$, we have (b) $\Psi + p/\rho + \frac{1}{2}h^2 = C$. If the only forces which act are pressures within the fluid, we may set $\Psi = 0$ and conclude that the velocity of the particles increases as they pass from places of higher pressure to places of lower pressure, and inversely.

For a motion for which there is a velocity-potential, the equation of continuity is (c) $\nabla^2\phi = 0$.

As an example of such a motion, we will consider a sphere at rest in an infinitely extended fluid. The particles of the fluid which

* Helmholtz, *Crelle's Journal*, Bd. 55, S. 25, 1858.

are at a great distance from the centre of the sphere move with equal velocities in the same direction.

Let the sphere be placed so that its centre is at the origin of coordinates O, and let the radius of the sphere be R. The particles whose distance r from O is infinite are supposed to move in a direction parallel to the positive z-axis with the velocity w_0. We set the velocity-potential (d) $\phi = V - w_0 z$, in which $V \doteq 0$ when $r = \infty$. Then (e) $u = -\partial V/\partial x$, $v = -\partial V/\partial y$, $w = -\partial V/\partial z + w_0$. Using equation (c) we have (f) $\nabla^2 V = 0$. If we set $V = 1/r$, or equal to a differential coefficient of $1/r$ taken with respect to x, y, or z, the equation (f) will be satisfied. Since the arrangement around the $\ast$ x-axis is symmetrical, we will consider if the assumption

(g) $$V = C \, . \, \partial(1/r)/\partial z = -Cz/r^3$$

will satisfy the given conditions.

The particles of the fluid move over the surface of the sphere, and hence the component of velocity in the direction of the radius is equal to zero, that is $(\partial\phi/\partial r)_{r=R} = 0$. If we set $z/r = \cos\gamma$, we have

$$\phi = -C\cos\gamma/r^2 - rw_0\cos\gamma \text{ and } d\phi/dr = 2C\cos\gamma/r^3 - w_0\cos\gamma.$$

From this it follows that (h) $C = \frac{1}{2}w_0 R^3$.

From equations (d), (g), and (h), it follows that

(i) $$\phi = -\tfrac{1}{2}w_0 R^3 z/r^3 - w_0 z = -w_0 z(1 + \tfrac{1}{2}R^3/r^3).$$

Using equations (e) we obtain

$$u = -\tfrac{3}{2}w_0 R^3 zx/r^5, \quad v = -\tfrac{3}{2}w_0 R^3 zy/r^5, \quad w = -\tfrac{1}{2}w_0 R^3(3z^2/r^5 - 1/r^3) + w_0.$$

If we set $u^2 + v^2 = s^2$ and $x^2 + y^2 = q^2$, we will have $s = -\frac{3}{2}w_0 R^3 qz/r^5$. If q and z are the coordinates of the path of a particle, the equation of the path will be (k) $dq/dz = s/w$.

Remembering that $r^2 = q^2 + z^2$, we may integrate equation (k) and obtain $q^2(1 - R^3/r^3) = c$. If c is constant, this is the equation of a stream line. If $c = 0$, we will have either $r = R$ or $q = 0$; in the first case, we get the equation of a great circle, in the second, the equation of the z-axis.

The pressure p may be determined by the help of equation (b). Since $\Psi = 0$, we have $p = \rho(C - \frac{1}{2}h^2)$. Now $h^2 = u^2 + v^2 + w^2$, and hence for a point on the surface of the sphere we have $h = \frac{3}{2}w_0 q/R$.

Hence the pressure p on the part of the sphere which lies toward the positive side of the z-axis, is as great as that on that part of the sphere which lies on the negative side of the z-axis; the moving mass of fluid will therefore impart no motion to the sphere. And further, a sphere which moves with constant velocity in any direction in an infinite mass of fluid experiences no resistance during its motion. This result, which is at first sight so startling, is explained by the fact that the resistance offered by friction is not taken into account.

SECTION XLV. LAGRANGE'S EQUATIONS OF MOTION.

Suppose a particle P of a fluid to be originally situated at the point whose coordinates are a, b, c, and after the lapse of the time dt, to have reached the point x, y, z. The general coordinates x, y, z, are functions of t, a, b, c; if t alone varies in these functions, we obtain the path of a particular particle. If, on the other hand, we give to the coordinates a, b, c, all possible values, and keep t constant, we have the positions of all the particles of the fluid at the same time. If the pressure is designated by p, and the density of the fluid by ρ, and if we set $U=\dot{x}$, $V=\dot{y}$, $W=\dot{z}$, we obtain from XLI. (b)

$$\text{(a)}\qquad \ddot{x}=X-1/\rho\,.\,\partial p/\partial x,\quad \ddot{y}=Y-1/\rho\,.\,\partial p/\partial y,\quad \ddot{z}=Z-1/\rho\,.\,\partial p/\partial z.$$

In order to eliminate the differential coefficients with respect to x, y, z, we multiply these equations respectively by $\partial x/\partial a$, $\partial y/\partial a$, $\partial z/\partial a$, by $\partial x/\partial b$, $\partial y/\partial b$, $\partial z/\partial b$, and finally by $\partial x/\partial c$, $\partial y/\partial c$, $\partial z/\partial c$.

By addition we then get the following equations:

$$\text{(b)}\quad \begin{cases} (\ddot{x}-X)\,.\,\partial x/\partial a+(\ddot{y}-Y)\,.\,\partial y/\partial a+(\ddot{z}-Z)\,.\,\partial z/\partial a+1/\rho\,.\,\partial p/\partial a=0,\\ (\ddot{x}-X)\,.\,\partial x/\partial b+(\ddot{y}-Y)\,.\,\partial y/\partial b+(\ddot{z}-Z)\,.\,\partial z/\partial b+1/\rho\,.\,\partial p/\partial b=0,\\ (\ddot{x}-X)\,.\,\partial x/\partial c+(\ddot{y}-Y)\,.\,\partial y/\partial c+(\ddot{z}-Z)\,.\,\partial z/\partial c+1/\rho\,.\,\partial p/\partial c=0,\end{cases}$$

These equations are due to Lagrange.

To these equations there must be added a relation which expresses the fact that the volume of the fluid does not change. The particles originally situated in a rectangular parallelepiped with the edges da, db, dc, are at the time t contained in a parallelepiped, the projections of whose edges are

$$\begin{matrix} \partial x/\partial a\,.\,da, & \partial y/\partial a\,.\,da, & \partial z/\partial a\,.\,da\,;\\ \partial x/\partial b\,.\,db, & \partial y/\partial b\,.\,db, & \partial z/\partial b\,.\,db\,;\\ \partial x/\partial c\,.\,dc, & \partial y/\partial c\,.\,dc, & \partial z/\partial c\,.\,dc.\end{matrix}$$

The volume of the parallelepiped at the time t will therefore be

$$\begin{vmatrix} \partial x/\partial a, & \partial y/\partial a, & \partial z/\partial a\\ \partial x/\partial b, & \partial y/\partial b, & \partial z/\partial b\\ \partial x/\partial c, & \partial y/\partial c, & \partial z/\partial c\end{vmatrix}.\,da\,db\,dc.$$

Since the fluid is assumed incompressible, the *equation of continuity* is

(c) $$\begin{vmatrix} \partial x/\partial a, & \partial y/\partial a, & \partial z/\partial a \\ \partial x/\partial b, & \partial y/\partial b, & \partial z/\partial b \\ \partial x/\partial c, & \partial y/\partial c, & \partial z/\partial c \end{vmatrix} = 1.$$

To apply Lagrange's equations, we will consider a fluid mass, which turns with a constant angular velocity ω about the z-axis, directed vertically downward. We then have $X=0$, $Y=0$, $Z=g$, and we set $z=c$, $a=r\cos\phi$, $b=r\sin\phi$, and further,

$$x = r\cos(\phi+\omega t) = a\cos\omega t - b\sin\omega t,$$
$$y = r\sin(\phi+\omega t) = b\cos\omega t + a\sin\omega t.$$

From these relations it follows that

$$\partial x/\partial a = \cos\omega t,\quad \partial x/\partial b = -\sin\omega t,\quad \partial x/\partial c = 0,$$
$$\partial y/\partial a = \sin\omega t,\quad \partial y/\partial b = \cos\omega t,\quad \partial y/\partial c = 0;$$
$$\partial z/\partial a = 0,\quad \partial z/\partial b = 0,\quad \partial z/\partial c = 1.$$
$$\ddot{x} = -\omega^2 x,\quad \ddot{y} = -\omega^2 y,\quad \ddot{z} = 0.$$

The equation of continuity (c) is satisfied and the equations of motion (b) are $\partial p/\partial a = \rho\omega^2 a$, $\partial p/\partial b = \rho\omega^2 b$, $\partial p/\partial c = g\rho$. Hence we obtain by integration $p = C + \rho\left(\tfrac{1}{2}\omega^2(a^2+b^2) + gc\right)$. This solution agrees with that given in XL.

SECTION XLVI. WAVE MOTIONS.

Lagrange's equations may be used to advantage in investigations on wave motions in a fluid acted on by gravity. All the particles of the fluid may be assumed to move in plane curves parallel to the xz-plane; let the x-axis be horizontal, and the z-axis be directed perpendicularly downward. Then, if we set $y=b$, we obtain

$$\partial x/\partial b = 0,\quad \partial y/\partial a = 0,\quad \partial y/\partial b = 1,\quad \partial y/\partial c = 0,\quad \partial z/\partial b = 0 \text{ and } \ddot{y} = 0.$$

If we further set $p = \rho P$, the equations of motion given in XLV. (b) become

(a) $$\begin{cases} \ddot{x}\,.\,\partial x/\partial a + (\ddot{z}-g)\,.\,\partial z/\partial a + \partial P/\partial a = 0, \\ \ddot{x}\,.\,\partial x/\partial c + (\ddot{z}-g)\,.\,\partial z/\partial c + \partial P/\partial c = 0, \end{cases}$$

and the equation of continuity XLV. (c) takes the form

(b) $$\partial x/\partial a\,.\,\partial z/\partial c - \partial z/\partial a\,.\,\partial x/\partial c = 1.$$

Suppose the particle B (Fig. 47), having, while in its position of equilibrium, the coordinates $OA=a$ and $AB=c$, to move in a circle

DFE, whose centre is at *C*. Let *D* be the position of the particle, represent the angle between *CD* and the perpendicular *CE* by Θ, and set $BC = s$, $CD = r$. We then have

(c) $$x = a + r\sin\Theta,\quad z = c + s + r\cos\Theta,\quad \Theta = mt + na.$$

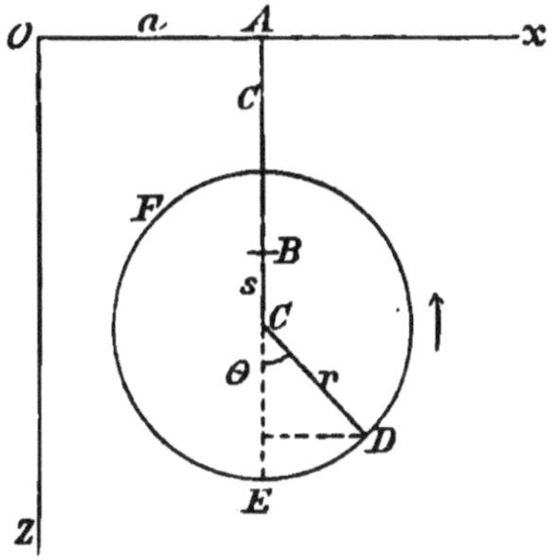

Fig. 47.

Here m and n are constants and r and s are functions of c. We therefore obtain

$$\partial x/\partial a = 1 + nr\cos\Theta,\quad \partial z/\partial a = -nr\sin\Theta,\quad \partial x/\partial c = \partial r/\partial c\,.\sin\Theta,$$
$$\partial z/\partial c = 1 + \partial s/\partial c + \partial r/\partial c\,.\cos\Theta.$$

By these relations, equation (b) takes the form

$$\partial s/\partial c + nr\,.\,\partial r/\partial c + \{nr(1 + \partial s/\partial c) + \partial r/\partial c\}\cos\Theta = 0.$$

Since this equation holds for all values of t or Θ, we have

(d) $$\partial s/\partial c + nr\,.\,\partial r/\partial c = 0 \text{ and } \partial r/\partial c + nr(1 + \partial s/\partial c) = 0.$$

We obtain further from equations (a) the relations

(e) $$-(m^2 - gn)r\sin\Theta + \partial P/\partial a = 0,$$

$$-m^2r\,.\,\partial r/\partial c - g(1 + \partial s/\partial c) - m^2r(1 + \partial s/\partial c)\cos\Theta - g\,.\,\partial r/\partial c\,.\cos\Theta + \partial P/dc = 0,$$

the last of which is transformed by the help of equations (d) into

(f) $$-(m^2 - gn)r\{\partial r/\partial c + (1 + \partial s/\partial c)\cos\Theta\} - g + \partial P/\partial c = 0.$$

If the pressure depends on c only, it follows from (e) and (f) that (g) (h) $m^2 = gn$, $P = gc$, if the constant is set equal to zero; the pressure therefore disappears for $c = 0$. This condition must hold at the free surface of the fluid.

The paths of the particles are circles. If the time required by the particle to traverse its path is T, that is, if T is the period of oscillation, we have $m = 2\pi/T$ and $\Theta = 2\pi/T\,.\,(t + 2\pi a/Tg)$.

If λ is the wave length and h the velocity of the wave, we will have $h = Tg/2\pi$ and $h = \lambda/T$, from which it follows that

(i) $$h = \sqrt{g\lambda/2\pi} \text{ and } n = 2\pi/\lambda.$$

The motion of the particle is such that it describes a circle whose centre lies a little above the position of equilibrium of the particle. From the first of equations (d) we have $s = -\frac{1}{2}nr^2$, where the constant disappears, since s and r vanish simultaneously. Hence also (k) $s = -\pi r^2/\lambda$. From the second of equations (d) it follows that $d \log r + nd(c+s) = 0$, and hence, by integration, $\log r + n(c+s) = k$, when k is a constant. For a particle on the surface we have $c = 0$; if the values of r and s for this particle are designated by R and S, we have $\log R + nS = k$. We have further $\log(r/R) + n(c+s-S) = 0$. The factor $c+s-S = H$ is the perpendicular distance between the centre of the path of the particle considered and the centre of the path of a particle in the surface. We have therefore (l) $r = Re^{-2\pi H/\lambda}$.

If ds/dc is eliminated from equations (d) we obtain

$$dr + nr(dc - nrdr) = 0,$$

and by integration $1/n \,.\, \log r + c - \frac{1}{2}nr^2 = k'$. For the particles on the surface we have $1/n \,.\, \log R - \frac{1}{2}nR^2 = k'$. Hence

(m) $$c = \lambda/2\pi \,.\, \log(R/r) - \pi/\lambda \,.\, (R^2 - r^2).$$

The free surface can be thought of as formed by the rolling of a circular cylinder on the under side of a horizontal surface AB (Fig. 48),

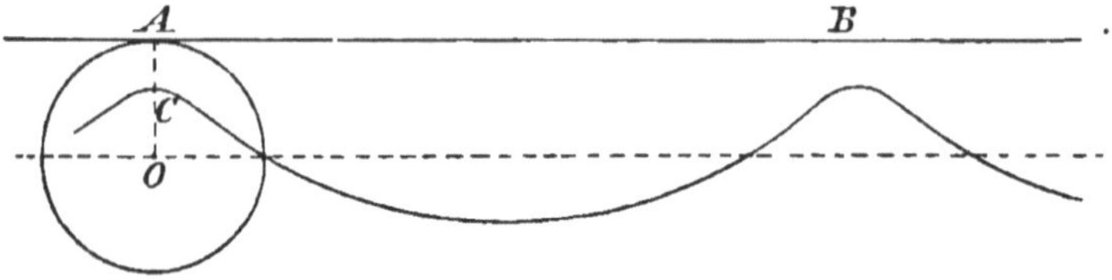

FIG. 48.

which lies at the height $OA = \lambda/2\pi$ over the centres of the paths which the particles in the surface describe. The free surface is then represented by a straight line whose distance from the axis of the cylinder is R.

CHAPTER V.

INTERNAL FRICTION.

SECTION XLVII. INTERNAL FORCES.

IN the discussion of the motion of fluids, the friction among the fluid particles has not been considered. Friction is excited in different degrees between the particles of the fluid when they move among themselves at different rates. In consequence of friction the viscosity of the fluid is more or less great. We will try to determine the friction caused by the motion of the fluid.

We will suppose that the particles of a fluid mass are moving in a direction parallel to the x-axis, and that those situated at the same distance from the xz-plane have the same velocity. The velocity increases in proportion to the distance from the xy-plane. One sheet of the fluid glides over another and thereby gives rise to a definite frictional resistance, which, according to Newton, may be assumed proportional to the rate of change of velocity with respect to the distance from the xz-plane, so that $du/dy = \epsilon$. The friction between two contiguous sheets is then proportional to the difference of their velocities, and inversely proportional to the distance between them. We therefore set the velocity $u = u_0 + \epsilon y$. Let OO' (Fig. 49) be a part of the fluid mass; on each unit of surface of $O'B$ there acts, in the direction Ox, a tangential force

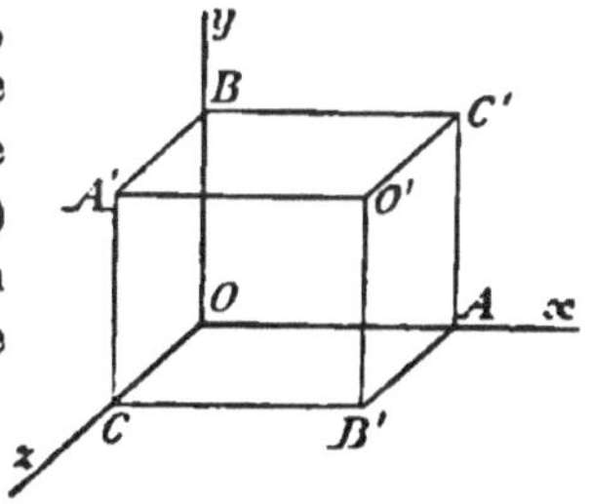

FIG. 49.

(a) $$T = \mu \,.\, du/dy = \mu\epsilon,$$

where μ is the *coefficient of friction*. A force $-T$ acts on OB' in the direction Ox. Further, the tangential forces T [XXV. (d)] must act on $O'A$ and OA', of which the one acting on $O'A$ is in the direction Oy, and the one acting on OA' is in the direction yO.

If the fluid mass moves in the direction of the y-axis with a velocity $v = v_0 + \epsilon' x$, the tangential force necessary to produce this motion is $T' = \mu . dv/dx$. If both motions exist simultaneously, a tangential force X_y acts on the fluid, such that we have

$$X_y = T + T' = \mu(du/dy + dv/dx).$$

In order to examine the physical meaning of this expression, we will consider a fluid particle, originally situated at the point x, y, z, which has moved through an infinitely small distance, whose projections on the axes are ξ, η, ζ. We then have $u = \partial\xi/\partial t$, $v = \partial\eta/\partial t$, and $X = \mu . \partial/\partial t(\partial\xi/\partial y + \partial\eta/\partial x) = 2\mu . \partial x_y/\partial t$. This expression gives the tangential force which arises from motion in a fluid in terms of the friction. We may therefore set

$$\text{(b)} \qquad \begin{cases} Z_y = 2\mu . \partial z_y/\partial t = \mu(\partial w/\partial y + \partial v/\partial z), \\ X_z = 2\mu . \partial x_z/\partial t = \mu(\partial u/\partial z + \partial w/\partial x), \\ Y_x = 2\mu . \partial y_x/\partial t = \mu(\partial v/\partial x + \partial u/\partial y). \end{cases}$$

We know by experiment that μ is independent of the pressure. The meaning of the other quantities in (b) is clear without explanation.

By the help of the formulas given in (b), we can determine the tangential forces which must act in the fluid to overcome the frictional resistances. We will now determine the magnitudes of the normal forces which are necessary for the extension of a viscous fluid in a given direction. Let the fluid move in a direction parallel to AB (Fig. 50), and let the velocity of a particle situated at the distance y from this line be equal to u. As before, we may set $u = v_0 + \epsilon y$.

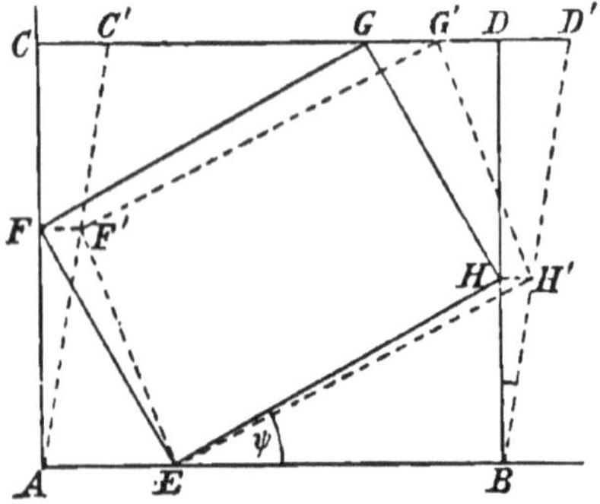

FIG. 50.

After the lapse of the time dt, A has traversed the distance $u_0 dt$, and C the distance $(u_0 + \epsilon AC)dt$. CC' represents the motion of the point C relative to A, and we have $CC' = \epsilon . AC . dt$. If we designate the angle CAC' by $d\phi$, we have (c) $d\phi = \epsilon . dt$.

The rectangle $EFGH$, described in the rectangle $ABCD$, transforms into the parallelogram $EF'G'H'$, and we determine the increments which the sides EH and EF receive by this transformation. Representing the angle HEB by ψ, and noticing that HH' and FF' are parallel to AB, we have $EH' = EH + HH' \cos\psi$, $EF' = EF - FF' \sin\psi$. We further have

$$HH' = BH \,.\, d\phi = EH \sin\psi \,.\, d\phi, \quad FF' = AF \,.\, d\phi = EF \cos\psi \,.\, d\phi,$$

and hence

$$(EH' - EH)/EH = \sin\psi \cos\psi d\phi;$$

$$(EF' - EF)/EF = -\sin\psi \cos\psi d\phi.$$

If the increment of length per unit-length of EH is designated by ds, we have (d) $ds = \sin\psi \cos\psi d\phi$; ds is also the diminution of length per unit-length of EF.

To bring about the deformation considered, a tangential force T must act on $ABCD$, which is, from (a), (e) $T = \mu\epsilon$. This force acts on the surface corresponding to CD in the direction CD, and on that corresponding to AB in the direction BA; on the other two surfaces the forces act in the directions CA and BD. To determine the normal force N acting on the surface EF, we set in XXIV. (a) $\alpha = \psi$, $\beta = \frac{1}{2}\pi - \psi$, $\gamma = \frac{1}{2}\pi$, and $X_y = Y_x = T$. Since all the other components of stress are equal to zero, we obtain

(f) $$N = P \cos\psi + Q \sin\psi = 2T \sin\psi \cos\psi.$$

From (c) and (d) we have $ds = \sin\psi \cos\psi \,.\, \epsilon \,.\, dt$, and from (e) and (f) $N = 2\mu\epsilon \sin\psi \cos\psi$. Hence, we have (g) $N = 2\mu \,.\, ds/dt$. The stress acting on the surface EH is $-N$. It has been shown that a unit of length, in the direction EF, is increased by $-ds$. If a normal stress were to act on the surface of $ABCD$, it would have no influence on the deformation; but a normal force $S + N$ would act on EF, and a normal force $S - N$ on EH.

If the normal stresses X_x, Y_y, Z_z act on a rectangular parallelepiped whose edges are parallel to the coordinate axes, when the fluid is in motion, they cause deformations and a change of volume. If, as in the theory of elasticity, we set the volume dilatation $\Theta = x_x + y_y + z_z$, then $x_x - \frac{1}{3}\Theta$ is the part of the increase in the direction of the x-axis which is here considered. Similarly, we set $3S = X_x + Y_y + Z_z$, and $X_x - S$ is the part of the normal force in the direction of the x-axis which causes the deformation. By the help of (g), we obtain

$$X_x - S = 2\mu(\partial x_x/\partial t - \tfrac{1}{3}\partial\Theta/\partial t).$$

If, finally, we set for $-S$ a quantity p, which may be considered a pressure, on account of its analogy with the pressure in ideal fluids and gases, and remember that $x_x = \partial\xi/\partial x$, $\partial x_x/\partial t = \partial u/\partial x$, etc., we will have (h) $X_x = -p + 2\mu \,.\, \partial u/\partial x - \frac{2}{3}\mu(\partial u/\partial x + \partial v/\partial y + \partial w/\partial z)$. Analogous expressions hold for Y_y and Z_z.

From equation (h) the dimensions of μ are $ML^{-1}T^{-1}$. The coefficient μ has been determined for many fluids and gases. It changes very much with the temperature. The following values hold for 0° C.:

Water, 0,01775; Alcohol, 0,01838; Air, 0,000182.

SECTION XLVIII. EQUATIONS OF MOTION OF A VISCOUS FLUID.

We will now present the equations of motion of a fluid exhibiting internal friction. From XXV. the components of stress act on the unit of volume in the direction of the x-axis with the force

$$(X) = \partial X_x/\partial x + \partial X_y/\partial y + \partial X_z/\partial z.$$

If U is the velocity of a single particle of the fluid in the direction of the x-axis, then $\rho\dot{U} = (X) + \rho X$, where, in the usual notation, X denotes the component of force in the direction of the x-axis. From XLVII. (b) and (h), we have

(a) $$\rho\dot{U} = \rho X - \partial p/\partial x + \mu\nabla^2 u + \tfrac{1}{3}\mu \,.\, \partial(\partial u/\partial x + \partial v/\partial y + \partial w/\partial z)/\partial x.$$

This equation, and those analogous to it, which hold for the velocities V and W in the directions of the y- and z-axes, are due to Stokes.* They hold in connection with the equation of continuity

(b) $$\partial\rho/\partial t + \partial(\rho u)/\partial x + \partial(\rho v)/\partial y + \partial(\rho w)/\partial z = 0.$$

We assume that the fluid is incompressible, and have

(c) $$\partial u/\partial x + \partial v/\partial y + \partial w/\partial z = 0.$$

(d) $$\left\{\begin{aligned} \rho(\dot{u} + u\,.\,\partial u/\partial x + v\,.\,\partial u/\partial y + w\,.\,\partial u/\partial z) &= \mu\nabla^2 u + \rho X - \partial p/\partial x, \\ \rho(\dot{v} + u\,.\,\partial v/\partial x + v\,.\,\partial v/\partial y + w\,.\,\partial v/\partial z) &= \mu\nabla^2 v + \rho Y - \partial p/\partial y, \\ \rho(\dot{w} + u\,.\,\partial w/\partial x + v\,.\,\partial w/\partial y + w\,.\,\partial w/\partial z) &= \mu\nabla^2 w + \rho Z - \partial p/\partial z. \end{aligned}\right.$$

The equations are simplified if the motion is steady, that is, if $\dot{u} = 0$, $\dot{v} = 0$, $\dot{w} = 0$. If the velocity is very small, the terms $u\,.\,\partial u/\partial x$, $v\,.\,\partial u/\partial y$, etc., may be neglected; we then have

(e) $\mu\nabla^2 u + \rho X - \partial p/\partial x = 0$, $\mu\nabla^2 v + \rho Y - \partial p/\partial y = 0$, $\mu\nabla^2 w + \rho Z - \partial p/\partial z = 0$.

If the forces have a potential Ψ, (f) $\nabla^2 p + \rho\nabla^2\Psi = 0$. If we introduce the components of rotation

$$\xi = \tfrac{1}{2}(\partial w/\partial y - \partial v/\partial z), \quad \eta = \tfrac{1}{2}(\partial u/\partial z - \partial w/\partial x), \quad \zeta = \tfrac{1}{2}(\partial v/\partial x - \partial u/\partial y)$$

and if the forces have a potential, we have from (e)

(g) $$\nabla^2\xi = 0, \quad \nabla^2\eta = 0, \quad \nabla^2\zeta = 0.$$

Further we have (h) $\partial\xi/\partial x + \partial\eta/\partial y + \partial\zeta/\partial z = 0$.

* Stokes, *Cambridge Phil. Tr.*, Vol. VIII., p. 297, 1845.

With reference to the boundary conditions, it is assumed that the particles of the fluid which are in contact with solid boundaries have no relative motion with respect to them; at the boundary of the fluid we have therefore $\bar{u}=0$, $\bar{v}=0$, $\bar{w}=0$, if $\bar{u}$, $\bar{v}$, and $\bar{w}$ represent the components of velocity at the bounding surface. If solids are present in the moving fluid, we may generally assume that each particle in the surface of the solid has the same velocity as the particle of the fluid which is in contact with it.

SECTION XLIX. FLOW THROUGH A TUBE OF CIRCULAR CROSS SECTION.

We consider a viscous fluid moving slowly through a narrow tube, which is set horizontal, so that gravity does not influence the motion. Let the axis of the tube be taken as the z-axis, and suppose that the particles of the fluid move parallel to it. We then have

$$u=0, \quad v=0.$$

Equations XLVIII. (e), (c) and (f) then become

(a) $\partial p/\partial x=0$, $\partial p/\partial y=0$, $\mu\nabla^2 w=\partial p/\partial z$; (b), (c) $\partial w/\partial z=0$; $\nabla^2 p=0$.

From (c) it follows that (d) $d^2p/dz^2=0$ and $p=fz+p_0$, where f and p_0 are constant.

It follows further from (a) that $\mu\nabla^2 w=f$. Since w depends on the distance r of the particle from the axis of the tube, we have, since $r^2=x^2+y^2$,

$$\nabla^2 w=d^2w/dr^2+1/r\,.\,dw/dr.$$

Hence we obtain $d^2w/dr^2+1/r\,.\,dw/dr=f/\mu$. By integration

$$w=c\log r+fr^2/4\mu+w_0.$$

Since w has a finite value for $r=0$, the constant c must equal 0. Therefore (e) $w=w_0+fr^2/4\mu$, where w_0 is the velocity in the axis of the tube. If the pressure is equal to p_0 when $z=0$ and to p_1 when $z=l$, we have from (d) $f=(p_1-p_0)/l$. If we substitute this value of f in equation (e), we have $w=w_0-r^2\,.\,(p_0-p_1)/4\mu l$. For all particles of the fluid which are in contact with the wall of the tube, we have $w=0$. Representing by R the radius of the tube, we will therefore have $0=w_0-R^2\,.\,(p_0-p_1)/4\mu l$. We obtain finally

$$w=(p_0-p_1)(R^2-r^2)/4\mu l.$$

The volume m of the fluid which flows in one second through a cross-section of the tube is given by

$$\text{(f)} \qquad m = \int_0^R 2\pi r dr \,.\, w = \pi(p_0 - p_1)R^4/8\mu l,$$

that is, *the volume of the fluid is directly proportional to the fourth power of the radius of the tube, inversely proportional to its length, and inversely proportional to the constant* μ.

Poiseuille was the first who investigated the flow of a fluid through narrow tubes; he was led to results which agree with the above formulas.*

* Among recent works on hydrodynamics are to be mentioned: Lamb, *Treatise on the Motion of Fluids.* Cambridge, 1879. Auerbach, *Die Theoretische Hydrodynamik.* Braunschweig, 1881.

CHAPTER VI.

CAPILLARITY.

SECTION L.—SURFACE ENERGY.

THE form of a fluid mass on which no external forces act is determined by the forces with which its particles act on one another. If the mass is very great, it will take the spherical form, in consequence of the gravitational attraction of its parts; if, on the other hand, the mass is small, the force of gravitation between the particles will have no perceptible influence. If the force of gravitation can be neglected, the force of cohesion, which acts in every fluid mass, tends to bring it into the same spherical form. From researches which have been made on the mode of action of this force, it appears that it acts only between particles which are at very small distances from one another. The law of its dependence on the distance between the particles is not yet known. We may nevertheless develop the laws of capillarity by the use of a method which does not require a knowledge of that law.

If the form of a fluid mass is originally a sphere, work must be done to change it into any other form. If the fluid offers no frictional resistance, this work can be due only to the fluid particles situated in or near the surface; since the only particles which can act on a particle at a greater distance from the surface are those which immediately surround it; these either remain in their positions or are replaced by others which act in the same way as those replaced. The work done is therefore expended in adding new particles to those already present in the surface, or, what is the same thing, in enlarging the surface. To increase the surface S of the fluid by the infinitely small quantity dS, the work CdS is necessary; C is constant and may be called the *capillary constant.*

Two bodies in general meet in a surface, and C depends on the character of these two bodies. In the case of a falling raindrop the two bodies in contact are water and air. At the surface of a drop of oil which floats in a mixture of water and alcohol, as in the well-known experiment of Plateau, two liquids are in contact. Even when a fluid is in contact with a solid, or when two solids are in contact, the common surface possesses a definite surface energy.

Let the capillary constant of two bodies a and b be C_{ab}, and let S be the surface in which the two bodies meet. The potential energy E_p of the surface S is (a) $E_p = C_{ab} . S$. Since $C_{ab} = E_p/S$, and since the dimensions of E_p and S are $L^2T^{-2}M$ and L^2 respectively, the dimensions of the capillary constant are $T^{-2}M$.

The surface of the fluid is under a definite tension somewhat analogous to that of an elastic membrane. If a rectangle $DEFG$ is described in a plane surface of a fluid, and if three sides of it retain their positions unchanged while the fourth side FG, along with the particles of the fluid present in it, is moved through the distance FH in the direction EF, the surface is enlarged by the area $FG . FH$, and the surface energy is increased by $C . FG . FH$, where C is the capillary constant. To produce the motion considered, a force K must act on FG; the work done is therefore equal to $K . FG . FH$. It follows that (b) $K = C$, or *the tension per unit-length of the fluid surface is numerically equal to the capillary constant.*

This tension existing in the surface exerts a pressure in the fluid. Let P (Fig. 51) be a point in the surface which is supposed convex in the neighbourhood of P. Suppose two plane sections erected at P, which contain the normal to the surface at that point. One of these planes cuts the surface in the curve PA, the other in the curve PB. PA and PB shall intersect at right angles and their radii of curvature shall be the principal radii of curvature of the surface at the point P. A third plane containing the normal to P cuts the surface in the curve PF, whose radius of curvature R is determined from Euler's theorem by the equation (c) $1/R = \cos^2\phi/R_1 + \sin^2\phi/R_2$, where ϕ is the angle between PA and PF. About the point P as a centre we suppose described a sphere of infinitely small radius which cuts the surface in the curve $AFBDE$. The element FG of this curve is acted on by the tension $C . FG$, proceeding from the adjacent parts of the surface. FG may be set equal to $r d\phi$ and the tension to $Cr d\phi$. Its direction makes an angle with the normal to the surface, whose cosine is r/R; hence the force acting in the direction of the normal

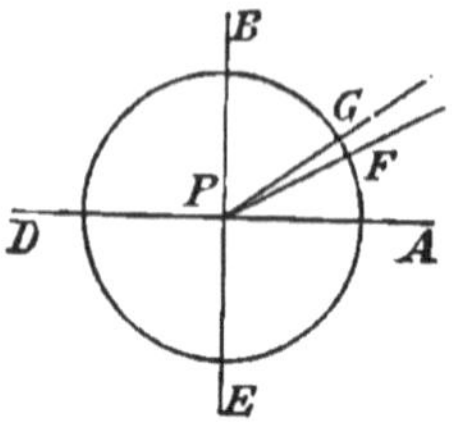

FIG. 51.

is $Cr^2/R \, . \, d\phi$. The surface tension therefore draws the surface-element $ABDE$ toward the interior of the fluid with a force

$$Cr^2 \int_0^{2\pi} d\phi/R.$$

Therefore if P represents the pressure on unit of surface we have

$$P\pi r^2 = Cr^2 \int_0^{2\pi} d\phi/R \quad \text{and} \quad P = C/\pi \, . \int_0^{2\pi} d\phi/R.$$

If for R we introduce the value given in (c), we obtain

(d) $$P = C(1/R_1 + 1/R_2).$$

It is probable that in addition to the pressure here found, which arises from the curvature of the surface, there also exists a constant pressure M which acts in the fluid when its surface is plane. The total pressure due to capillary forces is therefore $M + C(1/R_1 + 1/R_2)$, where M and C depend on the character of the two bodies which are in contact in the surface. Since the phenomena of capillarity do not permit of the measurement of this quantity M, it need not be further considered.

At 20° C. the value of C for the surface of contact between water and air is 81, between mercury and air 540, and between mercury and water 418.

Section LI. Conditions of Equilibrium.

Equilibrium exists in a fluid mass if its potential energy remains unchanged when the position and form of the mass are changed by an infinitely small amount. Since the energy depends on the extent of surface we must obtain an expression for the increment δS of the surface. Suppose a fluid A surrounded by another fluid B, the two fluids being such that they do not mix. If no external forces act on them, A will assume the spherical form.

Suppose the surface S to be concave toward A and to move toward B so that it undergoes an infinitely small change of form. Let s be the contour of the surface S, and S' represent that surface after the change of form has occurred. The contour of S' may be represented by s'. Erect at all points of s normals to S which cut the surface S' in a new curve σ, which may be supposed to lie within s'. If we designate the infinitely small distance between σ and s' by δl, the part of S' which lies between σ and s' will be given by (b) $\int \delta l \, . \, ds$.

We now erect at a point P in S the normal PP' which cuts S' at P'; set $PP' = \delta\nu$, and draw through P on the surface S two curves PE and PF, one of which corresponds to the maximum curvature of the surface at P, the other to the minimum. These principal curves and two others infinitely near them will bound a rectangle $PEQF$, whose sides $PE = a$ and $PF = b$ are infinitely small. If R_1 and R_2 are the principal radii of curvature, there will always be two angles α and β such that $a = R_1\alpha$, $b = R_2\beta$, and therefore $dS = a\,.\,b = R_1R_2\alpha\beta$. The normals to S erected at E, Q and F, intersect S' at E', Q', F'. We set $P'E' = a'$, $P'F' = b'$ and obtain $a' = (R_1 + \delta\nu)\alpha$, $b' = (R_2 + \delta\nu)\beta$ If S_1' is the part of S' bounded by σ, we will have

$$dS_1' = a'b' = (R_1R_2 + (R_1 + R_2)\delta\nu)\alpha\beta \text{ and } d(S_1' - S) = (1/R_1 + 1/R_2)\delta\nu dS.$$

We have therefore (c) $S_1' - S = \int(1/R_1 + 1/R_2)\delta\nu dS$. The total increment δS which S receives in consequence of the change of form is therefore (d) $\delta S = \int(1/R_1 + 1/R_2)\delta\nu dS + \int\delta l\,.\,ds$. This expression remains valid even if δl and $\delta\nu$ at particular points or at all points of the surface S are negative. If the fluid mass is bounded by a single surface, the contour s will be zero; if the contour is fixed we have $\delta l = 0$. In both cases the condition of equilibrium is

$$\text{(e)} \qquad \int(1/R_1 + 1/R_2)\delta\nu dS = 0.$$

Since the space occupied by the fluid mass is supposed constant, we have (f) $\int\delta\nu dS = 0$, since $\int\delta\nu dS$ represents the increment of volume.

From (e) and (f) it follows that (g) $1/R_1 + 1/R_2 = c$, where c is a constant. The same result is also given by L. (d) if we notice that the pressure in the fluid mass must be constant.

If three fluids which do not mix meet in a line, the three angles which the surfaces of the fluids make with one another may be determined. Such relations occur if a drop of oil lies on the surface of water. In this case the three fluids which meet are water, oil, and air. We shall designate these fluids, for greater generality, by a, b, and c; let the energy of a unit area of the surface separating a and b be C_{ab}; let C_{ac} and C_{bc} have similar meanings. It is sufficient to examine the case in which the edge is a straight line. The directions of the three surface tensions C_{ab}, C_{ac}, and C_{bc}, determine the inclination of the surfaces to each other; equilibrium exists when the three forces C_{ab}, C_{ac}, C_{bc} are in equilibrium. Let α, β, γ be the three angles sought, belonging respectively to the three fluids a, b, and c. We then have as the condition of equilibrium,

$$\text{(h)} \qquad C_{bc}/\sin\alpha = C_{ac}/\sin\beta = C_{ab}/\sin\gamma.$$

From these equations we may in general determine α, β, γ. If, however, one of the tensions, say C_{bc}, is greater than the sum of the other two, equilibrium cannot exist. In this case, the fluid spreads out into a very thin sheet which separates the fluids b and c; we have as an example the behaviour of a drop of oil of turpentine on water.

The theorem given in (h) may be also obtained by the following method: If the edge is displaced by an infinitely small distance from its original position, the sum of the surface energies

$$C_{bc} \,.\, S_1 + C_{ac} \,.\, S_2 + C_{ab} \,.\, S_3,$$

where S_1, S_2, and S_3 designate the three surfaces of separation, is increased by $C_{bc}\delta S_1 + C_{ac}\delta S_2 + C_{ab}\delta S_3$. This increment must be equal to zero, if equilibrium exists; we thus obtain equations (h).

If a solid c (Fig. 51) is in contact with two fluids a and b, the edges may be displaced infinitely little along the surface of the solid. In this case we have $\delta S_1 = -\delta S_2$, $\delta S_3 = -\delta S_1 \,.\, \cos \alpha$, and further, $C_{bc} - C_{ac} = C_{ab} \,.\, \cos \alpha$. Hence we have (i) $\cos \alpha = (C_{bc} - C_{ac})/C_{ab}$, where α is the so-called *contact angle.*

SECTION LII. CAPILLARY TUBES.

To make an application of the foregoing principles, we will consider a cylindrical tube c (Fig. 51 A) placed perpendicularly, the lower end of which is immersed in a fluid b; the upper part of the tube is surrounded by air, which may be represented by a. The bounding surfaces may be represented as before by S_1, S_2, and S_3. We may call the surface of contact between the fluid and the tube S_1, that between the air and the tube S_2, and that between the air and the fluid S_3. The fluid surface MM outside the tube may be infinitely great; it may also be considered as at rest, even when the surface in the tube is in motion.

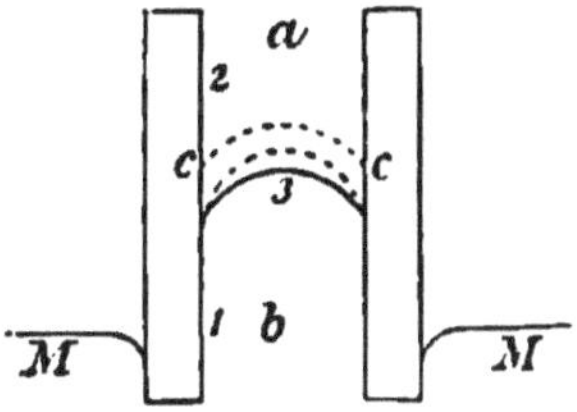

FIG. 51 A.

Take the surface MM for the xy-plane, and let the z-axis be directed perpendicularly upward. If g represents the acceleration of gravity, and ρ the density of the fluid, the potential energy of a particle of the fluid ρdv, will be $g \,.\, \rho dv \,.\, z$. Hence the potential energy of the fluid mass lying above the xy-plane is

$$\iiint g\rho z\,dx\,dy\,dz = \tfrac{1}{2}g\rho \iint z^2 dx\,dy,$$

if x, y, z are from now on considered as belonging to S_3. The part of the potential energy E_p whose variations are to be considered is

$$E_p = \tfrac{1}{2} g\rho \int\int z^2 dxdy + C_{ab}S_3 + C_{ac}S_2 + C_{bc}S_1.$$

In the case of equilibrium we have $\delta E_p = 0$, or

(a) $$0 = g\rho \int\int z\delta z dxdy + C_{ab}\delta S_3 + C_{ac}\delta S_2 + C_{bc}\delta S_1.$$

If s is the length of the line of section of the surface S_3 with the inner surface of the tube, if ϕ represents the angle between ds and the xy-plane, and if all points of the surface S_3 are elevated by the same infinitely small amount δz, where δz is constant, we will have $\delta S_3 = 0$, $\delta S_1 = -\delta S_2 = \int \cos\phi \delta z ds$. The equation (a) then becomes

(b) $$g\rho \int\int z dxdy = (C_{ac} - C_{bc}) \int \cos\phi ds.$$

Hence the fluid will be displaced by the difference of the tensions in the surfaces S_2 and S_1.

If, on the other hand, the line of section s retains its position, and if the only change is the change in the shape of the surface S_3, we will have from LI. (d) $\delta S_3 = \int (1/R_1 + 1/R_2)\delta\nu dS_3$ and $\delta S_1 = \delta S_2 = 0$, in which $\delta\nu$ is an element of a normal lying between the surfaces S and S'.

We may replace $\delta z dxdy$ by $\delta\nu dS_3$, and obtain from equation (a)

$$\int \{g\rho z + C_{ab}(1/R_1 + 1/R_2)\}\delta\nu dS_3 = 0.$$

Since $\delta\nu$ is arbitrary, we must have (c) $g\rho z + C_{ab}(1/R_1 + 1/R_2) = 0$. If the curvature of the surface is expressed by the differential coefficients of z with respect to x and y, we will obtain from (c) a differential equation for the determination of the form of the surface. If the contact angle is also given, the surface is completely determined.

If the cross-section of the tube is circular and very narrow, we may assume that approximately $R_1 = R_2 = r/\cos\alpha$, where r is the radius of the tube and α the contact angle. The height z to which the fluid rises is then $z = -2\cos\alpha / g\rho r \,.\, C_{ab}$. We obtain the same result from equation (b) if we set $\int\int z dxdy = \pi r^2 z$, $\int \cos\phi ds = 2\pi r$, and use equation LI. (i).

The theory of capillarity was discussed by Laplace in a supplement to the tenth book of the *Mécanique Céleste*. Poisson wrote a larger work on the subject, called *Nouvelle Théorie de l'Action Capillaire*: Paris, 1831. Finally Gauss made an epoch-making investigation on the theory of capillarity, published in the *Commentationes Soc. Scient.* Göttingensis. Vol. VII. 1830. (Works., Vol. V., p. 29.) The most elaborate recent publication on the subject is that of Mathieu, *Théorie de la Capillarité*: Paris, 1883.

CHAPTER VII.

ELECTROSTATICS.

SECTION LIII. FUNDAMENTAL PHENOMENA OF ELECTRICITY.

THE theory of electricity is founded upon the observation that amber and other bodies obtain by friction the property of attracting light bodies. Gray showed that this property may be transferred from one body to another. The conception was thus suggested that this property depends upon the presence of a *fluid* which is formed or set free by friction in the body, and which, under certain conditions, can pass from body to body. Dufay first showed that there are two so-called electrical conditions, or according to the conception just stated, two fluids, which Franklin named *positive* and *negative*, since they can completely neutralize each other.

The hypothesis of two fluids has had an extraordinary influence on the development of the theory of electricity. Poisson proceeded from this conception in his researches on electrical distribution, and W. Weber founded on it his theory of electrical currents.

In opposition to this theory, which, in its mathematical treatment, proceeds from the conception that electrical action is, like gravity, a force acting at a distance, Faraday adopted the view that the electrical forces propagate themselves from particle to particle, not immediately, therefore, but by the action of an intervening medium. On this view it is not possible to explain the phenomena of electricity completely without introducing a hypothetical medium, the *ether*, and we thus meet with peculiar difficulties, which have not yet been entirely overcome. Fruitful as Faraday's conceptions have been, we still cannot explain many phenomena, even by their help, and a complete systematic discussion of electricity cannot yet be given. In the following presentation it has not been possible to consider the whole subject from one point of view; we have only endeavoured to present the most important results which have been obtained.

The starting point of our study is Coulomb's investigation of the mechanical force which two electrified bodies exert on each other. Two bodies which carry the charges e_1 and e_2, measured in any manner and separated from each other by the distance r, will then act on each other, according to Coulomb's law, with a force $\mathfrak{F} = c \, . \, e_1 e_2 / r^2$, where c is a constant. According as the two charges are similar or dissimilar the bodies repel or attract each other. If the distribution of electricity on extended bodies is known, the mechanical force with which the bodies act on each other may be calculated. As a rule, however, the distribution of electricity on a body cannot be considered given. If electricity is developed by friction on a glass rod or a stick of sealing wax, that is, on relatively poor conductors, a slow discharge of the same occurs with lapse of time. The *distribution* of electricity on good conductors depends on their form, on the character of the body surrounding them, and on the charge. In determining the distribution, we start from the assumption that the same force acts between two quantities of electricity as between two bodies which are charged with these quantities of electricity. If a definite charge is imparted to an insulated conductor its separate parts will act on each other, and there will be a definite distribution.

Charged bodies excite an electrical distribution in neighbouring bodies. The attraction which a charged body exerts on an uncharged body is explained by the assumption that positive and negative electricities in equal quantities are present in the latter body, and that under the influence of the charged body a separation of the opposite electricities occurs, the force which proceeds from the charged body acting as an *electromotive* force. The distribution thus produced acts against the external electromotive force, and a condition of equilibrium is brought about if the electromotive force, which arises partly from the force acting from without which causes the distribution, and partly from the electricity separated in the body itself and therefore free, is everywhere zero within the conductor. In poor conductors also as, for example, the air, an electromotive force must arise under similar circumstances, whose action we will at present not consider.

Section LIV. Electrical Potential.

Suppose that electricity of density ρ is contained within the body L (Fig. 52), and that electricity of density σ is present on its surface. A volume-element $d\tau$ then contains the quantity of electricity $\rho \, . \, d\tau$, and a surface-element dS contains the quantity $\sigma . dS$. Suppose that unit quantity of electricity is present at the point P, whose coordinates are x, y, z, and that the charges within L and on its surface act upon it. Let the coordinates of any point in the body L be ξ, η, ζ, and let X, Y, Z be the components of the force which acts at the point P. We then have from Coulomb's law (cf. XII.),

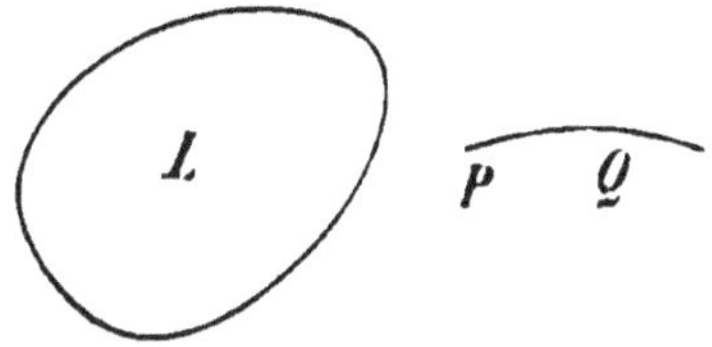

FIG. 52.

$$\text{(a)} \quad \begin{cases} X = \int (x-\xi)/r^3 . \rho d\tau + \int (x-\xi)/r^3 . \sigma dS, \\ Y = \int (y-\eta)/r^3 . \rho d\tau + \int (y-\eta)/r^3 . \sigma dS, \\ Z = \int (z-\zeta)/r^3 . \rho d\tau + \int (z-\zeta)/r^3 . \sigma dS, \end{cases}$$

where $r^2 = (x-\xi)^2 + (y-\eta)^2 + (z-\zeta)^2$.

If we set (b) $\Psi = \int \rho . d\tau/r + \int \sigma . dS/r$, it follows that

$$\text{(c)} \quad X = -\partial\Psi/\partial x, \quad Y = -\partial\Psi/\partial y, \quad Z = -\partial\Psi/\partial z.$$

Ψ is the *electrical potential*. If the unit charge moves from P to Q, and if r' is the distance of the point Q from the point (ξ, η, ζ) of the body L, we will have $\Psi' = \int \rho . d\tau/r' + \int \sigma . dS/r_1$. The work done by the electrical forces during the motion is

$$\int (Xdx + Ydy + Zdz) = \Psi - \Psi'.$$

If the point Q is so far distant from the body L that $\Psi' = 0$, the work done will equal Ψ. We can therefore say: *the electrical potential at a point is equal to the work done by the electrical forces when a unit of electricity moves from the point considered to a point at an infinite distance from the charged body.*

If the point P is within the body L, we describe a sphere K of radius R about the point P as centre, so small that the density ρ within it may be considered constant. The force due to the charge in K is then zero (cf. XIII.); the potential within L due to the electricity present in it cannot, therefore, be infinite.

The potential has the same value on both sides of a surface charged with electricity. If the surface density is σ, the potential Ψ which arises from the surface distribution is $\Psi = \int \sigma . dS/r$. We will consider the values Ψ_1 and Ψ_2 of the potential at two points P_1 and P_2 (Fig. 53) which lie on each side of the surface AB, and are separated by an infinitely small distance. Let the line P_1P_2 be a normal to the surface. We suppose an infinitely small portion of the surface cut off by a circular cylinder, whose axis is the line P_1P_2 and whose radius is R. The potential Ψ_1 is made up of two parts, one of which Ψ_1' arises from the part of the surface which is cut off by the cylinder, and the other $\Psi_1 - \Psi_1'$ from the remaining part of the surface. The latter value, $\Psi_1 - \Psi_1'$, is neither discontinuous nor infinite in the distance from P_1 to P_2. The value Ψ_2 is likewise made up of two parts. The radius of the infinitely small circle cut out of the surface is R. Let n be the distance of P_1 from the surface; we will then have

$$\Psi_1' = 2\pi\sigma . \int_0^R \frac{R\,.\,dR}{\sqrt{R^2+n^2}}, \quad \text{or} \quad \Psi_1' = 2\pi\sigma(\sqrt{R^2+n^2} - n).$$

It therefore appears that Ψ_1' vanishes if R and n are infinitely small.

It has been shown [XIV.] that

$$\partial^2(1/r)/\partial x^2 + \partial^2(1/r)/\partial y^2 + \partial^2(1/r)/\partial z^2 = 0.$$

We have therefore, from (b), for a point outside of L,

$$\text{(d)} \qquad \partial^2\Psi/\partial x^2 + \partial^2\Psi/\partial y^2 + \partial^2\Psi/\partial z^2 = \nabla^2\Psi = 0.$$

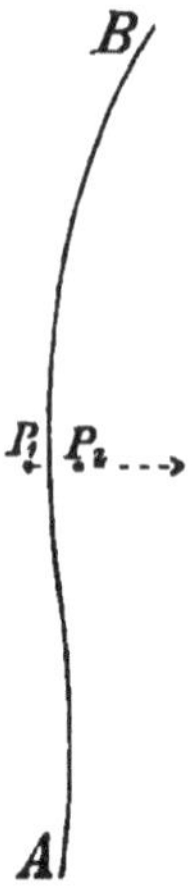

FIG. 53.

On the other hand, if P lies within the body we have [XIV. (h)] (e) $\nabla^2\Psi + 4\pi\rho = 0$. If we designate the normals to the surface drawn inward and outward by ν_i and ν_a, we have for the surface density σ [XIV. (l)], (f) $\overline{\partial\Psi_i}/\partial\nu_i + \overline{\partial\Psi_a}/\partial\nu_a + 4\pi\sigma = 0$. The horizontal lines over the differential coefficients indicate that their values are to be taken at the surface. *Hence if σ represents the electrical density on a surface, the sum of the forces acting in the direction of the normals drawn outward from both faces equals $4\pi\sigma$.*

These properties of the potential hold for every system of bodies charged in any way with electricity. If the distribution is given, the potential can be determined either from (b) or from (e) and (f). If the potential is given, the densities ρ and σ are determined from (e) and (f), while the components of the electrical force are given from (c). The force F, acting in any direction ds, is $F = -\partial\Psi/\partial s$.

SECTION LV. THE DISTRIBUTION OF ELECTRICITY ON A GOOD CONDUCTOR.

If a charge e is communicated to a good conductor, it distributes itself over the conductor. We will determine its volume density ρ

and the surface density σ. If Ψ_i is the potential inside, and Ψ_a the potential outside the conductor, we have from LIV. (d) and (e)

(a) $$\nabla^2\Psi_i + 4\pi\rho = 0 \text{ and } \nabla^2\Psi_a = 0.$$

After equilibrium is attained there is no electrical separation in the interior, that is, we have (b) $\partial\Psi_i/\partial x = 0$, $\partial\Psi_i/\partial y = 0$, $\partial\Psi_i/\partial z = 0$ for all points in the interior of the conductor. Hence (c) $\nabla^2\Psi_i = 0$, and therefore from (a) $\rho = 0$. Hence *the electricity is distributed only on the surface of the conductor.*

From (b) Ψ_i is constant in the interior of the conductor and equal to $\overline{\Psi}$, where $\overline{\Psi}$ is the value of the potential on the surface. We may determine Ψ_a from the equations $\nabla^2\Psi_a = 0$ and $\Psi_a = \overline{\Psi}$ for all points of the surface. The surface density is given by

$$\overline{\partial\Psi_i/\partial\nu_i + \partial\Psi_a/\partial\nu_a} + 4\pi\sigma = 0.$$

Since Ψ_i is constant we obtain (d) $4\pi\sigma = -\overline{\partial\Psi_a/\partial\nu_a}$.

If we represent the force acting outward at the surface of the conductor by F, we have (e) $F = -\overline{\partial\Psi_a/\partial\nu_a} = 4\pi\sigma$, that is, *the force acting at a point on the surface of the conductor in the direction of the normal is equal to the surface density σ at that point multiplied by 4π.*

If ds is an element of a curve drawn on the surface of the conductor, we have $\partial\Psi_a/\partial s = 0$, since Ψ_a is equal to $\overline{\Psi}$ everywhere on the surface, as has just been shown. Hence F has no components in the surface. The surface of the conductor is a *surface of constant potential,* and the direction of the force F is everywhere perpendicular to it.

To determine the potential of the conductor its charge e must be calculated; we have $e = \iint\sigma\,.\,dS$, and therefore

(f) $$e = 1/4\pi\,.\iint F\,.\,dS = -1/4\pi\,.\iint\overline{\partial\Psi_a/\partial\nu_a}\,.\,dS.$$

Since there is no charge in the interior of the conductor, we have $\Psi_a = \iint\sigma\,.\,dS/r$. If the electricity on the conductor, whose density is σ, is in equilibrium, it will remain so if the density becomes $n\sigma$, where n is a number. If one distribution which is in equilibrium is superposed on another, the new distribution is still in equilibrium. If the density is everywhere $n\sigma$, the potential has the value $n\Psi_a$. *The potential is therefore proportional to the charge.* If the charge C will bring the conductor to the potential 1, the charge (g) $Q = C\Psi$ must be imparted to the conductor in order to bring the potential from 0 to Ψ. We call C the *capacity* of the conductor. The capacity C is the ratio of the charge Q of a conductor to its potential Ψ. In order to give a means of representing the magnitude and direction of the electrical force in the region around the conductor, we determine the position of the surfaces of constant potential which surround it. Their equation is $\Psi = c$, where c is a constant. The first equi-potential surface is the surface of the conductor, for which $\Psi = \Psi$. At a distance which is very great in comparison with the dimensions of the conductor, the equi-potential surfaces will be spheres, since in that case the expression reduces to $\Psi = e/r$.

From VII. the *surfaces of constant potential* are perpendicular to the direction of the force. If two such surfaces are considered which are infinitely near each other, it appears from VII. that the force at every point in one of them is inversely proportional to the distance between the surfaces. If a line of force $PP_1P_2P_3$ (Fig. 54) is drawn from a point on the surface, and if the points P_1, P_2, P_3, etc., are so chosen that $F . PP_1 = F_1 . P_1P_2 = F_2 . P_2P_3$, etc., where F_1, F_2 are the electrical forces at the points P_1, P_2, we have a relation which may be carried out to any distance. If we draw the equi-potential surfaces in such a way that the potential, as we pass from one to the next, increases or diminishes regularly by the same amount, the product $F_n . (\overline{P_nP_{n+1}})$ will be constant. The further we pass from the acting quantity of electricity the greater will be the distance between the successive equi-potential surfaces. If the magnitude of the force at one of the equi-potential surfaces is given, its magnitude at another point of the figure may be calculated from the distance between the successive equi-potential surfaces, and if the magnitude of the electrical force F at the surface of the body is given, its magnitude at every other point of the figure may be determined.

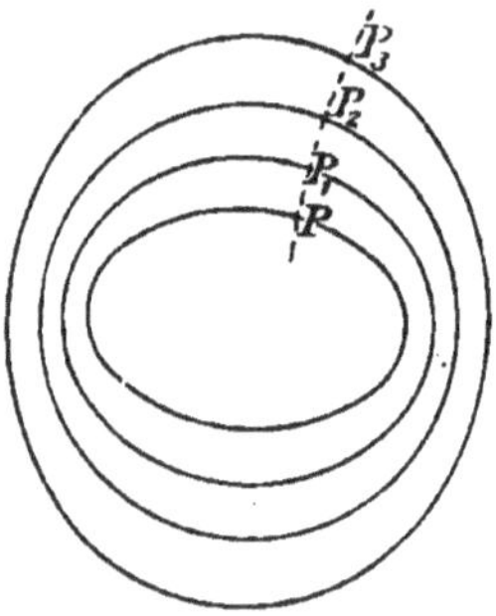

FIG. 54.

SECTION LVI. THE DISTRIBUTION OF ELECTRICITY ON A SPHERE AND ON AN ELLIPSOID.

1. *The Sphere.* Suppose the charge Q given to an insulated sphere of radius R. We are to determine the potential Ψ_a at a point outside the sphere. Let the centre of the sphere be taken as the origin of coordinates. We have $\nabla^2\Psi_a = 0$. Since Ψ_a is a function of the distance from the centre O, we have [XV. (1)] $1/r . d^2(r\Psi_a)/dr^2 = 0$, and hence $\Psi_a = c_1 + c_2/r$, where c_1 and c_2 are constants.

We assume that no charged bodies are present besides the sphere, so that $\Psi_a = 0$ when $r = \infty$, and hence $c_1 = 0$. The electrical force at the distance r from the centre is $F = -d\Psi_a/dr = c_2/r^2$.

From LV. (e) we have further

$$\sigma = 1/4\pi \,.\, F = 1/4\pi \,.\, c_2/R^2; \quad 4\pi R^2\sigma = Q = c_2.$$

The potential Ψ_a and the capacity C are therefore

$$\Psi_a = Q/r, \quad C = Q/\bar{\Psi} = R.$$

The dimensions of capacity are therefore those of a length.

2. *The Ellipsoid.* Represent the semi-axes of the ellipsoid by a, b, c, and its charge by Q. It is most natural to assume that the surfaces of constant potential are confocal ellipsoids. The equation of a system of such surfaces is

(a) $$E = x^2/(a^2+\lambda) + y^2/(b^2+\lambda) + z^2/(c^2+\lambda) = 1.$$

On our assumption the potential must be a function of λ, so that we will write $\Psi = f(\lambda)$. To find this function f we proceed from

$$\partial\Psi/\partial x = d\Psi/d\lambda \,.\, \partial\lambda/\partial x; \quad \partial^2\Psi/\partial x^2 = d^2\Psi/d\lambda^2 \,.\, (\partial\lambda/\partial x)^2 + d\Psi/d\lambda \,.\, \partial^2\lambda/\partial x^2$$

and the analogous expressions for y and z. These give

(b) $$\nabla^2\Psi = d^2\Psi/d\lambda^2 \,.\, \left((\partial\lambda/\partial x)^2 + (\partial\lambda/\partial y)^2 + (\partial\lambda/\partial z)^2\right) + d\Psi/d\lambda \,.\, \nabla^2\lambda.$$

We will set for brevity,

$$A = x^2/(a^2+\lambda)^2 + y^2/(b^2+\lambda)^2 + z^2/(c^2+\lambda)^2$$
$$B = x^2/(a^2+\lambda)^3 + y^2/(b^2+\lambda)^3 + z^2/(c^2+\lambda)^3.$$

We then have

(c) $$\partial E/\partial x = 2x/(a^2+\lambda) - A \,.\, \partial\lambda/\partial x;$$

(d) $$\partial A/\partial x = 2x/(a^2+\lambda)^2 - 2B \,.\, \partial\lambda/\partial x.$$

Analogous expressions hold for the differential coefficients with respect to y and z. Hence we have

$$\partial^2E/\partial x^2 = 2/(a^2+\lambda) - A \,.\, \partial^2\lambda/\partial x^2 - 2x/(a^2+\lambda)^2 \,.\, \partial\lambda/\partial x - \partial A/\partial x \,.\, \partial\lambda/\partial x = 0,$$

or, using equation (c),

(e) $$\partial^2E/\partial x^2 = 2/(a^2+\lambda) - A \,.\, \partial^2\lambda/\partial x^2 + 2B \,.\, (\partial\lambda/\partial x)^2 - 8/A \,.\, x^2/(a^2+\lambda)^3 = 0.$$

From equations (c) and (a) it follows that

(f) $$A^2\left((\partial\lambda/\partial x)^2 + (\partial\lambda/\partial y)^2 + (\partial\lambda/\partial z)^2\right) = 4A,$$

and from (e) and (f) that

(g) $$A \,.\, \nabla^2\lambda = 2/(a^2+\lambda) + 2/(b^2+\lambda) + 2/(c^2+\lambda).$$

By the help of equations (f) and (g) it follows from (b) that

(h) $$A \,.\, \nabla^2\Psi = 4d^2\Psi/d\lambda^2 + 2d\Psi/d\lambda \,.\, \left(1/(a^2+\lambda) + 1/(b^2+\lambda) + 1/(c^2+\lambda)\right).$$

Outside the ellipsoid we have $\nabla^2\Psi = 0$. If C is a constant, we have from (h), (i) $d\Psi/d\lambda = -C_1/\sqrt{(a^2+\lambda)(b^2+\lambda)(c^2+\lambda)}$ and

$$\Psi = -C_1\int d\lambda/\sqrt{(a^2+\lambda)(b^2+\lambda)(c^2+\lambda)} + C_2.$$

At an infinitely distant point, for which $x = y = z = \infty$, Ψ is assumed equal to zero; and for such a point equation (a) shows that $\lambda = \infty$. The potential Ψ at any point (x, y, z) is therefore given by

(k) $$\begin{cases} \Psi = C_1\int_\lambda^\infty d\lambda/\sqrt{(a^2+\lambda)(b^2+\lambda)(c^2+\lambda)}, \\ \text{where } \lambda \text{ is known from the equation} \\ x^2/(a^2+\lambda) + y^2/(b^2+\lambda) + z^2/(c^2+\lambda) = 1. \end{cases}$$

If the charge on the ellipsoid is Q, the potential at a point at a great distance from the ellipsoid is $Q/\sqrt{\lambda}$; by comparison with (k) we then obtain

(l) $$\Psi = Q/2\,.\int_\lambda^\infty d\lambda/\sqrt{(a^2+\lambda)(b^2+\lambda)(c^2+\lambda)}.$$

The electrical force F and its components X, Y, Z are determined from the equations

$$X = -d\Psi/d\lambda\,.\,\partial\lambda/\partial x, \quad Y = -d\Psi/d\lambda\,.\,\partial\lambda/\partial y, \quad Z = -d\Psi/d\lambda\,.\,\partial\lambda/\partial z.$$

$$F = -d\Psi/d\lambda\,.\,\sqrt{(\partial\lambda/\partial x)^2 + (\partial\lambda/\partial y)^2 + (\partial\lambda/\partial z)^2}.$$

From (f) and (l) we have

$$F = -2/\sqrt{A}\,.\,d\Psi/d\lambda = Q/\sqrt{A}\,.\,1/\sqrt{(a^2+\lambda)(b^2+\lambda)(c^2+\lambda)}.$$

If we represent the perpendicular let fall from the origin on the plane tangent to the ellipsoid at the point x, y, z of its surface by N, we have $\sqrt{A} = 1/N$, and hence $F = Q\,.\,N/\sqrt{(a^2+\lambda)(b^2+\lambda)(c^2+\lambda)}$.

The surface density σ on the ellipsoid itself is determined by the equation $4\pi\sigma = F$, and $\lambda = 0$ for this ellipsoid, so that (m) $\sigma = N\,.\,Q/4\pi abc$. *Hence the electrical density at a point on the ellipsoid is proportional to the perpendicular let fall from the centre on the plane tangent to the ellipsoid at that point.*

We will now consider several special cases. In the case of an *ellipsoid of rotation* $a = b$, and therefore from (l), if Ψ_0 is the potential of the ellipsoid, we have

$$\Psi_0 = Q/2\,.\int_0^\infty d\lambda/(a^2+\lambda)\sqrt{c^2+\lambda}.$$

Hence for $a > c$, (n) $\Psi_0 = Q/\sqrt{a^2-c^2}\,.\,(\tfrac{1}{2}\pi - \operatorname{arctg} c/\sqrt{a^2-c^2})$; for $a = c$, (o) $\Psi_0 = Q/a$; and for $a < c$,

(p) $$\Psi_0 = Q/2\sqrt{c^2-a^2}\,.\,\log\left[(c+\sqrt{c^2-a^2})/(c-\sqrt{c^2-a^2})\right].$$

If in (n) we set $c=0$ the ellipsoid becomes a *circular plate*, and its capacity is $C=Q/\Psi_0=a/(\frac{1}{2}\pi)$. For an *ellipsoid of rotation* whose length is great in comparison with the equatorial diameter, we have from (p) $\Psi_0=Q/c\,.\log(2c/a)$ and $C=c/\log(2c/a)$.

The surface density σ, from equation (m), is

$$\sigma=Q/4\pi abc\,.\,1/\sqrt{x^2/a^4+y^2/b^4+z^2/c^4}.$$

If z is eliminated by the help of the equation of the ellipsoid, and if c is infinitely small, we have for the density on an elliptical plate whose semi-axes are a and b, $\sigma=Q/4\pi ab\,.\,1/\sqrt{1-x^2/a^2-y^2/b^2}$. If the plate is circular, that is, if $a=b$, and if we set $x^2+y^2=r^2$, we have $\sigma=Q/4\pi a\,.\,1/\sqrt{a^2-r^2}$. At a point whose distance u from the edge is very small, we have $\sigma=Q/4\pi a\,.\,1/\sqrt{2au}$. In this case, therefore, the density is inversely proportional to the square root of the distance of the point from the edge.

SECTION LVII. ELECTRICAL DISTRIBUTION.

If several charged conductors are present in a region, the distribution of electricity on the conductors is determined not only by their form and magnitude, but also by their mutual action. The determination of the conditions of electrical equilibrium is, as a rule, very difficult. The most important work on this subject has been done by Poisson and William Thomson. We will here make use of the method of *electrical images* given by Thomson.

(*a*) *Distribution on a Plane Surface.*—Suppose the quantity of electricity e present at the point O (Fig. 55); let AB be the plane surface of a very large conductor L, which is in conducting contact with the earth. The potential Ψ_1 of L is therefore zero, since we assume the potential of the earth equal to zero (cf. VII.). We are to determine the surface density σ of the distribution on the surface. Let the potential at an arbitrary point in space due to the conductor L be Ψ_e, so that Ψ_e is the work done by the electrical forces of

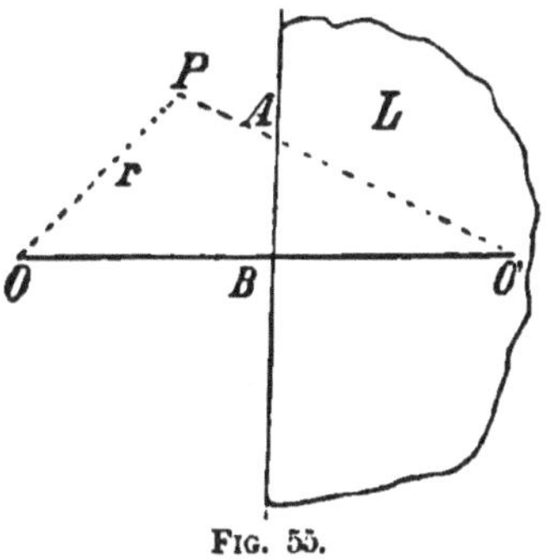

FIG. 55.

the conductor if unit quantity of electricity, which is supposed to be merely a test charge and to have no effect on the electrical distribution on L, is transferred from the point P to infinity. If $OP = r$, the potential Ψ at P will be $\Psi = e/r + \Psi_e$.

We now suppose a quantity of electricity $-e$ situated at the point O' (Fig. 55), which is the image of the point O with respect to the plane AB. This imaginary quantity at O' would act on all points lying on the same side of the plane AB as O, in the same way as the quantity of electricity which is distributed on AB; for the potential which arises from the quantities at O and O' satisfies Laplace's equation at all points which lie on the same side of the

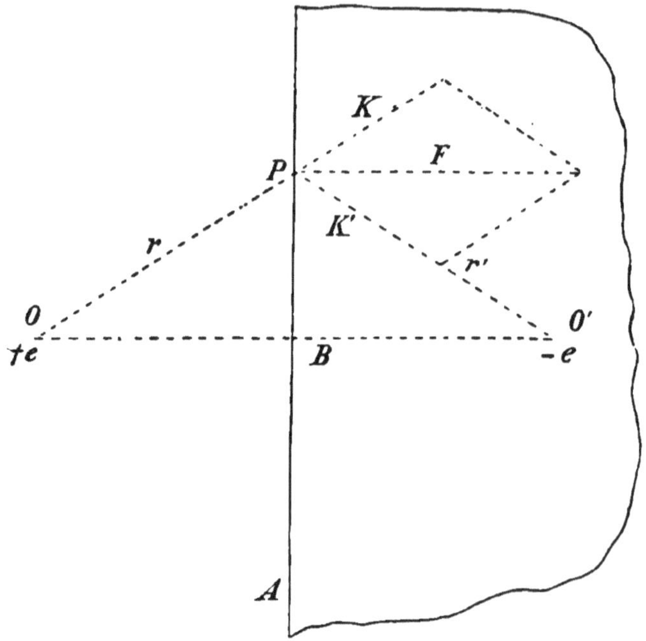

FIG. 56.

plane AB as O, except at the point O itself. Further, the potential vanishes at all points of the plane AB, since all points of that plane, which passes perpendicularly through the middle point of the line OO', are equally distant from the points O and O', and hence for all points of the plane AB we have $e/r - e/r' = 0$, where r and r' represent the distances of a point in the plane from O and O'. Now if a function satisfies Laplace's equation and assumes assigned values over a given surface, and if the function itself and its differential coefficients are continuous, it is single-valued and determinate. This theorem is known as Dirichlet's Principle.

It should be noticed that $\Psi_e = e/r'$, and that the potential Ψ at the point P is $\Psi = e/r - e/r'$.

A unit quantity of positive electricity lying at a point P (Fig. 56) in the plane AB is acted on by two forces, $K = e/r^2$ and $K' = e/r'^2$, whose directions coincide with OP and PO' respectively. Hence the direction of the resultant force F is parallel to OO' and equal to $F = -2eOB/OP^3$, if we consider the force positive when it is directed toward the region in which O lies. Now since $4\pi\sigma = F$ we obtain $\sigma = -e/2\pi \,.\, OB/OP^3$. *The surface density at the point P* (Fig. 56) *is therefore inversely proportional to the cube of the distance of the point P from the point O, at which the quantity of electricity $+e$ is situated.* The potential and the surface density are calculated in the same way when several points carrying charges are present in the region.

(*b*) *The Sphere.*—Suppose that the quantities of electricity e and e' are situated at the points O and O' (Fig. 57). The equi-potential

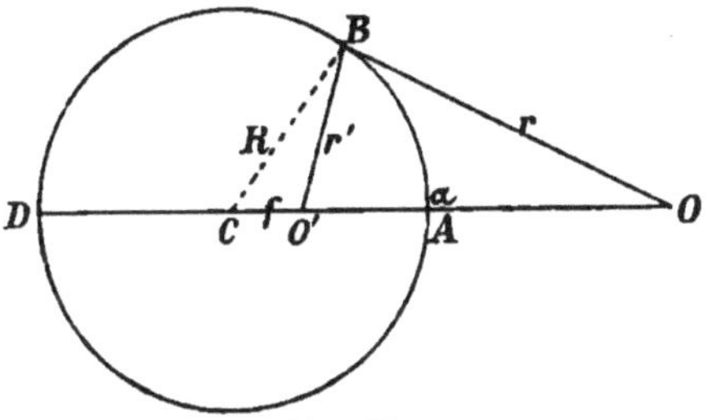

FIG. 57.

surface for which the potential vanishes is given by the equation $e/r + e'/r' = 0$ if r and r' represent the distances of a point on the equi-potential surface from O and O' respectively. If e and e' have the same signs, this equation represents a surface lying at infinity; if e and e' have opposite signs it represents a sphere and, in the limiting case, a plane.

The centre C of the sphere lies on the line OO'. and we have $OC : O'C = e^2 : e'^2$ and $CO'/CB = CB/CO = e'/e$. The triangles $CO'B$ and CBO are similar, and the radius CB of the sphere is the mean proportional between the distances of the points O and O' from the centre of the sphere.

If a hollow sphere of very thin sheet metal, in conducting connection with the earth, is brought into the place occupied by this spherical surface of zero-potential, the potential of points in the region will not be changed, either within or without the sphere; the electrical action depends only on the quantities of electricity e and e'.

If the sphere remains in conducting contact with the earth, and if we remove the quantity of electricity e' from the interior of the sphere, the potential within the sphere will become zero, while outside the sphere it retains its former value, since the quantity of electricity e does not change its position and the potential of the sphere still remains zero. Hence, the quantity e lying outside the sphere, kept at potential zero, together with the electricity induced on the sphere, exerts on points outside the sphere the same action as if the induced electricity were replaced by the mass e' lying inside the sphere. We call the point O', where the mass e' is situated, the *electrical image* of the point O. The quantity of electricity e' at that point will exert the same action as the quantity of electricity actually present on the sphere. In optics a point which appears to emit light from behind a mirror or lens which, if it were self-luminous, would emit rays in the same direction as those which proceed from the mirror or lens, is called a *virtual image*. Hence, we may consider O' the *electrical image*.

If we set $CO'=f$, $CO=a$, and $CB=R$, we have

$$e'/e=O'B/OB=f/R=R/a;$$

and hence $e'=Re/a$. We set $OB=r$ and $O'B=r'$. The force e/r^2 acts at B in the direction OB, and the force e'/r'^2 acts at the same point in the direction BO'. The former of these may be resolved into the components $e/r^2 . a/r$ along OC and $e/r^2 . R/r$ along CB, the latter into $-e'/r'^2 . f/r'$ along OC and $-e'/r'^2 . R/r$ along CB. The two components in the direction OC are equal but oppositely directed, and therefore annul each other. The other two combine to give the force $eR/r^3 - e'R/r'^3 = -(a^2-R^2)/R . e/r^3$, which acts in the direction CB. *The sphere is therefore an equi-potential surface, since the direction of the force, at any point of its surface, coincides with the direction of the normal to that surface.*

The density σ, as obtained from $F=4\pi\sigma$, is $\sigma = -(a^2-R^2)/4\pi R . e/r^3$, *and hence is inversely proportional to the third power of the distance from the charged point O.* The quantity of electricity on the sphere is $-e'=-Re/a$, since this charge produces the same potential in the region as the actual charge on the sphere, and therefore [LV. (f)] must be equal to it.

The sphere is attracted by the point O with the force

$$ee'/(a-f)^2=Re^2/a(a-f)^2.$$

If the distance $CO=a$ is very great, this force becomes Re^2/a^3.

(c) If the sphere is originally insulated and uncharged, we can find the electrical distribution on it by assuming that, besides carrying a charge, distributed as above described, it also carries a uniformly distributed charge whose surface-density is $e'/4\pi R^2 = e/4\pi Ra$. The sum of these two charges or the charge of the sphere is equal to zero. The surface-density is then $\sigma = e/4\pi R.(1/a - (a^2 - R^2)/r^3)$. The surface-density σ is zero on a circle whose periphery is distant $r = a\sqrt[3]{1 - R^2/a^2}$ from the point O. The plane of this circle lies nearer to O than the centre of the sphere. In order to find the potential Ψ of the sphere, we determine it for the centre. Since the charge on the sphere is zero, the potential due to that charge is also zero; the potential at the centre is therefore $\Psi = e/a$. This follows from the remark that the induced charge $-e'$ and the charge e at O together have no effect on the potential of the sphere; the potential is due to the additional charge $+e'$, which makes the potential $e'/R = e/a$.

The force with which the sphere is attracted by O is in this case very much smaller than if the sphere were in conducting connection with the earth. It is

$$ee'/(a-f)^2 - ee'/a^2 = Re^2/a^3 . R^2(2a^2 - R^2)/(a^2 - R^2)^2.$$

When R is very small in comparison with a, the force is approximately $Re^2/a^3 . 2R^2/a^2$. In this case we have a simpler expression for the surface-density σ. Designating the angle BCO by Θ, we have

$$r^2 = a^2 - 2aR\cos\Theta + R^2.$$

If a is so great that the higher powers of R/a can be neglected, we have $r^{-3} = a^{-3}(1 + 3R/a . \cos\Theta)$. If we designate the inducing force e/a^2 which proceeds from O by X, we have $\sigma = -3\cos\Theta/4\pi . X$, if, in the formula for σ, we consider the radius R as infinitely small in comparison with a and substitute the value just given for r^{-3}.

SECTION LVIII. COMPLETE DISTRIBUTION.

If a charged body A (Fig. 58) is situated in the interior of a metallic shell BC, there will be a distribution of electricity on the shell. If A is charged positively, the inner surface B will be negatively, and the outer surface C positively electrified. Let the charge on A be e, that on B $-e'$, and that on C $+e'$. We will show that the quantity of induced electricity e' is equal to the quantity of inducing electricity e. Let us suppose a closed surface D drawn in the interior of BC. If Ψ is the potential in the shell BC, and ν the normal to the element dS of the surface D, we have from XIV. (c), $4\pi(e-e') = -\int(\partial\Psi/\partial\nu)dS$. Since the potential Ψ in the shell is constant the integral vanishes, and hence $e = e'$. The charge e on C can be conducted off, or another charge can be conducted to it without causing any change in the charges A and B. If the inducing body is entirely surrounded by the body in which electricity is induced, we may say that the *induction* is *complete; the inducing and induced quantities of electricity are equal.*

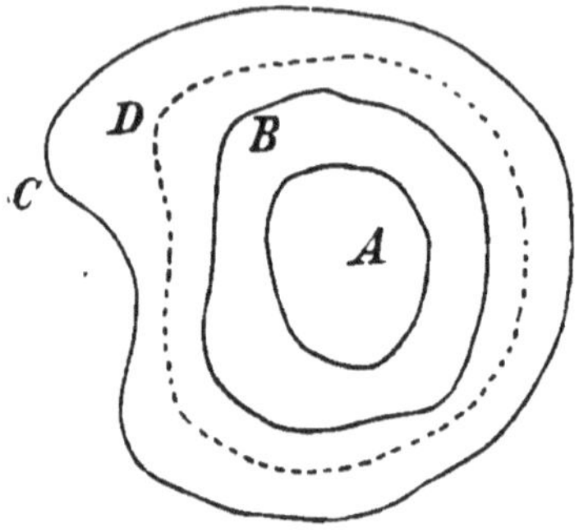

FIG. 58.

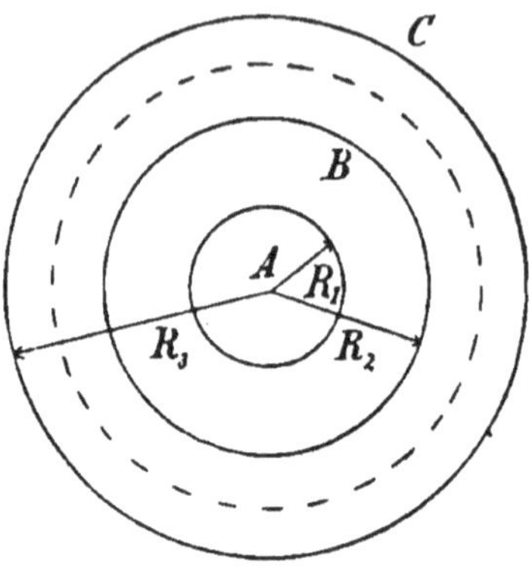

FIG. 59.

This theorem may be used in the comparison of the charges of different conductors. If the outer surface of BC is connected with an instrument which will measure potentials, and if the charged bodies to be tested are brought successively into the hollow within BC, the potentials indicated by the instrument are proportional to the magnitudes of the charges.

Suppose the sphere A, whose radius is R, to be charged with the quantity e (Fig. 59). Let BC be a spherical shell concentric with A, whose radii are R_2 and R_3. The inner surface of BC is then charged with the quantity $-e$. If there is no electricity on the outer surface C, the potential Ψ at A is $\Psi = e/R_1 - e/R_2$. The potential at a point within the shell BC or outside the surface C is zero, since both charges act with respect to an external point, as if they were concentrated at the common centre of the spheres. From the definition of the capacity C, we have [cf. LV. (g)]

(a) $$C = e/\Psi = R_1 R_2/(R_2 - R_1).$$

The induction can, in many cases, be almost complete, even when the one conductor is not completely surrounded by the other. Let ABC and DEF (Fig. 60) be two conductors whose surfaces BC and

DF lie very near each other. On the surface BC describe a closed curve GH, and from all points of it draw lines of force, which cut out on DF the curve KJ. Now draw a closed surface $G'GJJ'K'KHH'$ in such a manner that the two curves GH and JK lie in it. The surfaces bounded by the curves $G'H'$ and $J'K'$ lie inside the conductors and are congruent to the surfaces GH and JK respectively. Let dS be a surface-element of the closed surface $G'H'J'K'$, and let e and e' be the charges of the surfaces GH and JK. The potential is represented by Ψ and the normal by ν. We then have from LV.

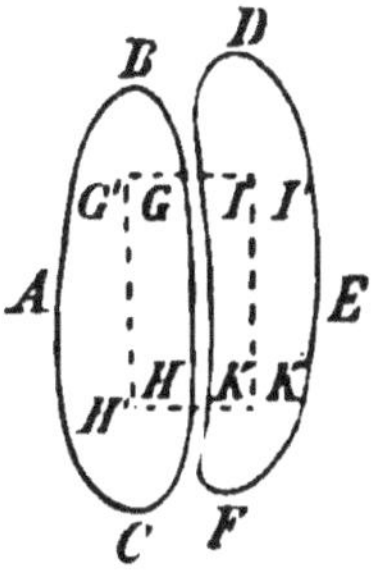

FIG. 60.

(b) $$4\pi(e+e') = -\iint \partial\Psi/\partial\nu \,.\, dS.$$

The integral vanishes, since Ψ is constant within the conductors, and since between the conductors the force is parallel to the closed surface. Hence we have $e = -e'$. If the surfaces BC and DF lie very near each other, we also have $\sigma = -\sigma'$, that is, the densities on the two surfaces are equal but of opposite sign.

If a is the distance between the surfaces BC and DF, and if Ψ_1 is the potential of ABC, and Ψ_2 the potential of DEF, the electrical force F in the intervening space is [VII. (e)]

$$\Psi_1 = \Psi_2 + Fa, \quad F = (\Psi_1 - \Psi_2)/a$$

The surface-density σ is [cf. LV. (e)] $\sigma = -\sigma' = (\Psi_1 - \Psi_2)/4\pi a$. The charge on the surface S is $e = (\Psi_1 - \Psi_2)/4\pi a \,.\, S$.

If the conductor DF is connected with the earth, that is, if $\Psi_2 = 0$, the capacity C will be $C = S/4\pi a$, that is, *the capacity is inversely proportional to the distance between the conductors.*

This formula is used in air-condensers, when a is very small.

SECTION LIX. MECHANICAL FORCE ACTING ON A CHARGED BODY.

If an element of volume contains the quantity of electricity $\rho d\tau$, it is acted on by a force whose components are $X\rho d\tau$, $Y\rho d\tau$, and $Z\rho d\tau$. From LV. (a) the x-component can be expressed by

$$1/4\pi \,.\, \partial\Psi/\partial x \,.\, \nabla^2\Psi d\tau = +1/4\pi \,.\, X(\partial X/\partial x + \partial Y/\partial y + \partial Z/\partial z)d\tau.$$

In the interior of a good conductor the force and electrical density are zero; but a force acts on each element of its surface, where the density is not zero. This force is determined in the following way:

Let AD (Fig. 61) be the electrified surface, and $BC=dS$ the surface-element, which is cut out by an infinitely small sphere described about the point P as centre, with the radius $PB=PC$. Let P_2 lie on the normal PP_2 to BC, and let P_2P be infinitely small in comparison with PB. A unit of electricity at the point P_2 is acted on by the force $2\pi\sigma$ arising from the distribution on the surface BC [cf. XIII. (3)]. At the corresponding point P_1 on the opposite side of BC the force acting is $-2\pi\sigma$. If l, m, n are the cosines of the angles which P_1P_2 makes with the axes, and X, Y, Z are the components of force which arise from all the electricity present except that on BC, we have

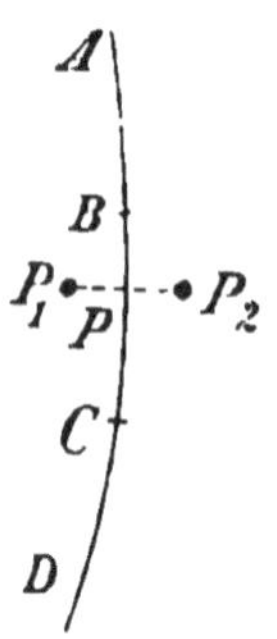

FIG. 61.

$$X_2 = -\partial\Psi_2/\partial x = X + 2\pi\sigma l, \quad X_1 = -\partial\Psi_1/\partial x = X - 2\pi\sigma l.$$

X_2, X_1 and Ψ_2, Ψ_1 are the components of force and the potentials at the points P_2 and P_1 respectively. We have (c) $X=\frac{1}{2}(X_2+X_1)$. Analogous expressions hold for the other components of force. X represents the force which acts on a unit of electricity on dS in the direction of the x-axis. The element dS is therefore moved in the direction of the x-axis by the force (d) $X\sigma dS=\frac{1}{2}(X_2+X_1)\sigma dS$.

If the element dS is part of the surface of a good conductor, and if P_1 lies within the conductor, the force at P_1 is equal to zero. If we represent the force acting at P_2 by F, the force which acts on dS is, from (c) and (d), (e) $\frac{1}{2}F\sigma dS$. Since, from LV. (e), we have $F=+4\pi\sigma$, the force sought will be (f) $2\pi\sigma^2 dS=1/8\pi\,.\,F^2\,.\,dS$.

W. Thomson has made an interesting application of this equation in the construction of his *absolute electrometer*. This consists of an insulated metal plate EF and a smaller circular plate CD, which is parallel with EF. CD forms a part of the base of a metallic cylinder AB (Fig. 62). If the potential of EF is Ψ, and that of CD and AB is zero, CD will be attracted to EF by a force which is determined in the following way: Since AB and CD are almost like a single continuous body, there is no perceptible surface distribution on the inner surface of $ABCD$. Represent the density on the external surface of CD by σ, the distance between CD and EF by a, and let the surface $CD=S$. The force K which attracts CD toward EF is,

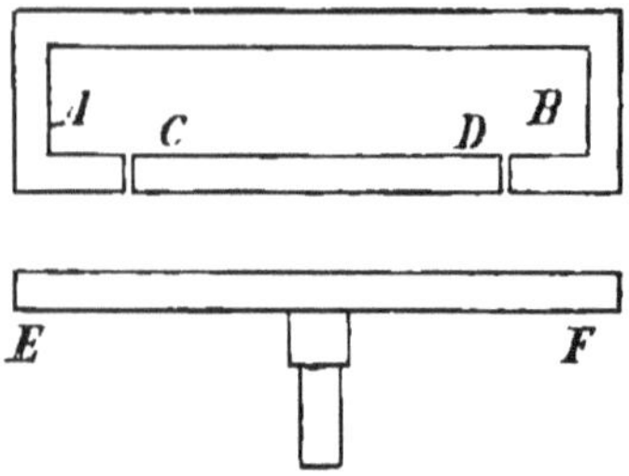

FIG. 62.

from (c), equal to $K = \frac{1}{2}F\sigma S$. We have $Fa = \Psi$ [cf. VII.], $4\pi\sigma = F$, and therefore $K = S\Psi^2/8\pi a^2$. We determine the weight which is necessary to counterbalance the electrical attraction. If M grams are necessary for this purpose, we have $\Psi = a\sqrt{8\pi Mg/S}$, where g is the acceleration of gravity.

Section LX. Lines of Electrical Force.

All actions by which electricity is produced, such as friction, induction, etc., produce equal quantities of positive and negative electricity; for this reason we are led to assume that in every unelectrified body equal quantities of positive and negative electricity are present. Let A and B (Fig. 63) be two bodies electrified by friction which are gradually separated further and further from each other, as in Fig. 64; during this separation they retain equal but opposite charges. Suppose A and B to be good conductors and to be insulated. Now, construct lines of force from all points of the contour of a surface-element dS on A. They will determine a surface-element dS' on B. The region bounded by the lines of force and the two elements dS and dS' may be called a *sphondyloid*. If the density is σ on dS and σ' on dS', we may prove, as in the discussion following LVIII. (b), that $\sigma dS - \sigma' dS' = 0$. The surface-densities on the two charged conductors A and B are therefore inversely proportional to the surfaces limited by the sphondyloid.

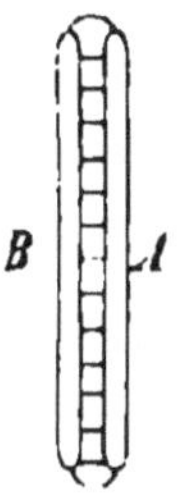

Fig. 63.

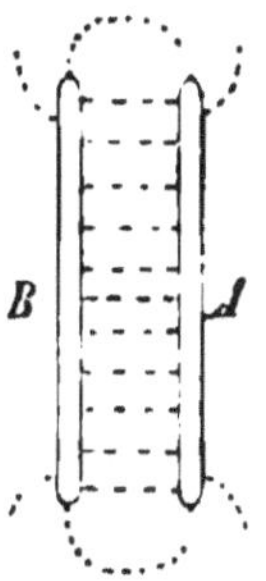

Fig. 64.

If the conductor A is charged with a quantity Q of positive electricity, and if the surface of A is cut into Q parts, each one of which is charged with unit quantity, the lines of force drawn from the contours of the Q parts on A cut the surface B also into Q parts, each one of which is charged with unit quantity of negative electricity.

If a conductor $ABCD$ is charged in any manner to the potential Ψ, the equi-potential surfaces about it are spheres, at all distances which are great in comparison with its dimensions. If we construct the equi-potential surface, whose potential is $(\Psi - 1)$, it lies nearest the conductor at the points A, B, and D (Fig. 65). At these points [LV.] the electrical force and the surface-density are greatest. The lines of force also lie in closest proximity to one another at these points. If the conductor has edges or points projecting outward, the density on them is very great; it will be infinitely great on a perfectly sharp edge. On this depends the so-called *action of points*; the density of the electricity is greatest at these points, and therefore the electricity flows out from them with especial ease.

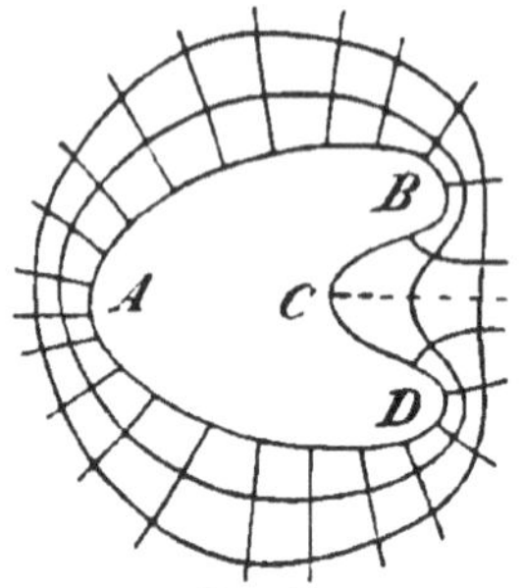

FIG. 65.

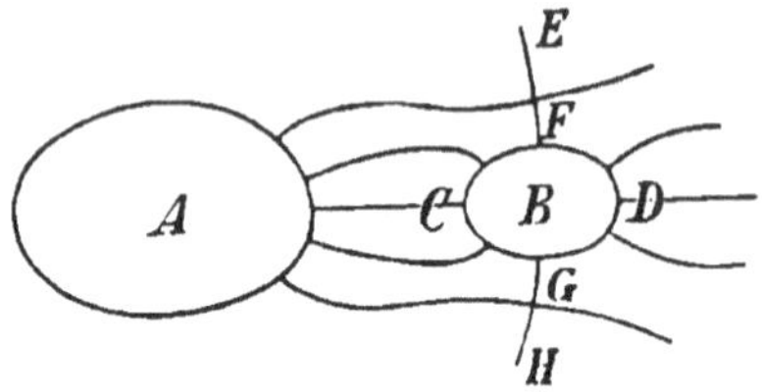

FIG. 66.

If A (Fig. 66) is a conductor charged with positive electricity, and B an insulated conductor without charge, negative electricity will be present at all parts of B which are met by the lines of force proceeding from A. Lines of force also proceed from the other points of B; the number of the lines which fall upon B is equal to the number of those proceeding from B. Hence, the surface of B is divided into two parts with opposite charges. The parts are separated by a curve encircling the body B, along which the surface-density σ and therefore also the electrical force are zero. This curve is the line of intersection of B and an equi-potential surface around A.

Let ABC (Fig. 67) be a conductor on whose surface the potential is constant and equal to Ψ, and let $A'B'C'$ be an equi-potential surface at which the potential is Ψ_1. We suppose that the charge of each surface-element, for example of $AB = dS$, is moved outward in the direction of the lines of force and transferred to the surface $A'B'C'$. If this surface is a conductor, the electricity transferred to it is in equilibrium. If the potential within the surface $A'B'C'$ is Ψ_1, and if it retains its former values outside of that surface, the condition $\nabla^2\Psi_a = 0$ is fulfilled for all external points. If the electrical forces at AB and $A'B'$ are F and F' respectively, we have from LV. (f), since AB and

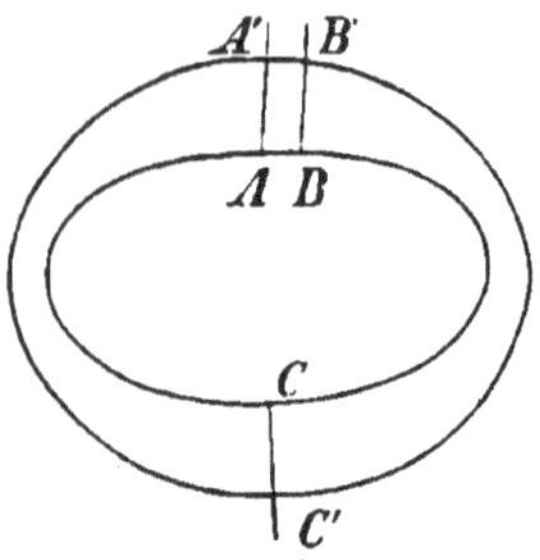

FIG. 67.

$A'B'$ are bounded by lines of force, $AB.F = A'B'.F'$. If σ and σ' represent the densities at AB and $A'B'$ respectively, we have further $AB.\sigma = A'B'.\sigma'$. We therefore obtain $F/\sigma = F'/\sigma'$. We have, however, $F = 4\pi\sigma$ and, therefore, also $F' = 4\pi\sigma'$.

Hence, if we divide the surface of the conductor into elements, each of which contains unit quantity of electricity, and if we draw lines of force outward from the boundary of the element, these lines bound a tube. The tube cuts the equi-potential surfaces surrounding the conductor in such a way that for all of them we have

$$F'/\sigma' = F''/\sigma'' = \ldots F/\sigma.$$

From the form of the tubes of force we obtain a representation of the distribution of electrical force in the region. And also from the distribution of the lines of force we may distinguish between the attractive and repulsive forces. Two lines of force proceeding in the same sense repel each other, so that the repulsion maintains equilibrium with the tension which acts along the lines of force [cf. XXVII].

Section LXI. Electrical Energy.

A conductor charged with the quantity of electricity e can do work in consequence of that charge; it possesses *electrical energy*. If the charged surface ABC (Fig. 67) is extended so that it assumes in succession the form of its equi-potential surfaces, the forces acting on the electrified surface do an amount of work which can be determined. If the surface of the conductor has the potential Ψ, and if it is extended until it coincides with the equi-potential surface whose potential is $\Psi + d\Psi$, the work done is calculated in the following way: A surface-element dS carrying the charge σdS is acted on by the force $\frac{1}{2}.F\sigma dS$. We represent by $d\nu$ the distance of the equi-potential surface $\Psi + d\Psi$ from the conductor. The work done on the element dS during its motion is $\frac{1}{2}F\sigma dS d\nu$. The total work done is therefore $\frac{1}{2}\int\int F\sigma dS d\nu$.

From the definition of the equi-potential surface [cf. LV. (e)] we have $Fd\nu = -d\Psi$; we therefore have for the total work done

$$-\tfrac{1}{2}\int\int \sigma d\Psi dS = -\tfrac{1}{2}\int e d\Psi.$$

If the surface of the body is extended until it coincides with the equi-potential surface at which the potential is zero, the work W done is

(a) $$W = -\tfrac{1}{2}\int_{\Psi}^{0} e d\Psi = \tfrac{1}{2}e\Psi.$$

All the work which can be done on the given conditions is represented by W, and hence it is called the *potential energy* of the conductor.

The electrical energy of the conductor can be geometrically represented in the following way: About the conductor L (Fig. 68), whose potential is Ψ, we construct the equi-potential surfaces whose potentials are successively $\Psi-1$, $\Psi-2$, $\Psi-3$, etc., and we divide the surface of the body L in such a way that each part carries unit charge. The space surrounding the body L is divided into $e\Psi$ parts by the lines of force starting from the boundaries of the separate parts and by the equi-potential surfaces; the number of these parts is double the value of the electrical energy.

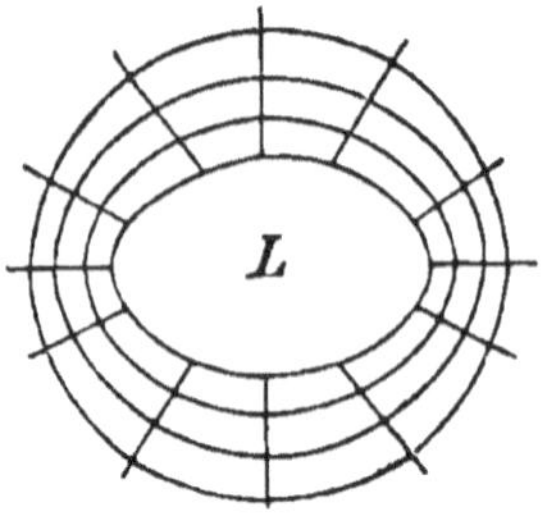

FIG. 68.

If we designate the electrical energy by W and the capacity by C, we have $e=C\Psi$, and thus (b) $W=\frac{1}{2}e\Psi=\frac{1}{2}C\Psi^2=\frac{1}{2}e^2/C$. We have seen before that the energy is also given by

$$W=\tfrac{1}{2}\int\int F\sigma dS d\nu = 1/8\pi \,.\int\int F^2 dS d\nu.$$

If X, Y, Z are the components of the electrical force, and if $dxdydz$ is a volume-element, we will have

(c) $$W=1/8\pi\,.\int\int\int(X^2+Y^2+Z^2)dxdydz.$$

If the two conductors ABC and $A'B'C'$ (Fig. 67) are charged with the quantities of electricity $+e$ and $-e$ respectively, and have the potentials Ψ_1 and Ψ_2, we obtain their potential energy in the same way by supposing the body ABC carrying the charge $+e$ to be gradually extended so as to coincide with the equi-potential surfaces which surround it. In this way the charge e is finally transferred to $A'B'C'$. The integral in (a) then becomes

(d) $$W=-\tfrac{1}{2}\int_{\Psi_1}^{\Psi_2} e\,d\Psi=\tfrac{1}{2}e(\Psi_1-\Psi_2).$$

This method of treatment may be applied in all cases. A system of conductors whose charges are e_1, e_2, e_3, ..., and whose potentials are Ψ_1, Ψ_2, Ψ_3, ..., have, with respect to a conductor whose potential is Ψ_0, the potential energy

(e) $$W=\tfrac{1}{2}e_1(\Psi_1-\Psi_0)+\tfrac{1}{2}e_2(\Psi_2-\Psi_0)+\ldots.$$

If the sum of all the charges is zero, that is, if $e_1+e_2+\ldots=0$ we have (f) $W=\frac{1}{2}e_1\Psi_1+\frac{1}{2}e_2\Psi_2+\ldots$. *Therefore, if all charged conductors are brought to the same potential, the electrical energy is independent of the value of the common potential.*

The expression for the electrical energy may be derived in still another way. In a system of conductors the electrical distribution is determined if the density ρ is given at all points in terms of the coordinates x, y, z. The potential at any point is then, in the usual notation, $\Psi = \int \rho d\tau / r$. The potential increases and diminishes proportionally to ρ. If the density of the electricity is doubled at all points, the value of the potential also is doubled.

If the charge $1/n \,.\, \rho d\tau$ is removed from every volume-element, the potential becomes $(n-1)/n \,.\, \Psi$. In order to transfer the quantity of electricity $1/n \,.\, \int \rho d\tau$ to a distant and very large body whose potential is Ψ_0, for instance to the earth, the work $1/n \,.\, \int \rho d\tau (\Psi - \Psi_0)$ must be done, if n is a very large number. If the quantity $1/n \,.\, \int \rho d\tau$ is again removed, the work required is

$$1/n \,.\, \int \rho d\tau \big((n-1)/n \,.\, \Psi - \Psi_0\big).$$

If the whole charge is at last transferred to the earth, the work W which is done is

$$W = 1/n \,.\, \int \rho d\tau \big(1 + (n-1)/n + (n-2)/n + \ldots + 1/n\big)\Psi - \Psi_0 \int \rho d\tau.$$

Now, we have $1 + (n-1)/n + (n-2)/n + \ldots + 1/n = n(n+1)/2n$, and if n is very great (g) $W = \frac{1}{2}\int \Psi \rho d\tau - \Psi_0 \,.\, \int \rho d\tau$. If the sum of the quantities of electricity present is zero, we will have (h) $W = \frac{1}{2}\int \Psi \rho d\tau$. Since $\nabla^2 \Psi = \partial^2\Psi/\partial x^2 + \partial^2\Psi/\partial y^2 + \partial^2\Psi/\partial z^2 = -4\pi\rho$, we have

$$W = -1/8\pi \,.\, \iiint \Psi \,.\, (\partial^2\Psi/\partial x^2 + \partial^2\Psi/\partial y^2 + \partial^2\Psi/\partial z^2) dx dy dz.$$

Now,

$$\iiint \Psi \,.\, \partial^2\Psi/\partial x^2 \,.\, dx dy dz$$
$$= \iint (\Psi \,.\, \partial\Psi/\partial x) \,.\, dy dz - \iiint (\partial\Psi/\partial x)^2 \,.\, dx dy dz.$$

Hence, by partial integration extended over the whole volume, we obtain

(i) $$W = 1/8\pi \,.\, \int \big((\partial\Psi/\partial x)^2 + (\partial\Psi/\partial y)^2 + (\partial\Psi/\partial z)^2\big) d\tau = 1/8\pi \,.\, \int F^2 d\tau.$$

This result has already been derived for the energy in a good conductor.

SECTION LXII. A SYSTEM OF CONDUCTORS.

If several insulated conductors A_1, A_2, A_3 are given, and if a unit of electricity is imparted to one of them, say to A_1, while the others have no charge, then the potential of A_1 becomes p_{11}, while the potentials of A_2 and A_3, etc., become p_{12} and p_{13} respectively. If A_2 were to become charged with unit quantity while the other conductors were to remain uncharged, the potential of A_2 would equal p_{22}, and the potentials of A_1, A_3, A_4, ... would be p_{21}, p_{23}, p_{24}, ... respectively. Now, if the conductor A_1 is charged with the quantity e_1, the conductor A_2 with the quantity e_2, etc., the potentials Ψ_1, Ψ_2, ... of the conductors A_1, A_2, A_3, ... respectively will be expressed by

$$\text{(a)} \quad \begin{cases} \Psi_1 = p_{11}e_1 + p_{21}e_2 + p_{31}e_3 + \dots \\ \Psi_2 = p_{12}e_1 + p_{22}e_2 + p_{32}e_3 + \dots \\ \Psi_3 = p_{13}e_1 + p_{23}e_2 + p_{33}e_3 + \dots . \end{cases}$$

The total energy of the electrical system is

$$\text{(b)} \quad W = \tfrac{1}{2}(\Psi_1 e_1 + \Psi_2 e_2 + \Psi_3 e_3 + \dots).$$

Hence,

$$\text{(c)} \quad \begin{cases} 2W = p_{11}e_1^2 + p_{22}e_2^2 + p_{33}e_3^2 + \dots + (p_{12} + p_{21})e_1 e_2 \\ \qquad + (p_{13} + p_{31})e_1 e_3 + (p_{23} + p_{32})e_2 e_3 + \dots . \end{cases}$$

If an infinitely small quantity of electricity δe_1 is communicated to one of the conductors, for example to A_1, the energy of the system will be increased by $\delta W = \Psi_1 \delta e_1$. This increment of the energy may also be expressed by the help of (c), since we have

$$\text{(d)} \quad \delta W = \left(p_{11}e_1 + \tfrac{1}{2}(p_{12} + p_{21})e_2 + \tfrac{1}{2}(p_{13} + p_{31})e_3 + \dots\right)\delta e_1.$$

Now, since δW is in this case equal to $\Psi_1 \delta e_1$, it follows from the first of equations (a) that (e) $\delta W = (p_{11}e_1 + p_{21}e_2 + p_{31}e_3 + \dots)\delta e_1$. Comparing the formulas (d) and (e), we obtain

$$2p_{21} = p_{21} + p_{12}, \quad 2p_{31} = p_{31} + p_{13}, \dots .$$

Therefore, (f) $p_{21} = p_{12}$, $p_{31} = p_{13}$, and in general $p_{mn} = p_{nm}$.

The electrical energy W_e of the system may therefore be expressed as a homogeneous quadratic function of the charges by

$$\text{(g)} \quad W_e = \tfrac{1}{2}p_{11}e_1^2 + \tfrac{1}{2}p_{22}e_2^2 + \tfrac{1}{2}p_{33}e_3^2 + \dots + p_{12}e_1e_2 + p_{13}e_1e_3 + p_{23}e_2e_3 + \dots ,$$

where the coefficients p_{mn} are called *coefficients of potential.*

If equations (a) are solved for e_1, e_2, e_3, we obtain

$$\text{(h)} \quad \begin{cases} e_1 = q_{11}\Psi_1 + q_{21}\Psi_2 + q_{31}\Psi_3 + \dots \\ e_2 = q_{12}\Psi_1 + q_{22}\Psi_2 + q_{32}\Psi_3 + \dots \\ e_3 = q_{13}\Psi_1 + q_{23}\Psi_2 + q_{33}\Psi_3 + \dots . \end{cases}$$

If the charge δe_1 is communicated to the conductor A_1 and the charge δe_2 to the conductor A_2, etc., so that the potential Ψ_1 is increased by $\delta\Psi_1$, while the other potentials retain their original values, we have

$$\delta e_1 = q_{11}\delta\Psi_1, \quad \delta e_2 = q_{12}\delta\Psi_1, \quad \delta e_3 = q_{13}\delta\Psi_1, \dots .$$

The increment of the energy is therefore

$$\text{(i)} \quad \delta W = q_{11}\Psi_1\delta\Psi_1 + q_{12}\Psi_2\delta\Psi_1 + q_{13}\Psi_3\delta\Psi_1 + \dots .$$

From equation (h) the energy W is given by

$$\text{(k)}\quad \begin{cases} 2W = q_{11}\Psi_1{}^2 + q_{22}\Psi_2{}^2 + q_{33}\Psi_3{}^2 + \ldots + (q_{12}+q_{21})\Psi_1\Psi_2 \\ \qquad + (q_{13}+q_{31})\Psi_1\Psi_3 + (q_{23}+q_{32})\Psi_2\Psi_3 + \ldots \end{cases}$$

If Ψ_1 is increased by $\delta\Psi_1$, we have

$$\text{(l)}\qquad \delta W = [q_{11}\Psi_1 + \tfrac{1}{2}(q_{12}+q_{21})\Psi_2 + \tfrac{1}{2}(q_{13}+q_{31})\Psi_3 \cdots +]\delta\Psi_1.$$

Comparing equations (i) and (l), it follows that (m) $q_{mn} = q_{nm}$. The energy W_Ψ, expressed in terms of the potentials, is therefore given by

$$\text{(n)}\quad \begin{cases} W_\Psi = \tfrac{1}{2}q_{11}\Psi_1{}^2 + \tfrac{1}{2}q_{22}\Psi_2{}^2 + \tfrac{1}{2}q_{33}\Psi_3{}^2 + \ldots + q_{12}\Psi_1\Psi_2 \\ \qquad + q_{13}\Psi_1\Psi_3 + \ldots + q_{23}\Psi_2\Psi_3 + \ldots . \end{cases}$$

The coefficients q_{nn}, in which the indices are the same, are the *capacities* of the different conductors; the coefficients q_{mn}, in which the indices differ from each other, are the *coefficients of induction*. The energy can therefore be expressed by the charges as well as by the potentials; in the former case it is represented by W_e, in the latter by W_Ψ.

The significance of the coefficients q_{11}, q_{22}, ... is shown as follows: If A_1 (Fig. 69) is an insulated conductor having the charge e_1, and if A_2, A_3, etc., are connected with the earth, we have

$$e_1 = q_{11}\Psi_1, \quad e_2 = q_{12}\Psi_1, \quad e_3 = q_{13}\Psi_1, \ldots,$$

since Ψ_2, Ψ_3, etc., are equal to zero. The coefficient q_{11} is the capacity of the conductor A_1 under these conditions. *Hence, the capacity of a conductor is the quantity of electricity which it must contain in order that its potential shall be unity, while the potential of all other conductors is zero.* The quantity of electricity induced on the conductors A_2, A_3 when connected with the earth is given by q_{12}, q_{13}, ..., if A_1 is electrified to unit potential. The coefficients q_{12}, q_{13} are negative. The lines of force proceeding from A_1 may either pass to the earth or terminate on the conductors A_2, A_3, etc. Since a positive charge is present at the points at which they leave A_1, the points at which they fall upon the other conductors must have a negative charge.

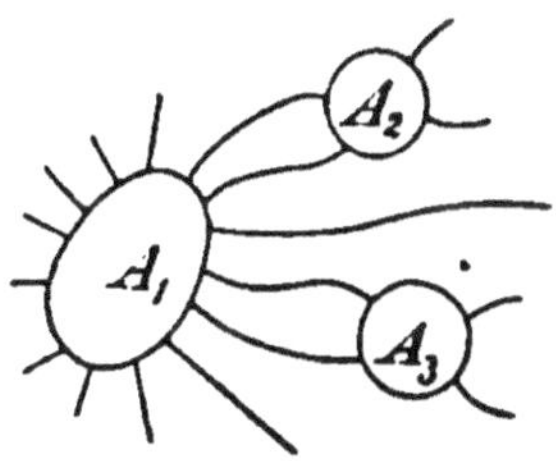

FIG. 69.

On the other hand, the coefficients p_{12}, $p_{13} \cdots p_{mn}$ are positive. If the charges of the conductors A_2, A_3, ... are $e_2 = e_3 = e_4 = \ldots = 0$, we have $\Psi_1 = p_{11}e_1$, $\Psi_2 = p_{12}e_1$, ... As many lines of force enter the uncharged conductors A_2, A_3 as pass out from them. Since lines of force pass from points of higher to points of lower potential, the potential of an uncharged conductor in an electrical field cannot be a maximum;

it lies between the greatest and least values of the potential in the field. If the conductor A_1 is charged with the unit of electricity, its potential is p_{11}. At a point infinitely distant the potential is zero. Hence, we have $p_{11} > p_{12}$, in general $p_{nn} > p_{nm}$ and $p_{mm} > p_{nm}$. Further, the value of p_{nm} lies between p_{nn} and zero, and since p_{nn} is positive, p_{nm} is also positive. The potentials of the two conductors are equal only when the charged conductor encloses the uncharged conductor. If one conductor does not enclose the other, we will always have

$$p_{nn} > p_{mn} \text{ and } p_{mm} > p_{mn}.$$

SECTION LXIII. MECHANICAL FORCES.

Let us suppose a set of insulated conductors; their charges will remain unchanged in quantity when the conductors are displaced. Their potentials depend on the charges in the manner given in LXII. The forces acting on the charged surfaces tend to set the conductors in motion. We assume that all the conductors except A_1 retain their relative positions; that A_1 can move in the direction of the x-axis; and we then determine the force which tends to move A_1 in this direction. Let the displacement of A_1 be δx. The energy W_e of the system will be diminished in consequence of this displacement by $X\delta x$. At the end of the motion the energy is $W_e + \delta W_e$, and hence we have $W_e - X.\delta x = W_e + \delta W_e$, and (a) $X = -\delta W_e/\delta x$. Now, from LXII. (g) we have

(b) $$X = \tfrac{1}{2}e_1^2 \delta p_{11}/\delta x + \tfrac{1}{2}e_2^2 \delta p_{22}/\delta x + \ldots + e_1 e_2 \delta p_{12}/\delta x + \ldots,$$

because the charges do not change during the motion, and are independent of the displacement δx. This method may be always applied if the motion of the conductor is one for which the mechanical work done by the system may be represented in the form $X\delta x$.

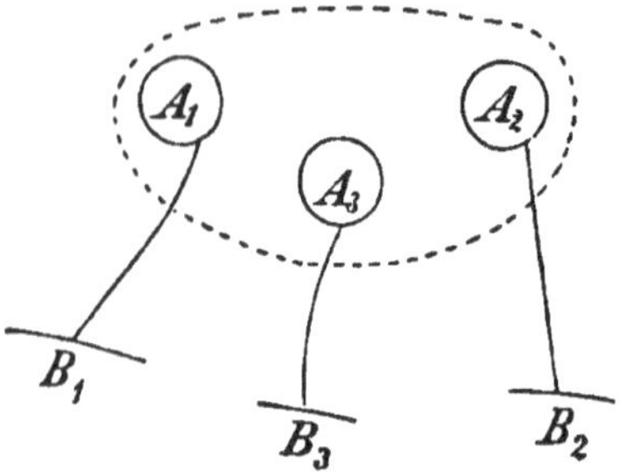

FIG. 70.

We now determine the force with which one of the conductors will move in the direction of the x-axis if the potentials remain constant. Let A_1, A_2, A_3, ... be the given conductors (Fig. 70). They are supposed to be connected by very thin wires with the very large conductors B_1, B_2, B_3, ..., whose potentials are Ψ_1, Ψ_2, Ψ_3, ... respectively, and which are so remote from the system of conductors A

that they have no influence upon it by induction. If the conductor A_1 is displaced by δx, the charges $e_1, e_2, e_3, \ldots$ increase by $\delta e_1, \delta e_2, \delta e_3, \ldots$, and we have from LXII. (h)

$$\delta e_1 = \Psi_1 \delta q_{11} + \Psi_2 \delta q_{21} + \Psi_3 \delta q_{31} + \ldots$$
$$\delta e_2 = \Psi_1 \delta q_{12} + \Psi_2 \delta q_{22} + \Psi_3 \delta q_{32} + \ldots$$
$$\delta e_3 = \Psi_1 \delta q_{13} + \Psi_2 \delta q_{23} + \Psi_3 \delta q_{33} + \ldots$$
$$\cdot \quad \cdot \quad \cdot \quad \cdot \quad \cdot$$

The electrical energy of the system thus increases by

$$\delta W = \Psi_1 \delta e_1 + \Psi_2 \delta e_2 + \Psi_3 \delta e_3 + \ldots,$$

or

$$\delta W = \Psi_1^2 \delta q_{11} + \Psi_2^2 \delta q_{22} + \Psi_3^2 \delta q_{33} + \ldots$$
$$+ 2\Psi_1 \Psi_2 \delta q_{12} + 2\Psi_1 \Psi_3 \delta q_{13} + 2\Psi_2 \Psi_3 \delta q_{23} + \ldots.$$

But from LXII. (n) the energy is equal to $W_\Psi + \delta W_\Psi$ in the new configuration of the system, where

$$\delta W_\Psi = \tfrac{1}{2}\Psi_1^2 \delta q_{11} + \tfrac{1}{2}\Psi_2^2 \delta q_{22} + \tfrac{1}{2}\Psi_3^2 \delta q_{33} + \ldots$$
$$+ \Psi_1 \Psi_2 \delta q_{12} + \Psi_1 \Psi_3 \delta q_{13} + \ldots.$$

The work doné is $X\delta x$. The sum of the energy W_Ψ originally present and the energy δW supplied is equal to the sum of the energy in the new position and the work done. We therefore have

$$W_\Psi + \delta W = W_\Psi + \delta W_\Psi + X \,.\, \delta x, \quad X \,.\, \delta x = \delta W - \delta W_\Psi.$$

If we substitute the expressions found for δW and δW_Ψ, we have

(c) $$\begin{cases} X \,.\, \delta x = \tfrac{1}{2}\Psi_1^2 \delta q_{11} + \tfrac{1}{2}\Psi_2^2 dq_{22} + \tfrac{1}{2}\Psi_3^2 dq_{33} + \ldots \\ \quad + \Psi_1 \Psi_2 \delta q_{12} + \Psi_1 \Psi_3 \delta q_{13} + \Psi_2 \Psi_3 \delta q_{23} + \ldots. \end{cases}$$

Hence, we obtain (d) $X \,.\, \delta x = \delta W_\Psi$, or $X = \delta W_\Psi / \delta x$, and further $\delta W = 2X \,.\, \delta x$.

The electrical energy supplied to the system is therefore twice as great as the mechanical work done. Now, if during the displacement of the conductors their potentials do not change, energy must flow from B_1, B_2, B_3, etc., to the conductors A_1, A_2, A_3, etc. One-half of the energy δW supplied is expended in doing the mechanical work, the other half in increasing the electrical energy.

Section LXIV. The Condenser and Electrometer.

1. *Parallel Plates.*

If two bodies at different potentials are placed near each other, a relatively great quantity of electricity can be collected on the surfaces which face each other. If A and B are two such bodies, whose potentials are Ψ_1 and Ψ_2 respectively, and if the opposing surfaces of the bodies are planes, the electrical force in the intervening space is everywhere constant, except near the edges of the plane surfaces. If a represents the distance between the planes, and if Ψ_1 is greater than Ψ_2, so that this force is directed from A to B, we have from VII. (c), (a) $\Psi_1 = \Psi_2 + Fa$ and $F = (\Psi_1 - \Psi_2)/a$. The surface-density on A_1 is determined from

$$4\pi\sigma = F \text{ or } \sigma = (\Psi_1 - \Psi_2)/4\pi a.$$

If S represents the surface of the conductor A which faces B, we have for the charge e_1 on S, (b) $e_1 = S\sigma = (\Psi_1 - \Psi_2) . S/4\pi a$. The charge e_2 on B is equal to $-e_1$. The electrical energy W_Ψ of the system is

$$W_\Psi = \tfrac{1}{2}(\Psi_1 e_1 + \Psi_2 e_2) = \tfrac{1}{2}(\Psi_1 - \Psi_2)e_1,$$

or (c) $W_\Psi = 1/8\pi . (\Psi_1 - \Psi_2)^2 . S/a$, where the energy is expressed in terms of the potentials. From (b) and (c) it follows that

(d) $$W_e = 2\pi a e_1^2/S.$$

If the x-axis is perpendicular to the plane surfaces of the conductors, and if it is directed from A to B, we have, representing the x-coordinate of the plane face of A by x_1, and that of the plane face of B by x_2, $a = x_2 - x_1$, and

$$W_\Psi = 1/8\pi . (\Psi_1 - \Psi_2)^2 . S/(x_2 - x_1)\,; \quad W_e = 2\pi e_1^2(x_2 - x_1)/S.$$

The mechanical force which acts on A is [LXIII.]

$$X_1 = -\delta W_e/\delta x_1 = 2\pi e_1^2/S\,; \quad X_1 = 2\pi\sigma e_1 = \tfrac{1}{2}Fe_1.$$

This corresponds to LIX. (e). We further have [LXIII. (d)]

$$X_1 = \delta W_\Psi/\delta x_1 = 1/8\pi . (\Psi_1 - \Psi_2)^2 . S/(x_2 - x_1)^2 = 1/8\pi . F^2 . S.$$

This agrees with LIX. (f). From the expressions which have been given for W_Ψ, we also have $W_\Psi = 1/8\pi . F^2 . S(x_2 - x_1)$, which agrees with LXI. (i).

The capacity C is $C = e_1/(\Psi_1 - \Psi_2)$; if $\Psi_2 = 0$, we have $C = S/4\pi a$.

2. *Concentric Spherical Surfaces.*

If a sphere A_1, whose radius is R, is enclosed by the concentric spherical shell, whose internal and external radii are R_2 and R_3, and if A_1 is given the charge e_1, and A_2 the charge e_2, the inner surface of A_2 will take the charge $-e_1$, while its outer surface has the charge $e_1 + e_2$.

The potentials within the sphere A_1 and within the spherical shell A_2 are therefore $\Psi_1 = e_1/R_1 - e_1/R_2 + (e_1+e_2)/R_3$; $\Psi_2 = (e_2+e_1)/R_3$; and hence $e_1 = R_2R_1/(R_2-R_1) . \Psi_1 - R_2 . R_1/(R_2-R_1) . \Psi_2$ and

$$e_2 = -R_2R_1/(R_2-R_1) . \Psi_1 + \big(R_3 + R_2 . R_1/(R_2-R_1)\big) . \Psi_2.$$

These equations agree with those given in LXII. (h). The potential Ψ_i in the space between the two spheres is

$$\Psi_i = e_1/r - e_1/R_2 + (e_1+e_2)/R_3,$$

where r is the distance of the point considered from the common centre of the spheres. The potential outside the spherical shell is $\Psi_a = (e_1+e_2)/r$. The capacity C of the inner sphere is determined by $e_1 = C\Psi_1$, when Ψ_2 is set equal to zero; we have therefore $C = R_1R_2/a$, if we represent by a the distance between the surface of the inner sphere and the inner surface of the spherical shell.

3. *Coaxial Cylinders.*

Suppose two coaxial cylindrical surfaces, A_1 and A_2, confronting each other. Let their potentials be Ψ_1 and Ψ_2, and their radii R_1 and R_2 respectively. Let a point in the space between A_1 and A_2 be at the distance r from the common axis of the cylinders. The potential Ψ must satisfy the equation $\nabla^2\Psi = 0$ for this space. Since the equi-potential surfaces in the space considered are cylindrical surfaces coaxial with A_1, the equation $\nabla^2\Psi = 0$ may be given the form [cf. XV.] $d^2\Psi/dr^2 + 2/r . d\Psi/dr = 0$, and we obtain by integration $\Psi = c \log r + c_1$. For $r = R_1$ we have $\Psi = \Psi_1$, and for $r = R_2$, $\Psi = \Psi_2$, therefore

$$\Psi_i = \Psi_1 . (\log R_2 - \log r)/(\log R_2 - \log R_1)$$
$$+ \Psi_2 . (\log r - \log R_1)/(\log R_2 - \log R_1).$$

For a point outside the outer cylinder, the potential is

$$\Psi_a = \Psi_2 + c(\log r - \log R_2),$$

where r is the distance of the point considered from the axis of the cylinder. The constant c cannot be determined from the potentials alone. The electrical force F in the intervening space is $F = -d\Psi_i/dr = -(\Psi_2 - \Psi_1)/(\log R_2 - \log R_1) . 1/r$. The surface-densities σ_1 and σ_2 on A_1 and A_2 are [LV. (e)]

$$\sigma_1 = (\Psi_1 - \Psi_2)/(\log R_2 - \log R_1) . 1/4\pi R_1,$$
$$\sigma_2 = (\Psi_2 - \Psi_1)/(\log R_2 - \log R_1) . 1/4\pi R_2.$$

The charge on a portion of the cylinder A_1, whose length is l, is $e_1 = 2\pi R_1 l \sigma_1$, or $e_1 = l/2 . (\Psi_1 - \Psi_2)/(\log R_2 - \log R_1)$. The charge on the inner surface of A_2 is equal and opposite to this. The capacity C of a portion of the inner cylinder, whose length is l, is

$$C=\tfrac{1}{2}l/(\log R_2-\log R_1).$$

4. *The Quadrant Electrometer.*

Suppose A_1A_1 to be two metal plates whose potential is Ψ_1, and A_2A_2 to be two similar plates whose potential is Ψ_2 (Fig. 71). In the middle between the two pairs of plates there is placed a plate A_3 whose potential is Ψ_3. We assume that $\Psi_1<\Psi_2<\Psi_3$. If A_3 is displaced by the distance δx in the direction from A_2 to A_1, and therefore in the direction of its length, the area $b\delta x$ is displaced from right to left, if b represents the width of the

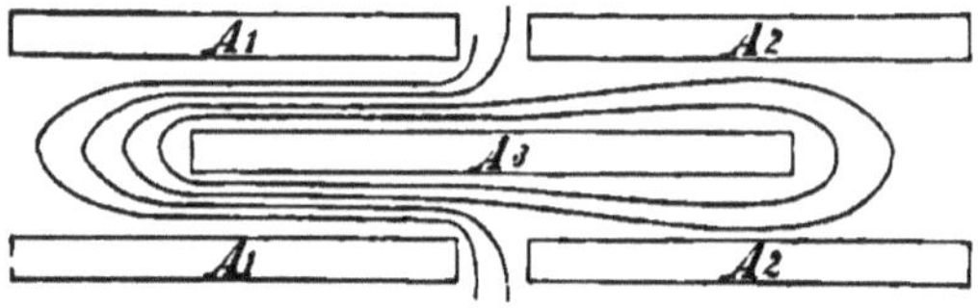

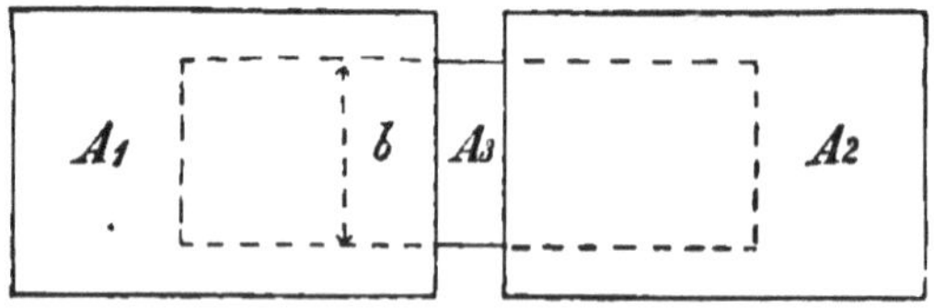

FIG. 71.

plate A_3. If the distance of A_3 from A_1 and A_2 is a, the force acting between A_1 and A_3 is $F_1=(\Psi_3-\Psi_1)/a$, and that acting between A_2 and A_3 is $F_1=(\Psi_3-\Psi_2)/a$, on the assumption that the points considered do not lie near the edges of A_3, or in the space between A_1 and A_2. From LXI. (i) the electrical energy is $W=1/8\pi\,.\int F^2d\tau$. During the motion considered, the energy on the left side of the pair of plates is increased by

$$1/8\pi\,.\,(\Psi_3-\Psi_1)^2/a^2\,.\,2ab\delta x\,;$$

that on the right side is diminished by $1/8\pi\,.\,(\Psi_3-\Psi_2)^2/a^2\,.\,2ab\delta x$. The gain of energy is therefore

$$\delta W_\Psi=b\delta x\big((\Psi_3-\Psi_1)^2-(\Psi_3-\Psi_2)^2\big)/4\pi a.$$

This expression does not fully represent the gain of energy, since no account is taken of the relations at the edges. We therefore set

$$\delta W_\Psi=\tfrac{1}{2}k\delta x\big((\Psi_3-\Psi_1)^2-(\Psi_3-\Psi_2)^2\big).$$

The force X which tends to move A_3 from right to left is then [LXIII. (d)] $X=k(\Psi_2-\Psi_1).(\Psi_3-\frac{1}{2}(\Psi_1+\Psi_2))$. We may apply this result to the quadrant electrometer. If, in this instrument, the movable aluminium plate turns through the angle Θ, we may set approximately

$$\Theta=a(\Psi_2-\Psi_1).(\Psi_3-\tfrac{1}{2}(\Psi_1+\Psi_2)),$$

where Ψ_1 and Ψ_2 are the potentials of the quadrants, Ψ_3 is the potential of the aluminium plate, and a is a constant whose value depends on the form and dimensions of the apparatus.

SECTION LXV. THE DIELECTRIC.

We have assumed until now that the bodies considered were either good conductors or perfect insulators, on which the charge was immovable. Experiment shows, however, that there are no perfect insulators.

Electricity on insulators is often lost by conduction, which for the most part is due to the film of fluid deposited on them by the air. But even if this film is removed by careful drying, conduction still persists. If a charge of electricity is communicated to one part of an insulator, it is distributed after a considerable time in the insulator in the same way as it would be in a good conductor. Besides this, another action also exists which is instantaneous. When a movable insulator is brought into the neighbourhood of a charged conductor, the insulator sets itself in the same way as a good conductor, from which it follows that an instantaneous distribution of electricity takes place in it. According to Faraday, insulators consist of very small conductors which are separated by an insulating medium. The capacity of a condenser is increased by replacing the air which serves as the insulator between its surfaces by other insulators, such as glass, shellac, calc spar, etc.

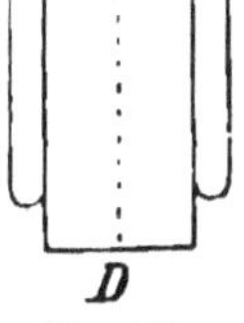

FIG. 72.

Let A and B (Fig. 72) be two conducting plates which are separated by the insulator CD. Let A be brought to the potential Ψ_1, and B to the potential Ψ_2; let the surface-density on A be σ, that on B is then $-\sigma$.

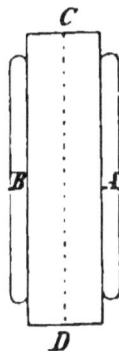

In order to explain this, Faraday assumed that a peculiar electrification exists in the insulator CD, by which each of the conducting particles contained in it acquires negative electricity on its right and positive electricity on its left. Just as a mechanical force can give rise to an elastic displacement, the forces proceeding from the plates of the condenser produce an electromotive action by setting up a current of electricity in the particles of the dielectric. The positive electricity flows in the particles toward the left, the negative toward the right.

By this process, which we may call *dielectric displacement*, there arises a polarization of all the particles.

The condition in the dielectric can be compared with the polarity of the particles of a permanent magnet.

The quantity $\mathfrak{D}$ which flows through a unit of area parallel to A and B must be equal to σ. A unit of area of the surface of the insulator at A receives the charge $-\sigma$, and that at B the charge $+\sigma$. Let a be the distance between A and B, whose difference of potential is $\Psi_1 - \Psi_2$. The force in the intervening space is (a) $F = (\Psi_1 - \Psi_2)/a$. If the quantity of electricity $\mathfrak{D}$ which flows through the unit of area is proportional to the force acting in the insulator, we can set

(b) $$\mathfrak{D} = K/4\pi \, . \, F,$$

where K is a constant. Hence, the surface of A, whose area is S, will have the charge (c) $S\mathfrak{D} = K/4\pi \, . \, (\Psi_1 - \Psi_2)/a \, . \, S$. A comparison of this equation with LXIV. (b) shows how many times greater the capacity of the condenser becomes if another insulator is used in place of air. We call K the *dielectric constant*; for air, which is chosen as the standard medium, we set $K = 1$.

The dielectric constant of an insulating medium is the ratio of the capacity of a condenser having that medium as an insulator to the capacity of the same condenser when air is the insulator. It is, however, more correct to set $K = 1$ for a vacuum; it has been shown that K for gases is a little greater than 1. As examples of the values of the dielectric constant, we have for

Glass, - - -	$K = 5,83—6,34$,
Paraffin, - - -	$K = 2—2,32$,
Sulphur, - - -	$K = 3,84$,
Shellac, - - -	$K = 3—3,7$,
Bi-sulphide of carbon,	$K = 2,6$,
Oil of turpentine, -	$K = 2,2$.

On the whole, the results obtained by different observers are not very consistent.

Let us, as in LVIII., suppose that the body A is brought into the space enclosed by the metallic shell BC. Suppose that

B receives the charge $-Q$, and C the charge Q. The quantity Q of positive electricity flows outward through the closed surface D taken within BC. Hence, when the quantity Q is introduced through the closed surface D, the same quantity flows out through the same surface. We are therefore justified in assuming that the quantity enclosed by the surface D is always zero. This holds for the closed surface E, which may be drawn in the insulator surrounding BC, and thus we obtain the general theorem *that the total quantity of electricity contained within a closed surface is equal to zero.*

If the quantity of electricity $\mathfrak{D}$, which flows through a unit of area perpendicular to the direction of electrical force, is proportional to that force at every point, we will have (d) $\mathfrak{D} = K/4\pi \,.\, F$. If f, g, and h are the quantities which pass through three units of area in an isotropic body taken perpendicular to the three coordinate axes, that is, if they are the rectangular *components of the displacement* $\mathfrak{D}$, and if X, Y, Z are the components of the electromotive force F, we have (e) $f = K/4\pi \,.\, X$, $g = K/4\pi \,.\, Y$, $h = K/4\pi \,.\, Z$. These expressions are consistent with the relations

(f) $$X = -\partial\Psi/\partial x, \quad Y = -\partial\Psi/\partial y, \quad Z = -\partial\Psi/\partial z.$$

SECTION LXVI. CONDITIONS OF EQUILIBRIUM.

Suppose the closed surface S to enclose a portion of the electrical system, passing partly through the dielectric, partly through the conductors. We will make use of our previous conclusion, that the total quantity of electricity within S is zero. If we represent by e the total quantity enclosed by S, and by $\mathfrak{D}'$ the quantity which flows out through a unit of area in consequence of the displacement in the dielectric, we have $e = \int \mathfrak{D}' \,.\, dS$. The normal to the surface S directed outward makes angles with the axes whose cosines are l, m, n. We have $\mathfrak{D}' = K/4\pi \,.\, F\cos\epsilon$, if ϵ is the angle between the normal to dS and the direction of the electromotive force F. Since $F\cos\epsilon = Xl + Ym + Zn$, we obtain

(b) (c) $$e = 1/4\pi \,.\, \int K(Xl + Ym + Zn)dS = \int (fl + gm + hn)dS.$$

If we apply equation (c) to an infinitely small parallelepiped, whose edges are dx, dy, and dz, we find that the quantity which enters it is $f dydz + g dxdz + h dxdy$, and that which leaves it

$$\left(f + \frac{\partial f}{\partial x}dx\right)dydz + \left(g + \frac{\partial g}{\partial y}dy\right)dxdz + \left(h + \frac{\partial h}{\partial z}dz\right)dxdy;$$

if ρ is the volume-density, the charge contained in the parallelepiped is $\rho dxdydz$; now the total quantity of electricity within it equals zero, and therefore

$$fdydz + gdxdz + hdxdy + \rho dxdydz$$
$$= \left(f + \frac{\partial f}{\partial x}dx\right)dydz + \left(g + \frac{\partial g}{\partial y}dy\right)dxdz + \left(h + \frac{\partial h}{\partial z}dz\right)dxdy.$$

From this it follows that the volume-density ρ of the free electricity within the body [cf. LXVIII.] is given by (d) $\rho = \partial f/\partial x + \partial g/\partial y + \partial h/\partial z$, and from LXV. (e) we obtain

(e) $$\partial(KX)/\partial x + \partial(KY)/\partial y + \partial(KZ)/\partial z = 4\pi\rho.$$

In terms of the potential, this becomes

(f) $$\partial(K.\partial\Psi/\partial x)/\partial x + \partial(K.\partial\Psi/\partial y)/\partial y + \partial(K.\partial\Psi/\partial z)/\partial z + 4\pi\rho = 0.$$

If we consider a surface on which the surface-density is σ, we obtain by the same method as that used in LIV. (g) $\sigma = \mathfrak{D}_1 + \mathfrak{D}_2$, if $\mathfrak{D}_1$ and $\mathfrak{D}_2$ are the polarizations in the directions of the normals to the surface drawn outward. If the forces along the same normals are N_1 and N_2, we have (h) $\sigma = K_1/4\pi . N_1 + K_2/4\pi . N_2$, where K_1 and K_2 are the dielectric constants on the opposite sides of the surface.

By means of this equation questions on electrical distribution and on the relations between density and potential may be solved. If K is constant, it follows from (f) that $K\nabla^2\Psi + 4\pi\rho = 0$. In a region where the dielectric constant is K, the potential which arises from a given charge ρ is equal to only the Kth part of that which arises from the same charge in a region where $K = 1$. In the latter case the potential Ψ' is determined by $\nabla^2\Psi' + 4\pi\rho = 0$, so that $\Psi = \Psi'/K$.

The electrical force is also diminished in the same ratio. If the charges e_1 and e_2 are placed at the points A and B respectively, and if the distance $AB = r$, the charges repel each other with the force R,

$$R = 1/K . e_1e_2/r^2.$$

SECTION LXVII. MECHANICAL FORCE AND ELECTRICAL ENERGY IN THE DIELECTRIC.

Suppose that the dielectric constant is constant in the region considered. The forces acting in the directions of the axes on the parallelepiped $dxdydz = d\tau$, in which the density is ρ, are

(a) $$(X)d\tau = \rho Xd\tau, \quad (Y)d\tau = \rho Yd\tau, \quad (Z)d\tau = \rho Zd\tau.$$

Using LXVI. (e), we have

$$(X) = 1/4\pi\big(X\,.\,\partial(KX)/\partial x + X\,.\,\partial(KY)/\partial y + X\,.\,\partial(KZ)/\partial z\big).$$

Since [LXV. (f)] the forces have a potential, we obtain from XXVII. (b)

$$(X) = K/8\pi\,.\,\big(\partial(X^2 - Y^2 - Z^2)/\partial x + 2\partial(XY)/\partial y + 2\partial(XZ)/\partial z\big).$$

The force which acts on the volume-element $d\tau$ may be considered as due to the stresses X_x, Y_y, etc. [cf. XXVII. (c)], where

(b) $$\begin{cases} X_x = K/8\pi\,.\,(X^2 - Y^2 - Z^2), & Y_z = Z_y = K/4\pi\,.\,YZ, \\ Y_y = K/8\pi\,.\,(Y^2 - X^2 - Z^2), & Z_x = X_z = K/4\pi\,.\,XZ, \\ Z_z = K/8\pi\,.\,(Z^2 - X^2 - Y^2), & X_y = Y_x = K/4\pi\,.\,YX. \end{cases}$$

If the direction of the x-axis is that of the electrical force, we have

(c) $$X_x = KX^2/8\pi, \quad Y_y = -KX^2/8\pi, \quad Z_z = -KX^2/8\pi,$$

and the tangential components vanish. In the direction of the electromotive force F, there is a tension S, and in all directions perpendicular to the force F, a pressure T, such that (d) $S = T = KF^2/8\pi$.

If A and B (Fig. 73) are two conducting surfaces which are separated by an insulator whose dielectric constant is K, and if AB is a line of force, a tension S acts along that line, and a pressure $T = S$ acts perpendicularly to it. A surface-element at A is under a tension

$$1/8\pi\,.\,KF^2,$$

acting in the direction of the normal drawn outward. When $K = 1$ this reduces to the result reached in LIX.

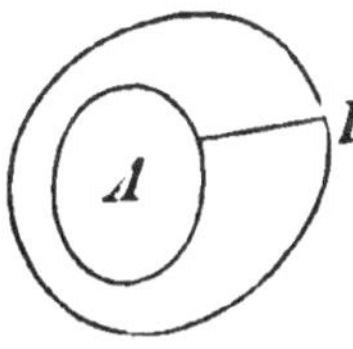

FIG. 73.

For example, let AB be a hollow sphere of glass whose inner and outer radii are r_1 and r_2 respectively. Let the potential of the surface A be Ψ_1, and that of B be 0. The potential in the interior of the spherical shell is determined from LXVI. (f); when $\rho = 0$, we have $\nabla^2\Psi = 0$. Since the potential depends only on the distance r from the centre, we have from XV. $d^2\Psi/dr^2 + 2/r\,.\,d\Psi/dr = 0$, and hence $\Psi = A + B/r$. Having regard to the boundary conditions, we obtain $\Psi = r_1(r_2 - r)/(r_2 - r_1)\,.\,\Psi_1/r$. The forces F_1 and F_2 which act at the inner and outer surfaces are

$$F_1 = +\Psi_1/r_1\,.\,r_2/(r_2 - r_1); \quad F_2 = +\Psi_1/r_2\,.\,r_1/(r_2 - r_1).$$

Representing the stresses on these surfaces by p_1 and p_2, we obtain

$$p_1 = K/8\pi\,.\,\big(\Psi_1/r_1\,.\,r_2/(r_2 - r_1)\big)^2; \quad p_2 = K/8\pi\,.\,\big(\Psi_1/r_2\,.\,r_1/(r_2 - r_1)\big)^2.$$

These stresses may be regarded as pressures which act on the surfaces.

If $d\phi/dr$ represents the increment of r which results from the pressure on the surface, we have from XXXI. (h) $d\phi/dr = \frac{1}{3}ar + b/r^2$, where $a = 3/(3\lambda+2\mu).(p_1r_1^3 - p_2r_2^3)/(r_2^3 - r_1^3)$; $b = 1/4\mu.(p_1 - p_2)r_1^3.r_2^3/(r_2^3 - r_1^3)$. From this it follows that

$$p_1r_1^3 - p_2r_2^3 = K\Psi_1^2/8\pi.r_1r_2/(r_2 - r_1);$$
$$(p_1 - p_2)r_1^3r_2^3 = K\Psi_1^2/8\pi.r_1r_2(r_2^4 - r_1^4)/(r_2 - r_1)^2.$$

If we set $K\Psi_1^2/8\pi.r_1r_2/[(r_2 - r_1)(r_2^3 - r_1^3)] = N$, we will have

$$d\phi/dr = \big(r/(3\lambda + 2\mu) + (r_2^4 - r_1^4)/(r_2 - r_1).1/4\mu r^2\big)N.$$

The volume contained by the hollow sphere will be increased by the action of the electrical force.

If we represent by Θ_0 the increment of the unit of volume, we have $4\pi/3.(r + d\phi/dr)^3 = 4\pi/3.r^3(1 + \Theta_0)$, so that $\Theta_0 = 3d\phi/rdr$. For the volume of the sphere, for which $r = r_1$, we obtain

$$\Theta_0 = 3\big(1/(3\lambda + 2\mu) + (r_2^4 - r_1^4)/(r_2 - r_1).1/4\mu r_1^3\big)N.$$

If we set $r_2 - r_1 = \delta$, and if δ is very small in comparison with r_1, we have $\Theta_0 = 9N.(\lambda + \mu)/\mu(3\lambda + 2\mu)$, where $N = K\Psi_1^2/8\pi.1/3\delta^2$. It thus follows, using XXIX. (d), that (e) $\Theta_0 = 3/E\delta^2.K\Psi_1^2/8\pi$. The increase of volume here considered has been observed for various condensers.

If a region, in which the dielectric constant K is a function of the coordinates, contains electrical charges, whose density is ρ, and which give rise to the potential Ψ, the energy W is determined as in LXI. by (f) $W = \frac{1}{2}\int\rho\Psi d\tau$.

If ρ is expressed in terms of the potential by

$$\partial(K.\partial\Psi/\partial x)/\partial x + \partial(K.\partial\Psi/\partial y)/\partial y + \partial(K.\partial\Psi/\partial z)/\partial z + 4\pi\rho = 0,$$

we have for the energy W,

$$W = -1/8\pi.\iiint\Psi\big(\partial(K.\partial\Psi/\partial x)/\partial x + \partial(K.\partial\Psi/\partial y)/\partial y + \partial(K.\partial\Psi/\partial z)/\partial z\big)dxdydz.$$

By integration by parts, we obtain

$$W = 1/8\pi.\iiint K\big((\partial\Psi/\partial x)^2 + (\partial\Psi/\partial y)^2 + (\partial\Psi/\partial z)^2\big)dxdydz,$$

where the integration is extended over the entire region, and it is assumed that the force and the potential vanish at infinity. If F represents electrical force, we have (g) $W = 1/8\pi.\iiint KF^2dxdydz$.

Electrical Double Sheets. We have seen in LVIII. that two conductors which are very near each other, and are kept at the two different potentials Ψ_1 and Ψ_2, are oppositely charged. The surface-densities σ and $-\sigma$ of their charges are given by $\sigma = (\Psi_1 - \Psi_2)/4\pi a$, where a

is the distance between the surfaces. If a is taken infinitely small, there is still a finite difference of potential between the surfaces; calling this difference V, we have (a) $\sigma = V/4\pi a$.

Now, there are several ways by which such finite potential differences may be established across a surface; for example, it may be done by friction or, what amounts to the same thing, by contact. This being so, we must necessarily assume, as was first remarked by v. Helmholtz, that a *double sheet* of electricity is formed on the two surfaces which are near each other, in which, if the distance a is extremely small, the density σ must be extremely great if the potential difference V is to be finite. When two such bodies are separated from each other, provided they still retain the electricity thus disposed on them, they will both be very strongly charged. It is usually not possible to separate them without discharging them, but if one or both of them are insulators a very considerable charge remains. V. Helmholtz thus explains the action of the rubber of the frictional electrical machine.

In the same way v. Helmholtz explained several remarkable phenomena, for example the phenomena of electrical convection, studied by Quincke and Wiedemann. Let us consider a capillary tube, of circular cross-section, whose inner radius is R and whose length is l. A liquid is supposed to be flowing through this tube. If there is a difference of potential V between the liquid and the wall of the tube, a layer of electricity forms around the liquid, whose density σ is determined by equation (a). Setting the radius of this cylindrical electrical layer equal to r, so that $a = R - r$, we have (b) $\sigma = V/4\pi(R-r)$. Let the velocity of the flow at the place at which this layer is present be ω. The quantity of electricity which is carried on by the current in unit time through any cross-section of the tube is

$$Q = 2\pi r\sigma\omega = \frac{rV\omega}{2(R-r)}.$$

Now, we have found [XLIX.] that

$$\omega = \frac{(p_0 - p_1)(R^2 - r^2)}{4\mu l}, \text{ and hence } Q = \frac{rV(p_0 - p_1)(R + r)}{8\mu l}.$$

We may in this equation set $R = r$, and obtain

$$\text{(c)} \qquad Q = \frac{R^2 V(p_0 - p_1)}{4\mu l} = \frac{SV(p_0 - p_1)}{4\pi\mu l},$$

where S is the area of the cross-section of the tube.

If the pressure at the two ends of the tube is the same, while a difference of potential $\Psi_1 - \Psi_2$ exists between them, an electrical force $F=\frac{\Psi_1-\Psi_2}{l}$ will act within the tube, and therefore a force $F\sigma$ will act on every unit of area of the electrical layer. The liquid will thus be set in motion. The velocity increases from the wall of the tube, where it is 0, to the layer σ, where it may be called u. By the definition of internal friction [XLVII.], we then have $F\sigma=\mu\frac{u}{a}$. If σ is expressed in terms of the potential difference, we have

(d) $$u=\frac{VF}{4\pi\mu}.$$

CHAPTER VIII.

MAGNETISM.

SECTION LXVIII. GENERAL PROPERTIES OF MAGNETS.

IT was very early known to the Greeks that in the city of Magnesia, in Asia Minor, stones were to be found which had the power of attracting iron. If such a stone, which consisted mostly of magnetic oxide of iron, were thrown into iron filings, they would adhere with special strength to it at certain points. A long magnet is especially active in the vicinity of its ends, which are called *poles*. If a bar magnet is suspended horizontally, so that it can turn about a vertical axis passing through its centre, it assumes of itself a definite direction which approximately coincides with the meridian of the place. The end of the bar which points toward the north is called the *north pole*, the other end the *south pole*. Following the indication of this experiment, we assume the existence of two magnetic fluids, which are separated in each particle of iron by the influence of a magnetic force. If a bar magnet is floated on a fluid at rest, it assumes a definite direction when exposed to the influence of a magnetic force, but the force does not set the floating magnet in motion if the dimensions of the magnet are very small in comparison with its distance from the seat of the magnetic force. We conclude from this that the north and south magnetic fluids are present in a magnet in equal quantities. Let us represent the quantity of one fluid by $+m$, that of the other by $-m$, the north magnetism being conventionally taken as positive. Coulomb proved that the poles of two magnets repel each other with a force F, which is given by (a) $F = m_1 m_2 / r^2$, where m_1 and m_2 are the quantities of magnetism at the poles and r is the distance between the poles.

If a magnet is broken into many small pieces, each part is still a magnet. Because of this fact, we conclude that every magnet is made up of a very great number of very small magnets.

If a magnet is broken, positive magnetism appears on one, and negative magnetism on the other, of the surfaces formed by the fracture, and their quantities are equal unless the magnetization is changed by the jar given to the magnet when it is broken. We will represent by $+\sigma$ the quantity of magnetism that is present on a unit of area of one of the newly formed faces. For each point on this face σ has a definite value, dependent on the position of the point on the face and of the face in the original magnet. Construct a normal to this face drawn outward; σ then depends on the coordinates x, y, z of the point and on the direction of the normal, which makes angles with the axes whose cosines are l, m, n. Let $ABC = dS$ (Fig. 74) be an element of the positive face and O a point in the magnet infinitely near it. A, B, C are points in which the surface dS is met by the lines Ox, Oy, Oz parallel to the axes drawn through O. The surface OBC, one of the surfaces of the tetrahedron $OABC$, may be considered as a negative face. Suppose that the magnet is magnetized with the components A, B, C in the directions OA, OB, and OC, the surface ABC of the tetrahedron exhibits positive magnetism, the surfaces OBC, etc., negative magnetism. Represent the surface-density perpendicular to the x-axis by A; each unit of surface of OBC then exhibits the quantity of magnetism $-A$. The unit of surface of OAC and OBA will, in a similar notation, exhibit the quantities of magnetism $-B$ and $-C$ respectively. The position of the surface ABC is determined by the cosines l, m, n of the angles which the normal to the surface makes with the axes; we have $OBC = l\,.\,dS$, $OAC = m\,.\,dS$, $OBA = n\,.\,dS$. If h represents the perpendicular let fall from O to the surface ABC, and if the quantity of magnetism in the unit of volume is represented by ρ, the total quantity of magnetism contained in the tetrahedron is

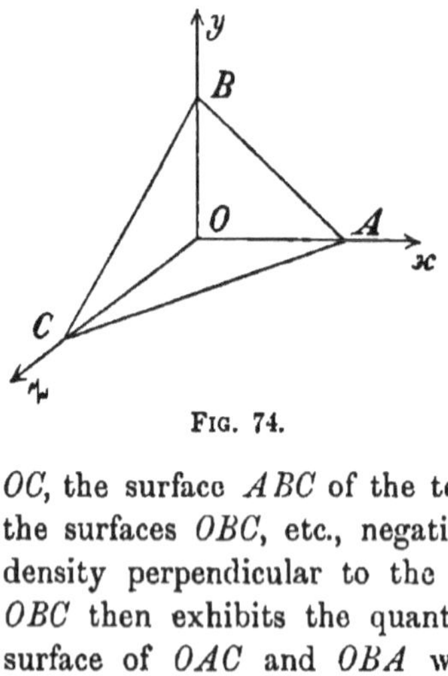

FIG. 74.

$$(\sigma - lA - mB - nC)dS + \tfrac{1}{3}h\rho dS.$$

Now, since the total quantity of magnetism in a magnet is zero, and the altitude h of the tetrahedron is assumed to be infinitely small, we have (b) $\sigma = Al + Bm + Cn$. Hence, if the surface-density on three perpendicular surface-elements passed through a point is known, the density on any other surface passed through the same point is determined from (b). If we set

(c) $$J^2 = A^2 + B^2 + C^2 \text{ and } A = J\lambda,\ B = J\mu,\ C = J\nu,$$

where $\lambda^2 + \mu^2 + \nu^2 = 1$, it follows from (b) that $\sigma = J(l\lambda + m\mu + n\nu)$. J is the *intensity* or *strength of magnetization*. The direction of the intensity makes angles with the coordinate axes, whose cosines are λ, μ, ν.

If we set $l\lambda + m\mu + n\nu = \cos\epsilon$, where ϵ is the angle between the intensity of magnetization and the normal to the surface-element, we have $\sigma = J\cos\epsilon$, that is, σ is the component of the intensity of magnetization along the normal to the surface-element. The greatest value of σ is reached when $\epsilon = 0$, that is, when the direction of the intensity of magnetization coincides with the normal to the surface-element on which the density is σ. A surface may be passed through any point in a magnet for which the surface-density is a maximum. The direction of the intensity of magnetization J lies in the normal to this surface. For such a surface, whose normal coincides with the direction of magnetization, $\sigma = J$; that is, the quantity of magnetism on the unit of area of this surface is J. We construct a parallelepiped, one of whose ends, dS, lies in the surface considered, and whose edges perpendicular to the surface are ds; $J.dS.ds$ is called the *magnetic moment* of this parallelepiped. *The intensity of magnetization J is equal to the ratio of the magnetic moment of the magnet to its volume.*

The magnetic condition of a magnet is defined by the components of magnetization A, B, C. The density σ on the surface of the magnet is determined by (b) from these components. Free magnetism may be present in the interior of the magnet also. If OO' (Fig. 75) is a rectangular parallelepiped, whose edges are dx, dy, dz, taken within the magnet, and if A, B, C are the components of the intensity of magnetization at the point O, OA' will contain the quantity of magnetism $-A\,dydz$, and $O'A$ the quantity $(A + \partial A/\partial x\,.\,dx)dydz$. Analogous expressions hold for the other surfaces. Representing by ρ the quantity of magnetism contained in the unit of volume, we have

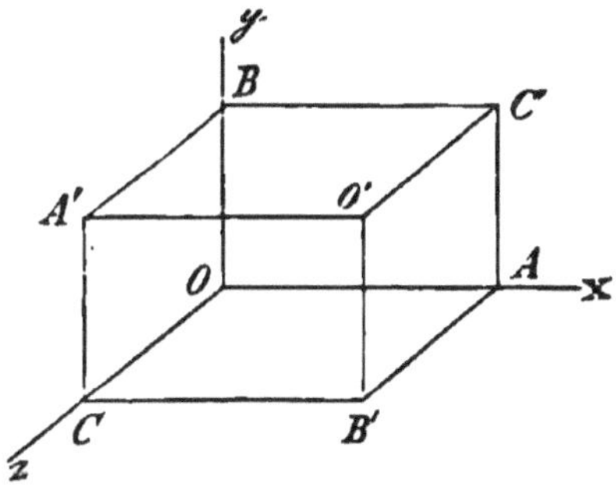

FIG. 75.

$$(\partial A/\partial x + \partial B/\partial y + \partial C/\partial z + \rho)dxdydz = 0,$$

since the total magnetism in the parallelepiped is zero. Hence (c) $\rho = -(\partial A/\partial x + \partial B/\partial y + \partial C/\partial z)$. While we may consider A, B, C as the natural and direct expressions for the condition of a magnet, we determine its free magnetism from the derived magnitudes ρ and σ.

SECTION LXIX. THE MAGNETIC POTENTIAL.

The force with which a magnet acts on a pole at which unit quantity of magnetism is concentrated is called the *magnetic force.* Its components are represented by α, β, γ. These are determined from the potential in the same way as the components of the electrical force. Let the quantities of magnetism m, m', m'', ... be present at given points; let the pole P, for which the potential is to be determined, be distant r, r', r'', ... respectively from those points. The potential is then given by

$$V = m/r + m'/r' + m''/r'' + \ldots, \text{ and } m + m' + m'' + \ldots = 0.$$

If N is a north pole with magnetism $+m$ (Fig. 76) and S a south pole with magnetism $-m$, and if P is the point for which the potential is to be determined, and whose distances from the north and south poles are r and r' respectively, we have

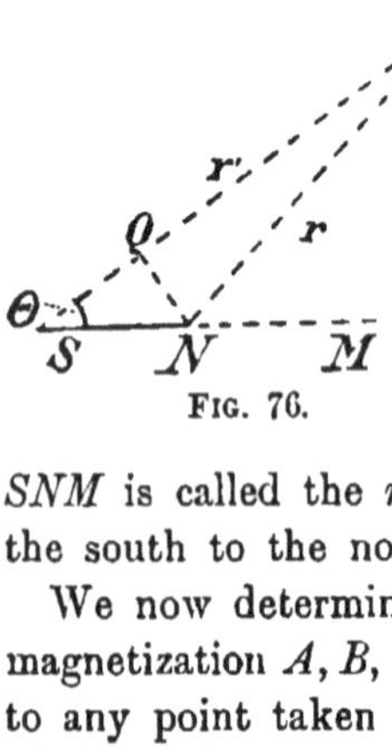

FIG. 76.

$$V = m/r - m/r' = m(r' - r)/rr'.$$

If the length l of the magnet is very small in comparison with r and r', and if we set $lm = \mathfrak{M}\cos\Theta/r^2$, we have (a) $V = \mathfrak{M}\cos\Theta/r^2$. $\mathfrak{M} = lm$ is called the *moment* of the magnet; SNM is called the *magnetic axis.* Its positive direction is that from the south to the north pole.

We now determine the *potential of a magnet* whose components of magnetization A, B, and C are given. Let the coordinates with respect to any point taken as origin be ξ, η, ζ; A, B, C are then functions of these three coordinates. A parallelepiped whose edges are $d\xi$, $d\eta$, $d\zeta$ will have the magnetic moment $A\,d\eta d\zeta . d\xi$, if, for the present, we consider only the magnetization determined by A. If the coordinates of the point P, for which the potential is to be determined, are x, y, z, and if its distance from the point ξ, η, ζ is r, we have

$$r^2 = (x-\xi)^2 + (y-\eta)^2 + (z-\zeta)^2, \quad \cos\Theta = (x-\xi)/r.$$

The potential due to the element $d\xi . d\eta . d\zeta$ arising from the component of magnetization A is, from (a), $A\,d\xi d\eta d\zeta/r^2 . (x-\xi)/r$. B and C

give rise to the potentials $Bd\xi d\eta d\zeta/r^2 . (y-\eta)/r$, $Cd\xi d\eta d\zeta/r^2 . (z-\zeta)/r$. The sum of these three potentials, integrated over the whole magnet, gives the total potential V,

(b) $$V = \iiint[A(x-\xi) + B(y-\eta) + C(z-\zeta)]d\xi d\eta d\zeta/r^3.$$

Since $r\partial r/\partial x = x-\xi$, $r\partial r/\partial y = y-\eta$, $r\partial r/\partial z = z-\zeta$, we have

(c) $$V = -\iiint\big(A . \partial(1/r)/\partial x + B . \partial(1/r)/\partial y + C . \partial(1/r)/\partial z\big)d\xi d\eta d\zeta.$$

If we set

$$\psi_1 = \iiint A/r . d\xi d\eta d\zeta,\quad \psi_2 = \iiint B/r . d\xi d\eta d\zeta,\quad \psi_3 = \iiint C/r . d\xi d\eta d\zeta,$$

we have (d) $V = -(\partial\psi_1/\partial x + \partial\psi_2/\partial y + \partial\psi_3/\partial z)$, since the components of magnetization A, B, C are independent of the coordinates x, y, z of the point P.

Another more general transformation may be made by help of the equations, $r . \partial r/\partial\xi = -(x-\xi)$, $r . \partial r/\partial\eta = -(y-\eta)$, $r . \partial r/\partial\zeta = -(z-\zeta)$, by the use of which we obtain from (b),

(e) $$V = \iiint[A . \partial(1/r)/\partial\xi + B . \partial(1/r)/\partial\eta + C . \partial(1/r)/\partial\zeta]d\xi d\eta d\zeta.$$

Let the normal to the surface-element make angles with the axes whose cosines are l, m, n. By integration by parts, it follows that

(f) $$V = \iint(Al + Bm + Cn)/r . dS - \iiint(\partial A/\partial\xi + \partial B/\partial\eta + \partial C/\partial\zeta) . d\xi d\eta d\zeta/r,$$

and using LXVIII. (b) and (e), (f) becomes

(g) $$V = \iint\sigma dS/r + \iiint\rho d\xi d\eta d\zeta/r.$$

The correctness of the last equation is immediately evident from the meaning of σ and ρ.

The components α, β, γ of the magnetic force are expressed, as in the theory of electricity, by

(h) $$\alpha = -\partial V/\partial x,\quad \beta = -\partial V/\partial y,\quad \gamma = -\partial V/\partial z.$$

The force N, which acts in the direction of the element of length $d\nu$, is $N = -\partial V/\partial\nu$. If the potential is V_i inside the magnet and V_a outside the magnet, we have at the surface $\overline{V}_i = \overline{V}_a$. If ν_i is the normal to a surface-element on which the density is σ drawn into the magnet, and ν_a the normal to the same element drawn outward, we have from the general laws of the potential [cf. XIV. (1)], which are applicable here, (i) $\partial V_i/\partial\nu_i + \partial V_a/\partial\nu_a + 4\pi\sigma = 0$. For every point within the magnet, we have (k) $\nabla^2 V_i + 4\pi\rho = 0$, or introducing the value for ρ given in LXVIII. (e), (l) $\nabla^2 V_i = 4\pi(\partial A/\partial x + \partial B/\partial y + \partial C/\partial z)$, if ρ and A, B, C are functions of x, y, and z. Outside the magnet we have, on the other hand, (m) $\nabla^2 V_a = 0$. If S is a closed surface, ν its normal drawn outward, and M the total magnetism enclosed by the surface, we have from XIV. (c) $4\pi M = -\iint \partial V/\partial\nu \,.\, dS$, or designating by $\mathfrak{H}_n$, the magnetic force in the direction of the normal to the surface,

(n) $$4\pi M = \iint \mathfrak{H}_n dS.$$

The equations (i) and (k) are special cases of this equation.

SECTION LXX. THE POTENTIAL OF A MAGNETIZED SPHERE.

If the components of magnetization are given functions of the coordinates, and ξ, η, ζ are the coordinates of a point within the magnet, we obtain the potential most easily by using the formula LXIX. (d). If A, B, C are constant, the problem is to determine the potential of a body of constant volume-density. Hence we set

(a) $$\psi = \iiint d\xi \,.\, d\eta \,.\, d\zeta/r,$$

and obtain (b) $V = -(A \,.\, \partial\psi/\partial x + B \,.\, \partial\psi/\partial y + C \,.\, \partial\psi/\partial z)$. The potential of a sphere whose components of magnetization are A, B, C is to be determined by means of this equation. Take the origin of the system of coordinates at the centre of the sphere. The potential ψ has different values, according as the point for which the potential is to be determined lies inside or outside the sphere. In the usual notation we have, from XIII. (c) and (d), $\psi_a = 4\pi R^3/3r$; $\psi_i = 2\pi(R^2 - r^2/3)$, where R is the radius of the sphere.

If we represent the magnetic potential for points outside the sphere by V_a, and for points inside it by V_i, we have

(c), (d) $$V_a = 4\pi/3 \,.\, R^3/r^3 \,.\, (Ax + By + Cz)\,;\quad V_i = 4\pi/3 \,.\, (Ax + By + Cz).$$

Let J be the intensity of magnetization, and let its direction make angles with the axes whose cosines are λ, μ, ν. Let Θ be the angle between the direction of J and the line r. We then have

$$\lambda x/r + \mu y/r + \nu z/r = \cos\Theta$$

and (e) $V_a = 4\pi/3 \,.\, R^3 J \cos\Theta/r^2$; $V_i = 4\pi/3 \,.\, Jr\cos\Theta$.

Hence, the potential outside the magnet is the same as that which is set up by an infinitely small magnet, whose magnetic moment is $\mathfrak{M} = 4\pi/3 \,.\, R^3 J$ [LXIX (a)].

If the x-axis lies in the direction of magnetization, the potential in the interior of the magnet is $V_i = 4\pi/3 \,.\, Jx$. The magnetic force $\mathfrak{H}$ inside the sphere is therefore constant, and is expressed by

(f) $$\mathfrak{H} = -4\pi/3 \,.\, J.$$

Outside the sphere we divide the force into two components, one of which, P, acts in the direction of the line r, the other, Q, perpendicularly to that line. We then have $P = -\partial V_a/\partial r$, $Q = -1/r \, . \, \partial V_a/\partial \Theta$, or

$$P = 8\pi/3 \, . \, R^3 J \cos\Theta/r^3 = 2\mathfrak{M}/r^3 \, . \cos\Theta;$$
$$Q = 4\pi/3 \, . \, R^3 J \sin\Theta/r^3 = \mathfrak{M}/r^3 \, . \sin\Theta.$$

From LXVIII. the surface-density is determined by $\sigma = J\cos\Theta$. The resultant force F is $F = \mathfrak{M}/r^3 \, . \sqrt{1 + 3\cos^2\Theta}$. We have further $tg\Theta = 2Q/P$. If ϕ is the angle between the direction of the force F and the direction of r, we have $tg\phi = Q/P = \frac{1}{2} tg\Theta$.

Section LXXI. The Forces which Act on a Magnet.

Let us suppose that the magnetic forces of a magnet, whose components we may represent by a, β, γ, are functions of the coordinates. Let us suppose also another magnet, whose components of magnetization are A, B, C. Its action on the first magnet is to be determined. Consider the infinitely small parallelepiped OO' within the second magnet (Fig. 77); on the face OA' there is a quantity of magnetism present equal to $-Adydz$, which is acted on by the force $-Adydza$ in the direction of the positive x-axis. The face AO', on which is the quantity of magnetism $Adydz$, is acted on by the force $(a + \partial a/\partial x \, . \, dx)Adydz$ in the same direction. The resultant of these two forces is $A \, . \, \partial a/\partial x \, . \, dxdydz$. On the surface-element OB' there is the quantity of magnetism $-Bdxdz$, and on BO' the quantity $Bdxdz$. The former is acted on by the force $-Bdxdz \, . \, a$ in the direction of the positive x-axis, and the latter by the force $+B(a + \partial a/\partial y \, . \, dy)dxdz$ in the same direction. The resultant of these two forces is $B \, . \, \partial a/\partial y \, . \, dxdydz$. For the surface-elements OC' and $O'C$, we obtain the resultant $C \, . \, \partial a/\partial z \, . \, dxdydz$. We form the sum of these three resultants, integrate over the whole volume occupied by the magnet, and obtain for the force X, which tends to move the magnet in the direction of the x-axis,

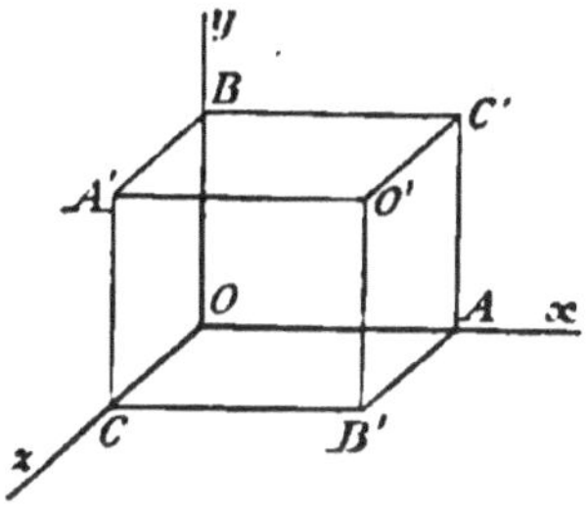

Fig. 77.

(a) $$X = \iiint (A \, . \, \partial a/\partial x + B \, . \, \partial a/\partial y + C \, . \, \partial a/\partial z) dxdydz.$$

Analogous expressions hold for the forces Y and Z.

If the magnetic force whose components are a, β, γ is due to a system of magnets which give rise to the potential V at the point

x, y, z, we have $\alpha = -\partial V/\partial x$, $\beta = -\partial V/\partial y$, $\gamma = -\partial V/\partial z$. We then have also $\partial\alpha/\partial y = \partial\beta/\partial x$, $\partial\alpha/\partial z = \partial\gamma/\partial x$, and hence

(b) $$X = \iiint (A\,.\,\partial\alpha/\partial x + B\,.\,\partial\beta/\partial x + C\,.\,\partial\gamma/\partial x)dxdydz.$$

We now determine the *moment of the forces* which tend to turn the magnet about one of the coordinate axis, say the x-axis, on the assumption that the magnetic forces are constant. Let the coordinates of the point O (Fig. 77) be x, y, z. The force which acts on the surface BO' in the direction of the z-axis has a moment with respect to the x-axis equal to $Bdxdz\,.\,\gamma\,.\,(y+dy)$. The force acting on OB' has a moment with respect to the same axis equal to $-Bdxdz\,.\,\gamma\,.\,y$. Neglecting small terms of higher order, the resultant moment is $B\gamma dxdz\,.\,dy$. The forces acting on the surfaces $O'C$ and OC' give rise to the moment $-C\beta dxdy\,.\,dz$. The moment L, which tends to turn the magnet about the x axis, is therefore

(c) $$L = \iiint (B\gamma - C\beta)dxdydz.$$

The moments of rotation M and N, with respect to the two other axes, are determined from analogous expressions.

If the magnet is subjected to the action of the earth's magnetism only, the magnetic force may be considered as constant both in magnitude and direction. The components α, β, γ are then independent of x, y, z, and therefore $X = Y = Z = 0$. *The centre of gravity of a magnet does not move under the action of the earth's magnetism.* The magnet is, however, acted on by a moment of rotation, which may be determined in the following way:

Let the magnetic moment of the magnet be $\mathfrak{M}$, and suppose its direction with respect to the coordinate axes to be determined by the angles whose cosines are l, m, n. We then have

$$\mathfrak{M}l = \iiint A\,.\,d\tau, \quad \mathfrak{M}m = \iiint B\,.\,d\tau, \quad \mathfrak{M}n = \iiint C\,.\,d\tau,$$

and hence $L = \mathfrak{M}(\gamma m - \beta n)$; $M = \mathfrak{M}(\alpha n - \gamma l)$; $N = \mathfrak{M}(\beta l - \alpha m)$. From these equations, it follows that $L\alpha + M\beta + N\gamma = 0$ and $Ll + Mm + Nn = 0$, that is, *the resultant moment is perpendicular to the magnetic force and also to the magnetic axis of the magnet.* If the direction of the force is parallel to the x-axis, and if the magnetic axis lies in the xy-plane and forms with the x-axis the angle Θ, we will have

(d) $$L = 0, \quad M = 0, \quad N = -\mathfrak{M}\alpha\,.\,\sin\Theta.$$

If a magnet can turn about a vertical axis, the moment which tends to increase the angle Θ between the magnetic axis and the magnetic meridian is $-\mathfrak{M}H\,.\,\sin\Theta$, where H denotes the horizontal component of the earth's magnetism. If ω is the angular velocity of the magnet and J its moment of inertia, we have, from XXII. (c),

$$d(J\omega) = -\mathfrak{M}H \,.\, \sin\Theta \,.\, dt, \quad \text{or since} \quad \omega = d\Theta/dt = \dot{\Theta},$$

(e) $$J\ddot{\Theta} = -\mathfrak{M}H \,.\, \sin\Theta.$$

If the angle Θ is very small, the period of oscillation of the magnet is given by XXII. (o), (f) $\tau = \pi\sqrt{J/\mathfrak{M}H}$.

Section LXXII. Potential Energy of a Magnet.

By the potential energy of a magnet is meant the work which is needed to transfer the magnet from a position in which no magnetic forces act on it to the position in which the magnetic potential is V. We will first consider an infinitely small parallelepiped (Fig. 78), whose components of magnetization are A, B, C. In order to bring the magnetic surface OA' to the position in which the potential is V, the work $-A \,.\, dydz \,.\, V$ must be done. The opposite surface $O'A$ is brought to the position in which the potential is $V + \partial V/\partial x \,.\, dx$, and the work done on it is $A \,.\, dydz \,.\, (V + \partial V/\partial x \,.\, dx)$. The work done on these two surfaces therefore amounts to $A \,.\, \partial V/\partial x \,.\, dxdydz$. If we obtain in like manner the work done on the two other pairs of surfaces, we find that the whole work W done in transporting the magnet is

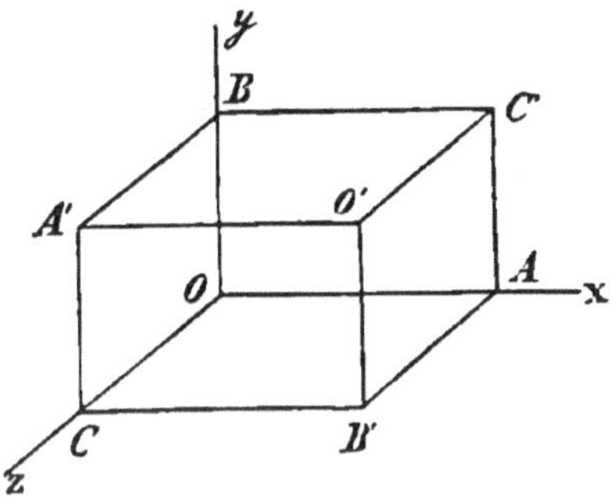

Fig. 78.

(a) $$W = \iiint (A \,.\, \partial V/\partial x + B \,.\, \partial V/\partial y + C \,.\, \partial V/\partial z)dxdydz,$$

or since α, β, γ, the components of the magnetic force, are

$$\alpha = -\partial V/\partial x, \quad \beta = -\partial V/\partial y, \quad \gamma = -\partial V/\partial z,$$

we have (b) $W = -\iiint (A\alpha + B\beta + C\gamma)dxdydz$. We will apply this equation to the case of a magnet subjected to the action of the earth's magnetism only. Let its magnetic moment be $\mathfrak{M}$, and let the direction of its magnetic axis make angles with the coordinate axes whose cosines are l, m, n. We then have

$$\iiint A \,.\, dxdydz = l\mathfrak{M}, \quad \iiint B \,.\, dxdydz = m\mathfrak{M}, \quad \iiint C \,.\, dxdydz = n\mathfrak{M},$$

$$W = -\mathfrak{M}(l\alpha + m\beta + n\gamma).$$

Representing the magnetic force by $\mathfrak{H}$, and supposing its direction to make angles with the coordinate axes whose cosines are λ, μ, ν, we obtain $W = -\mathfrak{M}\mathfrak{H}(l\lambda + m\mu + n\nu)$. Letting Θ represent the angle between the magnetic axis of the magnet and the magnetic force, we have (c) $W = -\mathfrak{M}\mathfrak{H}\,.\cos\Theta$. If the direction of the force is parallel to the x-axis, as in LXXI., and if the magnetic axis lies in the xy-plane, the work done in turning the magnet through the angle $d\Theta$ is $dW = +\mathfrak{M}\mathfrak{H}\,.\sin\Theta\,.\,d\Theta$. This agrees with LXXI. (d).

We will now consider a very small magnet situated near a very strong magnet. If the small magnet has sufficient freedom of motion, it will turn so that its magnetic axis is parallel to the direction of the magnetic force. In this case we have $\Theta = 0$ and its potential energy W is $W = -\mathfrak{M}\mathfrak{H}$.

Since the motion of the small magnet involves the loss of potential energy, it moves in such a way that W diminishes. This occurs by the last equation, when $\mathfrak{H}$ increases; the magnet therefore moves in the direction in which the magnetic force increases. A particle of a paramagnetic substance therefore tends to move towards the place where the magnetic force is greatest. On the other hand, diamagnetic bodies move toward the place where the magnetic force is a minimum.

In order to find the magnetic energy residing in a system of magnets, we proceed in the following way: The potential at every point within the system varies proportionally with the values of the components of magnetization. We assume that the components of magnetization change only in such a way that, in successive instants, they always increase by the same fraction of their final values. On these conditions the potential increases in the same proportion. If the components of magnetization are originally zero, the potential is also originally equal to zero. Let the final values of the components of magnetization be A, B, C. At a particular instant during the increase of the components of magnetization, let these be represented by nA, nB, nC, where n is a proper fraction. At the same time the potential at any point is equal to nV. If the components of magnetization increase by $A\,.\,dn$, $B\,.\,dn$, $C\,.\,dn$ respectively, the potential at the point considered increases by Vdn. If A, B, C increase by $A\,.\,dn$, $B\,.\,dn$, $C\,.\,dn$ respectively, the work needed to accomplish this is, by (a),

$$\iiint(A\,.\,dn\,.\,n\partial V/\partial x + B\,.\,dn\,.\,n\partial V/\partial y + C\,.\,dn\,.\,n\partial V/\partial z)dxdydz$$
$$= n\,.\,dn\iiint(A\,.\,\partial V/\partial x + B\,.\,\partial V/\partial y + C\,.\,\partial V/\partial z)dxdydz.$$

Now, if n increases from 0 to 1, we have $\int_0^1 ndn = \frac{1}{2}$, and the work done is

(d) $$W = \frac{1}{2}\iiint(A\,.\,\partial V/\partial x + B\,.\,\partial V/\partial y + C\,.\,\partial V/\partial z)dxdydz,$$

or, by introducing the components of the magnetic force,

(e) $$W = -\frac{1}{2}\iiint(A\alpha + B\beta + C\gamma)dxdydz.$$

The energy of a magnetic system may be expressed in another way. The same method by which we before obtained an expression for the energy, shows that (f) $W = \frac{1}{2}\int\sigma\,.\,V\,.\,dS + \frac{1}{2}\iiint\rho\,.\,V\,.\,dxdydz$, where σ and ρ represent the surface and volume-densities respectively.

If V and its first differential coefficient vary continuously, we have $\sigma = 0$ and $W = \frac{1}{2}\iiint\rho\,.\,V\,.\,dxdydz$. Now, $\nabla^2 V + 4\pi\rho = 0$, and hence

$$W = -1/8\pi\,.\iiint V(\partial^2 V/\partial x^2 + \partial^2 V/\partial y^2 + \partial^2 V/\partial z^2)dxdydz.$$

By integration by parts over the whole infinite region, we obtain

$$W = 1/8\pi\,.\iiint[(\partial V/\partial x)^2 + (\partial V/\partial y)^2 + \partial V/\partial z)^2]dxdydz,$$

or (g) $W = 1/8\pi\,.\iiint(\alpha^2 + \beta^2 + \gamma^2)dxdydz$. Similar expressions hold for dielectric polarization [cf. LXI.].

SECTION LXXIII. MAGNETIC DISTRIBUTION.

A piece of soft iron brought into a magnetic field becomes magnetized by induction. We assume that the intensity of magnetization at any point is a function of the total magnetic force acting at that point. We assume that the intensity of magnetization is proportional to the magnetic force, or that (a) $A = k\alpha$, $B = k\beta$, $C = k\gamma$, where k is a constant. The magnetizing force proceeds partly from the permanent magnets present in the field, and partly from the quantities of magnetism induced in the soft iron. The potential due to the former may be designated by V, that due to the latter by U, so that

$$A = -k\,.\,\partial(V+U)/\partial x,\quad B = -k\,.\,\partial(V+U)/\partial y,\quad C = -k\,.\,\partial(V+U)/\partial z.$$

Now, in the space not occupied by permanent magnets, we have $\nabla^2 V = 0$, and therefore $\partial A/\partial x + \partial B/\partial y + \partial C/\partial z = -k\nabla^2 U$, or since, from LXVIII. (e), $\rho = -(\partial A/\partial x + \partial B/\partial y + \partial C/\partial z)$, we have finally $\nabla^2 U - \rho/k = 0$.

Since the potential W is due to the components of magnetization A, B, C, the equation that holds within the soft iron is $\nabla^2 U + 4\pi\rho = 0$. From the last two equations we obtain (b) $(1 + 4\pi k)\rho = 0$, and hence $\rho = 0$; that is, *there is no free magnetism present within the soft iron.* The magnetism present is therefore situated on the surface of the iron. We will now determine the surface-density σ of this distribution.

For this purpose we use the equation

$$4\pi\sigma + \partial(V + U_i)/\partial\nu_i + \partial(V + U_a)/\partial\nu_a = 0,$$

where ν_a and ν_i are the normals drawn from any point on the surface of the iron inward and outward respectively. U_i and U_a are the values of the potential due to the induced magnetism inside and outside the iron mass. Now we have

$$\partial V/\partial\nu_i = -\partial V/\partial\nu_a, \text{ and hence } 4\pi\sigma + \partial U_i/\partial\nu_i + \partial U_a/\partial\nu_a = 0.$$

The magnetizing force just outside the surface of the soft iron is $-\partial(V + U_i)/\partial\nu_i$ in the direction of ν_i. The free magnetism on the corresponding surface-element is therefore $\sigma \,.\, dS = k \,.\, \partial(V + U_i)/\partial\nu_i \,.\, dS$. Hence we have

(c) $$4\pi k \,.\, \partial V/\partial\nu_i + (1 + 4\pi k)\partial U_i/\partial\nu_i + \partial U_a/\partial\nu_a = 0$$

The relation (c) in connection with the equations (d) $\nabla^2 U_i = 0$, $\nabla^2 U_a = 0$ serves to determine the potentials U_i and U_a.

As an example of the theory here presented, we will consider the magnetization of a sphere subjected to the action of a constant magnetizing force $\mathfrak{H}$ which acts in the direction of the x-axis. Let the intensity of magnetization of the sphere in the direction of the x-axis be A; the force due to the magnetization and acting in the direction of the x-axis is [LXX. (f)] equal to $-4\pi/3 \,.\, A$. From equation (a) we have, therefore, $A = k(\mathfrak{H} - 4\pi/3 \,.\, A)$, and hence $A = k\mathfrak{H}/(1 + 4\pi/3 \,.\, k)$. We have in this case [LXX. (e)]

$$U_i = 4\pi/3 \,.\, Ax; \quad U_a = 4\pi/3 \,.\, R^3 A \,.\, \cos\Theta/r^2.$$

These values satisfy equation (c), since $V = -\mathfrak{H}x$.

Section LXXIV. Lines of Magnetic Force.

If M represents the free magnetism within a closed surface, and $\mathfrak{H}_n$ the component of magnetic force in the direction of the normal to the surface, we have, from LXIX., (a) $4\pi \,.\, M = \iint \mathfrak{H}_n \,.\, dS$.

If we lay a sheet of paper on a magnet which lies horizontally, and scatter iron filings over it, they arrange themselves in curves which are called *lines of magnetic force*. Let DE (Fig. 79) be a small surface of area dS; lines of force proceed from its perimeter which bound a *tube of force*. If $D'E'$ represents another section cut through the tube of force, equation (a) can be applied to the part of the tube of force thus bounded. Since the direction of the magnetic force coincides with the direction of the lines of force, the normal force over the surface of the tube is everywhere zero except at the ends DE and $D'E'$. Let $\mathfrak{H}$ be the force acting at DE and $\mathfrak{H}'$ that acting at $D'E'$; let Θ and Θ' be the angles between the direction of the force and the directions of the normals to DE and $D'E'$ respectively. Since there is no free magnetism present in the interior of the tube, we have

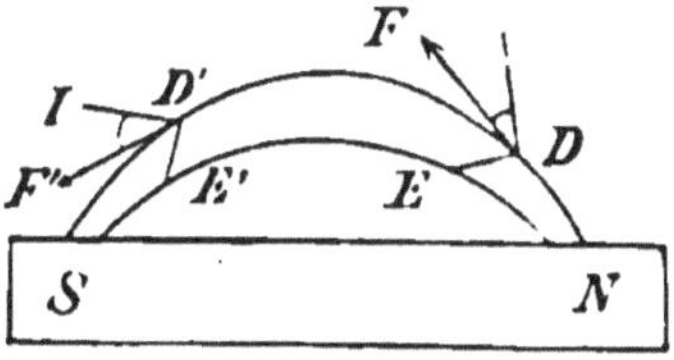

FIG. 79.

$$-\mathfrak{H} . dS . \cos\Theta + \mathfrak{H}' . dS' . \cos\Theta' = 0.$$

If the sections dS and dS' are perpendicular to the lines of force, we have $\mathfrak{H}/\mathfrak{H}' = dS'/dS$. *The force is therefore inversely proportional to the cross-section of the tube of force.* The lines of force may be so drawn that their distances from one another furnish a representation of the magnitude of the magnetic force in the field.

A tube of magnetic force cannot return into itself or form a hollow ring. If this were not so, the work which is done by the magnetic forces during the transfer of a unit quantity of magnetism from any point over a closed path back to the same point again would not be zero. If ds is an element of the tube of force, the work done would be $\int \mathfrak{H} . ds > 0$, if the direction of motion coincides with the direction of the force. If the magnetic potential is V, we have, however, $\mathfrak{H} = -dV/ds$, and therefore, for a closed path,

$$\int dV/ds . ds = V - V = 0,$$

since the potential is a single-valued function of the position of the point.

Any tube of magnetic force must begin and end on the surface of a magnet. If the tube ends with the cross-section PQ (Fig. 80), so that a magnetic force is present in the tube $TUQP$, while it is zero outside the tube at R and S, we may apply equation (a) to the region $TUQSRP$. Since a magnetic force acts at the surface TU, but not in the region $PQSR$, the surface integral taken over $TUQSRP$ cannot be zero. Magnetism must therefore be present within the closed surface, which contradicts our assumption. Therefore, any tube of magnetic force ends at the surface of a magnet.

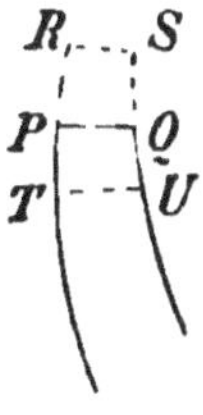

FIG. 80.

In order to represent the magnitude and direction of magnetic forces, Faraday used lines of magnetic force; *he assumed that the lines of force are continued in the body of the magnet.* His mode of representation has become of very great importance. If a magnet is broken, and the surfaces exposed by the fracture are placed so as to face each other and separated by only a small distance, a strong magnetic

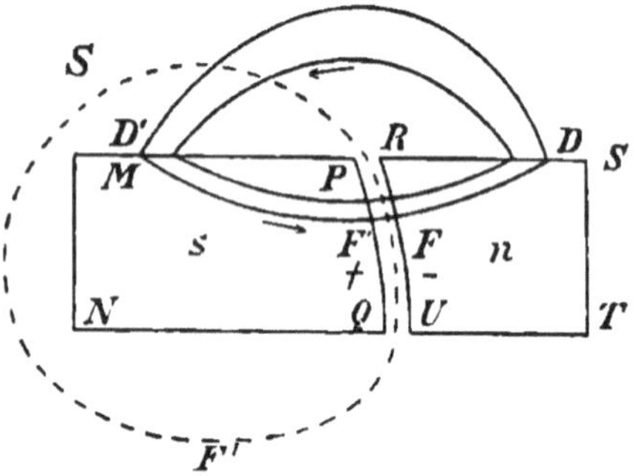

FIG. 81.

force acts in the region *PQRU* (Fig. 81). This force is due partly to the free magnetism in the interior of the magnet and on its original surface, and partly to the free magnetism on the newly-formed surfaces. The force due to the former cause is directed from the north pole *n* to the south pole *s*, that due to the latter from *s* to *n*. The latter force is in practice the stronger, so that we may say with a certain propriety that the magnetic tubes of force are produced through the interior of the magnet along the path *D'F'FD* (Fig. 81).

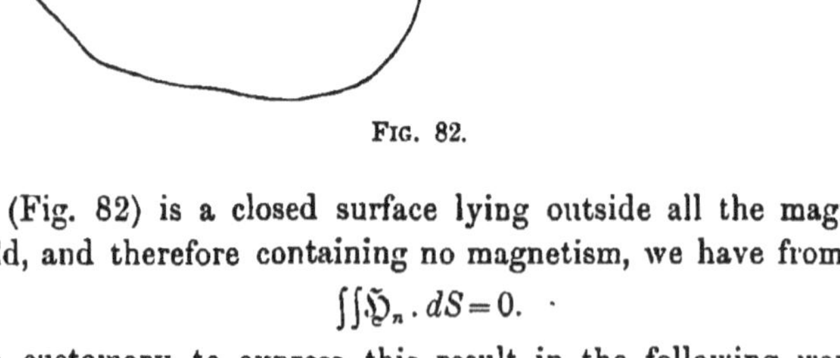

FIG. 82.

If S (Fig. 82) is a closed surface lying outside all the magnets in the field, and therefore containing no magnetism, we have from XIV.,

$$\iint \mathfrak{H}_n \, . \, dS = 0.$$

It is customary to express this result in the following way: The integral $\int \mathfrak{H}_n \, . \, dS$ which is extended over a part of the surface may be divided into the parts $\mathfrak{H}_{n1}.dS_1$, $\mathfrak{H}_{n2}.dS_2$, etc. Let them be so taken that they are all equal, and let their common value be taken as unity. Since the product $\mathfrak{H}_n.dS$ is constant for the same tube or line of force, the integral $\int\mathfrak{H}_n.dS$ gives the *number of lines of force which traverse the surface. If this integral is zero, as many lines of force enter the surface as leave it.*

This holds for a surface which contains one or more magnets, for the sum of the magnetism in every magnet is zero. On the other hand, the theorem does not hold if the surface cuts through a magnet. Nevertheless, if the magnet is divided into two parts, *MNQP* (Fig. 81) and *RSTU*, and if they are situated infinitely near each other, the theorem holds for either of them if the surface considered contains one part, but excludes the other. Now, if this mode of division of the magnet produces no disturbance in its magnetization, the theorem can be expressed in the following way:

We represent the components of the magnetic force by α, β, γ. In the part of the surface lying outside the cleft, no other magnetic force is acting; on the other hand, the free magnetism $+\sigma$ on the element dS of PQ (Fig. 81), and $-\sigma$ on the corresponding element of RU, produce a force which can be determined in the following way: From XIII. a surface on which the surface-density is σ exerts an attractive force $2\pi\sigma$ on a unit of mass lying very near it. In the case of magnetism, the force $2\pi\sigma$ is a repulsion. If there are two parallel surfaces, on one of which the density of the magnetic distribution is σ, while on the other it is $-\sigma$, the magnetic force acting between the surfaces is $4\pi\sigma$. If the normal to the surface-element dS directed outward makes angles with the axes whose cosines are l, m, n, we have, if A, B, C are the components of magnetization, $\sigma = lA + mB + nC$. The magnetic force in the direction of the normal is $l\alpha + m\beta + n\gamma$. Since, in the case considered, the surface-integral must be zero, we have $\iint(l\alpha + m\beta + n\gamma + 4\pi\sigma)dS = 0$, or, from LXVIII. (b),

$$\int[l(\alpha + 4\pi A) + m(\beta + 4\pi B) + n(\gamma + 4\pi C)]dS = 0.$$

We set (b) $a = \alpha + 4\pi A$, $b = \beta + 4\pi B$, $c = \gamma + 4\pi C$, and obtain

$$\text{(c)} \qquad \int(al + bm + cn)dS = 0.$$

The quantities a, b, c are the components of *magnetic induction.* If, therefore, the directions of the lines of force are determined by the directions of the resultants of the magnetic induction, it follows that *the lines of force may be considered as continued through the magnet itself, and that they therefore return into themselves.* Now, equation (c) shows that as many lines enter a surface as leave it. If we consider an arbitrary surface S so drawn as to pass through every point of a closed curve s, and determine the magnetic induction whose components are a, b, c, the magnitude $N=\int(al+bm+cn)dS$ is determined by the boundary s of the surface S. We may say, *the curve s encloses N lines of magnetic force.*

From equations (b) it follows that

$$\partial a/\partial x+\partial b/\partial y+\partial c/\partial z$$
$$=\partial \alpha/\partial x+\partial \beta/\partial y+\partial \gamma/\partial z+4\pi(\partial A/\partial x+\partial B/\partial y+\partial C/\partial z)=-\nabla^2 V-4\pi\rho.$$

Hence, we have $$\partial a/\partial x+\partial b/\partial y+\partial c/\partial z=0.$$

SECTION LXXV. THE EQUATION OF LINES OF FORCE.

We will develop the equation of the lines of force of a small straight magnet NS (Fig. 83) which is magnetized in the direction

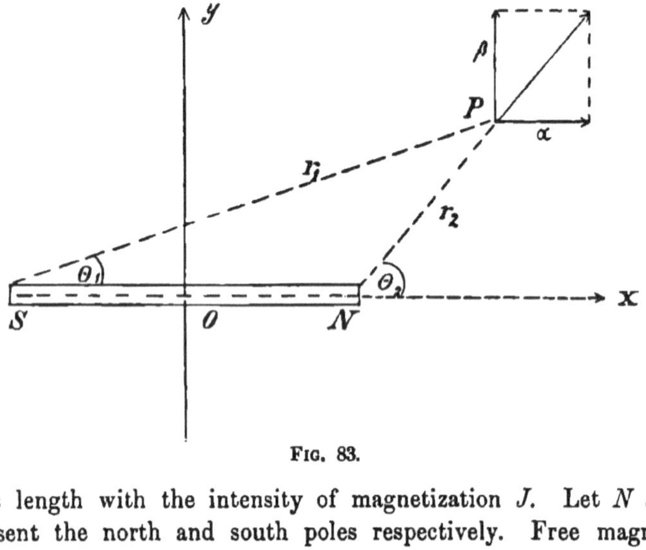

FIG. 83.

of its length with the intensity of magnetization J. Let N and S represent the north and south poles respectively. Free magnetism is present on the end surfaces S and N, supposed to be plane and to have the area dA; the quantity at the north end is $J.dA$, that at the south end $-J.dA$. Let the centre of the magnet be the origin of coordinates, and the x-axis coincide with the length of the magnet. We are to determine the components α and β of the magnetic force which acts at the point P, whose coordinates are x and y. If $2l$ is the length of the magnet, and if $PS=r_1$, $PN=r_2$, $J.dA=q$, we have

$$\alpha=q.(x-l)/r_2^3-q.(x+l)/r_1^3,\quad \beta=q.y/r_2^3-q.y/r_1^3.$$

If dx and dy are the projections of an element of the line of force, we have $dy/dx = \beta/\alpha$, or

(a) $$(x-l)/r_2^3 \,.\, dy - (x+l)/r_1^3 \,.\, dy = y/r_2^3 \,.\, dx - y/r_1^3 \,.\, dx.$$

If we set $\angle PSx = \Theta_1$, $\angle PNx = \Theta_2$, we have

$$\cos\Theta_1 = (x+l)/r_1 \text{ and } \cos\Theta_2 = (x-l)/r_2.$$

If x and y increase by dx and dy respectively, $\cos\Theta$ will increase by $d\cos\Theta$, or

$$d\cos\Theta_1 = \left(1/r_1 - (x+l)^2/r_1^3\right)dx - (x+l)y/r_1^3 . dy = y^2/r_1^3 . dx - (x+l)y/r_1^3 . dy.$$

In the same way $d\cos\Theta_2 = y^2/r_2^3 \,.\, dx - (x-l)y/r_2^3 \,.\, dy$. From equation (a) we obtain $d(\cos\Theta_1 - \cos\Theta_2) = 0$, or, if c is a constant, $\cos\Theta_1 - \cos\Theta_2 = c$. This is the equation of the lines of force.

SECTION LXXVI. MAGNETIC INDUCTION.

The components of magnetic induction are

$$a = \alpha + 4\pi A, \quad b = \beta + 4\pi B, \quad c = \gamma + 4\pi C.$$

If we consider a, b, c as components of flux, they have a property similar to that of the components of flow of an incompressible fluid [cf. XLI. (e)], for $\partial a/\partial x + \partial b/\partial y + \partial c/\partial z = 0$, or, what amounts to the same thing, $\int(al + bm + cn)dS = 0$. If $\mathfrak{B}$ represents the resultant of a, b, c, and ϵ the angle between $\mathfrak{B}$ and the normal to the surface-element dS of the closed surface S, we have $\int \mathfrak{B} \,.\, \cos\epsilon \,.\, dS = 0$.

Let $EF = dS$ (Fig. 84) be an element of the surface of the magnet, and let the quantity of magnetism $\sigma \,.\, dS$ be present on dS. Let the induction outside the surface EF, in the direction EE', be $\mathfrak{B}_a$, and inside that surface, in the same direction, be $\mathfrak{B}_i$. Suppose perpendiculars erected on the perimeter of the element EF and the surfaces $E'F'$ and $E''F''$ drawn parallel to EF. We then have

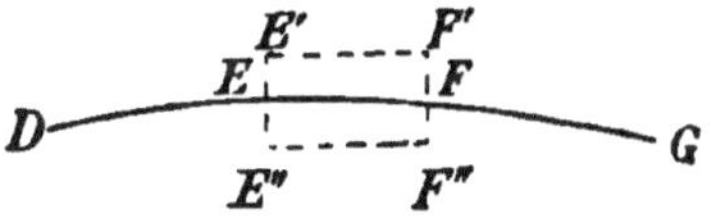

FIG. 84.

$$(\mathfrak{B}_i - \mathfrak{B}_a) \,.\, dS = 0, \text{ or } \mathfrak{B}_i = \mathfrak{B}_a.$$

Outside the surface, $\mathfrak{B}_a$ is the same as the magnetic force in the direction EE'; it will be $\mathfrak{B}_a = -\partial V_a/\partial \nu_a$, if the potential outside the surface is V_a and the normal EE' is equal to $d\nu_a$. Let V_i be the potential within the surface. The induction $\mathfrak{B}_i$ is then [LXXIV.]

$$\mathfrak{B}_i = \partial V_i/\partial \nu_i + 4\pi\sigma,$$

and we have, therefore, $\partial V_i/\partial \nu_i + \partial V_a/\partial \nu_a + 4\pi\sigma = 0$. This is the same equation as LXIX. (i). If the body considered is a mass of iron whose coefficient of magnetization is k, we have

$$A = k\alpha,\quad B = k\beta,\quad C = k\gamma,\quad \mu = 1 + 4\pi k,\quad a = \mu\alpha,\quad b = \mu\beta,\quad c = \mu\gamma.$$

It follows that $\partial a/\partial x + \partial b/\partial y + \partial c/\partial z = \mu . \nabla^2 V_i = 0$ within the mass of iron. Within that mass, therefore, no free magnetism is present. The magnetic induction perpendicular to the surface has the same value on both sides of it, that is [cf. LXXIII. (c)],

$$\mu\partial(U_i + V)/\partial\nu_i = -\partial(U_a + V)/\partial\nu_a.$$

The magnitude μ, which is the ratio of the magnetic induction to the magnetic force, may be called the *magnetic inductive capacity* (*magnetic permeability*). *The coefficient of induction* or *the permeability* μ is equal to unity in vacuo, where $k = 0$; in *paramagnetic* bodies $\mu > 1$, in *diamagnetic* bodies $\mu < 1$.

SECTION LXXVII. MAGNETIC SHELLS.

Suppose a thin steel plate to be magnetized so that one face is covered with north magnetism and the other with south magnetism. At any point A in the face N (Fig. 85) draw a normal to the plate which cuts the surface S at B. Let the plate be so magnetized

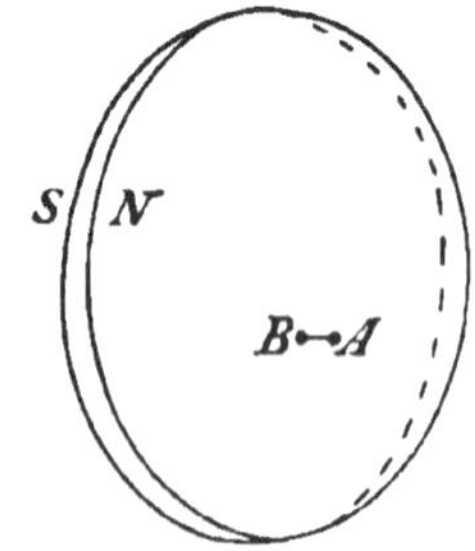

FIG. 85.

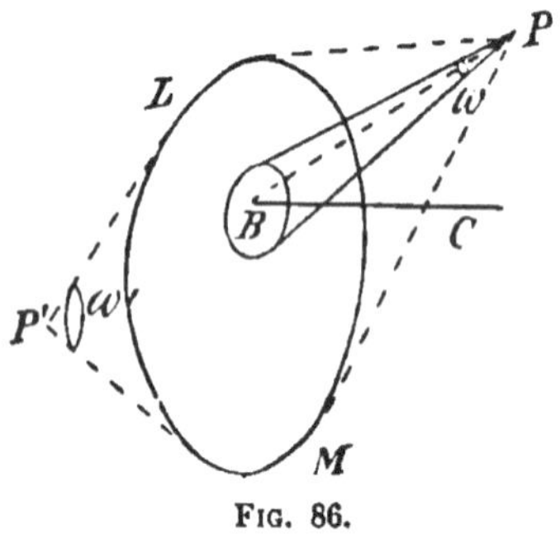

FIG. 86.

that $-\sigma$ represents the magnetic surface-density at B, and $+\sigma$ that at A. We set $AB = e$, and call $\sigma e = \Phi$ the *strength of the shell* * at the point under consideration. If the plate is infinitely thin and the surface-density infinitely great, Φ has a finite value. Such a plate is called a magnetic "*shell*."

The potential of such a shell may be expressed in the following way: Let LM (Fig. 86) be the shell, dS a surface-element on its positive face, BC the normal to this surface-element, and P the point

* In the original, *the moment of the surface*.—TR.

at which the potential is to be determined. We represent the angle between BC and BP by ϵ. The potential at the point P, due to that part of the shell whose end-surface is dS, is [LXIX.]

$$dV = \sigma . dS . e . \cos\epsilon/r^2.$$

Hence, if the strength of the shell is constant, (a) $V = \Phi . \iint \cos\epsilon . dS/r^2$, where the integral is to be extended over the whole surface. If the solid angle subtended by dS at the point P is called $d\omega$, and if we set $BP = r$, we have $dS . \cos\epsilon = r^2 d\omega$.

Therefore $dV = \sigma . e . d\omega = \Phi . d\omega$, and hence (b) $V = \Phi . \omega$, where ω is the *solid angle* subtended by the shell at the point P. We may call ω the apparent magnitude of the shell seen from the point P.

If the point for which the potential is to be determined lies on the opposite side of the surface, say at P', and if ω' is the solid angle subtended by the shell at the point P', we have $V' = -\Phi . \omega'$. If the points P and P' approach each other until they are infinitely near, but on opposite sides of the shell, we have (c) $V' = -\Phi . (4\pi - \omega)$, since 4π is the total solid angle about a point. Hence, finally, $V - V' = 4\pi . \Phi$.

If PQP' (Fig. 87) is a curve which does not cut the shell, and whose ends lie infinitely near each other on opposite sides of it, the work done by the magnetic forces in moving a unit magnet pole over the path PQP' is equal to $4\pi\Phi$. This theorem holds even if other magnets are present in the field. They act on the pole with forces which have a single-valued potential, and the work done by them during the motion of the unit pole in the curve PQP' is equal to zero; for this curve may be considered as a closed curve, since P and P' are infinitely near each other.

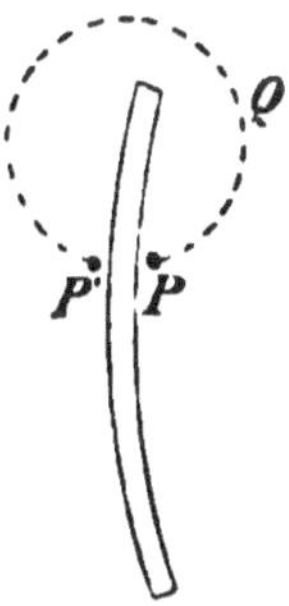

FIG. 87.

After obtaining an expression for the potential of a magnetic shell, we determine the force with which the shell acts on a magnet pole of unit strength. The normal to the shell makes angles with the axes whose cosines are l, m, n; let ξ, η, ζ be the coordinates of a point in the shell, and x, y, z the coordinates of the point outside the shell for which the potential is to be determined. We then have

$$\cos\epsilon = l . (x - \xi)/r + m . (y - \eta)/r + n . (z - \zeta)/r,$$

where $r^2 = (x - \xi)^2 + (y - \eta)^2 + (z - \zeta)^2$. From equation (a) the potential is

$$V = \Phi . \iint [(x - \xi) . l + (y - \eta)m + (z - \zeta)n]/r^3 . dS.$$

Since $\partial r^{-1}/\partial\xi = (x-\xi)/r^3$, we obtain

$$V = \Phi\iint(l\,.\,\partial r^{-1}/\partial\xi + m\,.\,\partial r^{-1}/\partial\eta + n\,.\,\partial r^{-1}/\partial\zeta)dS.$$

Represent the components of the magnetic force in the direction of the x-axis by α, we then have (d) $\alpha = -\partial V/\partial x$, and

$$\alpha = +\Phi\,.\iint(l\,.\,\partial^2 r^{-1}/\partial\xi^2 + m\,.\,\partial^2 r^{-1}/\partial\xi\partial\eta + n\,.\,\partial^2 r^{-1}/\partial\xi\partial\zeta)dS,$$

because $\partial r^{-1}/\partial x = -\partial r^{-1}/\partial\xi$. If the shell does not pass through the point x, y, z, r will never become zero, and we have

$$\partial^2 r^{-1}/\partial\xi^2 + \partial^2 r^{-1}/\partial\eta^2 + \partial^2 r^{-1}/\partial\zeta^2 = 0,$$

(e) $\alpha = +\Phi.\iint[m.\partial^2 r^{-1}/\partial\eta\partial\xi + n.\partial^2 r^{-1}/\partial\zeta\partial\xi - l(\partial^2 r^{-1}/\partial\eta^2 + \partial^2 r^{-1}/\partial\zeta^2)]dS.$

From the theorem of VI. (f), we have

$$\text{(f)}\quad \begin{cases} \int(X.d\xi/ds + Y.d\eta/ds + Z.d\zeta/ds)ds \\ = \iint[l(\partial Z/\partial\eta - \partial Y/\partial\zeta) + m(\partial X/\partial\zeta - \partial Z/\partial\xi) + n(\partial Y/\partial\xi - \partial X/\partial\eta)]dS. \end{cases}$$

We set $X=0$, $Y = +\Phi.\partial r^{-1}/\partial\zeta$, $Z = -\Phi.\partial r^{-1}/\partial\eta$, by which the right sides of equations (e) and (f) become identical, and then obtain

$$\text{(g)}\qquad \alpha = \Phi\,.\int(\partial r^{-1}/\partial\zeta\,.\,d\eta/ds - \partial r^{-1}/\partial\eta\,.\,d\zeta/ds)ds.$$

Analogous expressions hold for β, γ. By carrying out the differentiation, we obtain (h) $\alpha = \Phi\,.\int[(z-\zeta)/r^3\,.\,d\eta/ds - (y-\eta)/r^3\,.\,d\zeta/ds]ds.$

The force is therefore determined by the contour and the strength of the magnetic shell. This result follows from the fact that the potential is determined by the solid angle and the strength of the shell.

In order to find the geometrical meaning of equation (h), we use the following method: Let ξ, η, ζ be the coordinates of the point O (Fig. 88); Oy and Oz represent the directions of the y- and z-axes respectively. Let the element ds be parallel to the z-axis, and represented by $OA = d\zeta$; we then have $d\eta = 0$. Let the point P, for which the potential is to be determined, lie in the yz-plane, and let $OP = r$. We set $AOP = \Theta$, and have $y - \eta = r\,.\sin\Theta$. The magnetic force due to $ds = OA$ is, (i) $\alpha = -\Phi\,.\,ds\,.\sin\Theta/r^2$; it is perpendicular to the yz-plane. Its direction may be determined in the following way: *If the right hand is held so that the fingers point in the direction of ds, and the palm is turned toward the pole P, the thumb gives the direction of the force.*

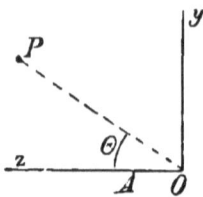

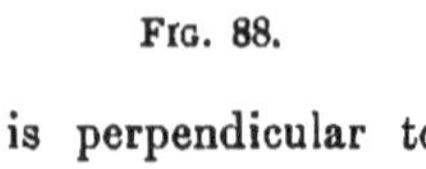

FIG. 88.

Finally, we determine the work which must be done to bring a magnetic shell from an infinite distance to a place where the magnetic potential is equal to V. Let the shell be divided into elements dS. In order to bring the surface-element which carries the quantity

$\sigma . dS$ of south magnetism to its final position, work equal to $-\sigma . dS . V$ must be done. In order to bring the corresponding surface-element carrying the same quantity of north magnetism to its place, work equal to $(V + dV/d\nu . e)\sigma . dS$ must be done, if ν represents the normal to the surface-element dS. Hence the total work done is

$$A = \iint dV/d\nu . e\sigma . dS = \Phi . \iint dV/d\nu . dS.$$

Now, since $dV/d\nu = -(l\alpha + m\beta + n\gamma)$, we obtain for the work done

$$A = -\Phi . \iint (l\alpha + m\beta + n\gamma) dS.$$

If N represents the number of lines of force contained by the contour of the shell, we have $A = -\Phi . N$.

CHAPTER IX.

ELECTRO-MAGNETISM.

SECTION LXXVIII. BIOT AND SAVART'S LAW.

OERSTED discovered that the electrical current exerts an action on magnets; the law of the magnetic force which is due to an electrical current was discovered by Biot and Savart. Let AB (Fig. 89) be a conductor traversed by a current which is measured by the quantity of electricity flowing in unit time through any cross-section. Let the quantity of magnetism μ be situated at the point P, and let the conductor AB be divided into infinitely small parts ds. If $CD = ds$ is an infinitely small part of the conductor $CP = r$, and Θ the angle between r and the direction of the current in CD, the direction of the force will be perpendicular to the plane determined by r and ds.

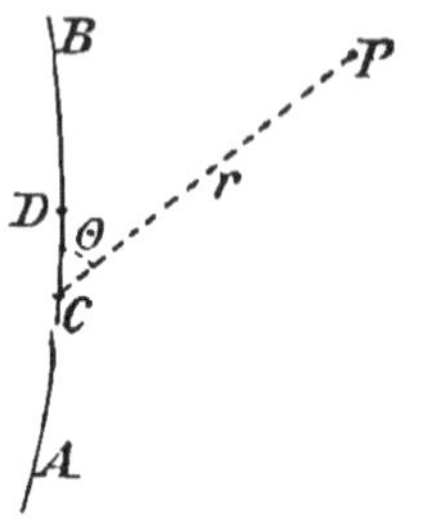

FIG. 89.

If the right hand points in the direction of the current and the palm is turned toward the magnet pole, the direction of the force exerted by the current-element on the pole is given by the direction of the thumb. The magnitude of the force K is (a) $K = \mu \, . \, i \, . \, ds/r^2 \, . \sin \Theta$.

The magnetic force which is due to any system of electrical currents, whose direction, strength, and position in space are known, may be calculated from (a).

If the current forms a *closed circuit*, and if the intensity of the current is the same at all points in the conductor, we may determine the force due to the current and also the potential which the current produces. The force due to any current-element is equal to that exerted by a line-element of the same length, which forms part of the contour of a magnetic shell, whose strength is equal to the current-strength. This follows from a comparison of equation (a) with LXXVII. (i).

From LXXVII. (b), *the potential V of a closed circuit* of strength i, at the point P, is (b) $V = i\omega$, where ω is the *solid angle* subtended by the circuit at P. If α, β, γ are the components of the magnetic force acting at P, we have, from LXXVII. (h),

$$\text{(c)} \quad \begin{cases} \alpha = i \,.\, \int \Big((z-\zeta)/r^3 \,.\, d\eta/ds - (y-\eta)/r^3 \,.\, d\zeta/ds \Big) ds, \\ \beta = i \,.\, \int \Big((x-\xi)/r^3 \,.\, d\zeta/ds - (z-\zeta)/r^3 \,.\, d\xi/ds \Big) ds, \\ \gamma = i \,.\, \int \Big((y-\eta)/r^3 \,.\, d\xi/ds - (x-\xi)/r^3 \,.\, d\eta/ds \Big) ds. \end{cases}$$

If ABC (Fig. 90) is a conductor through which a current i flows in the direction indicated by the arrow, and if a unit magnet pole moves around the current in the direction found by using the right hand in the manner before described, the work done by the magnetic forces during the movement over the path $DFED$ is, from (b), equal to $4\pi i$. If the path of the pole encircles several currents i, i', i'' etc., the magnetic forces due to these currents do upon it, during its motion, the work A, given by (d) $A = 4\pi(i + i' + i'' + \ldots)$, in which the currents which flow in one direction are to be reckoned positive, and those which flow in the opposite direction, negative. Hence, the potential which an electrical current produces at the point F is not determined only by the position of that point. If we bring a unit pole (Fig. 90) to F over the path GF from an infinite distance, the work which is done will be equal to V, the potential at the point F. If the pole then passes around the current over the path $FEDF$, the work $4\pi i$ will be done, and the potential at F becomes $V + 4\pi i$. If the pole passes n times around the current in the same way, the potential at F becomes $V + 4\pi ni$. Hence, the potential at the point F has an infinite number of values. The differential coefficients of the potential with respect to x, y, z are nevertheless completely determined.

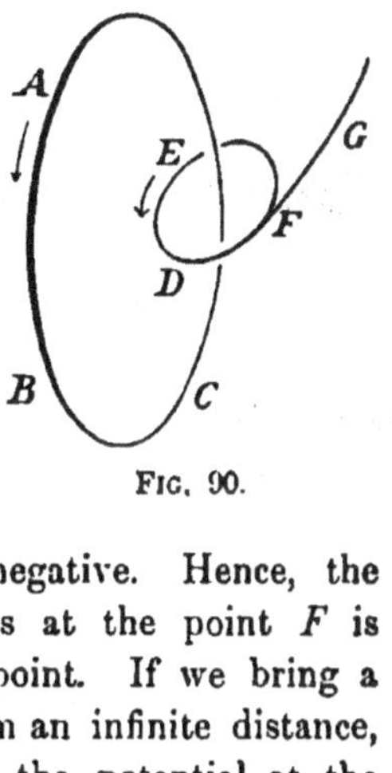

FIG. 90.

If a pole of strength μ passes once around the current, the work done on it is $4\pi i\mu$; if a magnet passes once around the current and returns to its original position, the work done is $4\pi i \Sigma\mu$, when $\Sigma\mu$ represents the sum of the quantities of magnetism in the magnet. But since for any magnet $\Sigma\mu = 0$, the work done is in this case equal to zero.

Since an electrical current may be replaced by a magnetic shell, we can obtain the magnetic moment of an infinitely small closed current. If i is the current-strength, the strength of the equivalent magnetic shell is $\sigma e = i$. If dS is the surface of the shell, we have $i.dS = \sigma e.dS$. The quantity of magnetism on one side of the shell is σdS, and the thickness of the shell is e. Hence, the magnetic moment of the current equals the product of the current-strength and the area enclosed by the circuit.

SECTION LXXIX. SYSTEMS OF CURRENTS.

Let a conductor be wound around a cylinder, so that the distances of the separate turns from each other are equal. We may approximately determine the magnetic action of the system in the following way: If L is the length of the cylinder, N the number of turns, and i the current-strength, the current flowing in unit length of the cylinder is Ni/L, and that flowing in the length dx is $Ni.dx/L$. A portion of the cylinder whose length is dx may be replaced by a magnetic shell, whose thickness is dx and whose surface-density is σ, if (a) $\sigma.dx = Ni.dx/L$ and $\sigma = Ni/L$. If this substitution is carried out for the whole length of the cylinder, the actions of the positive and negative faces of the substituted magnetic shells annul each other everywhere except on the ends of the cylinder. If the current flows in the way shown in Fig. 91, A exhibits negative and B positive magnetism. Such a system of currents is called a *solenoid.* Outside the cylinder, the only magnetic forces which act are those which proceed from the poles A and B. If the length of the solenoid is great in comparison with its diameter, the magnetic force outside of it vanishes near its middle point. The force in the interior of the solenoid may be determined in the following way: Let the line CD be parallel to the axis of the solenoid (Fig. 91), let CF and DE be perpendicular to its surface, and let the line FE be parallel to CD. We assume that the magnetic force γ is parallel to the axis of the solenoid in its interior, and that no magnetic force acts outside of it. Suppose a unit pole to traverse the closed path $CDEF$. The work done by the magnetic force is $\gamma.dx$, if we set $CD = dx$. From LXXVIII. (d), we have (b) $\gamma.dx = 4\pi.Ni.dx/L$,

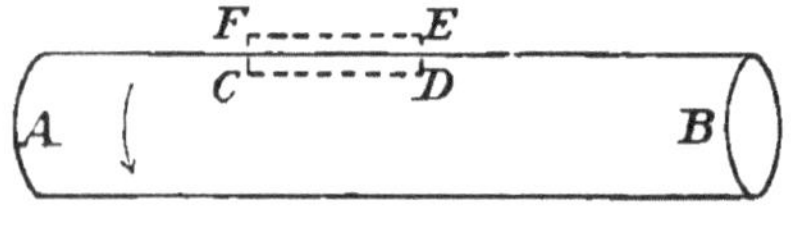

FIG. 91.

$\gamma = 4\pi Ni/L$. If we represent the number of turns in unit length by N_1 we have $\gamma = 4\pi N_1 i$, that is, *the magnetic force* (number of lines of force per square centimetre) *in the interior of the solenoid and near its middle point is given by the product of the current-strength i into the number N of turns per unit length of the solenoid.*

We will now determine the magnetic force of a sphere on whose surface a conductor is wound. Suppose that a conductor is wound on a sphere of radius R, so that the planes of the turns are parallel and separated from each other by the distance a. Suppose that $ABCD$ and $EFGH$ (Fig. 92) are two of the turns. If the current-strength is i, a single turn may be replaced by a magnetic shell whose surface-density is s, if $as = i$. For points outside the sphere, the action of the positive magnetism on the surface BC will be nearly annulled by the action of the negative magnetism on the surface EH; the only effective part of the two surfaces is the circular ring, whose width is BE. Suppose the magnetism on this ring to be distributed with the density σ, over the zone $BFGC$. We have then

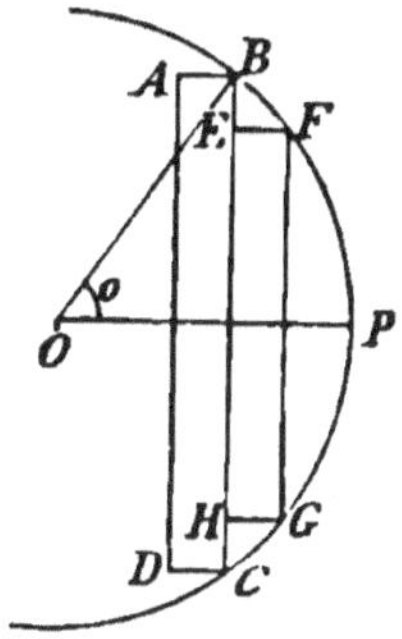

FIG. 92.

$$BE \, . \, s = BF \, . \, \sigma, \text{ or } \sigma/s = BE/BF = \cos\Theta,$$

if Θ is the angle between the radius OB, and the line OP perpendicular to the plane of the coils. Using the relation $as = i$, we obtain $\sigma = i/a \, . \cos\Theta$. From LXX. (e), (f) the magnetic potential for points outside the sphere is given by $V_a = 4\pi/3 \, . \, R^3 i/a \, . \cos\Theta/r^2$, since i/a is equivalent to the intensity of magnetization J.

In determining the magnetic force in the interior of the sphere, we must remember that the magnetic lines of force due to the currents are continuous, and that the magnetic force in this case is the same as the magnetic induction of the equivalent magnetized sphere [LXXIV.]. To find it, we suppose, as in that section, that the magnetized sphere is divided into two parts by cutting out an infinitely thin section perpendicular to the lines of magnetic force, and that the unit pole is placed in the opening between these two parts. It will then be subjected to the force $2\pi s$ directed toward the north end of the magnetized sphere due to the repulsion of the north magnetism exposed on one face of the cut, to the force $2\pi s$ due to the attraction of the south face of the cut and in the same direction, and to the force $-\frac{4}{3}\pi i/a$ [LXX. (f)] due to the distribution on the outside of the sphere and in the opposite direction. The total force F acting on the pole is therefore $F = 4\pi i/a - \frac{4}{3}\pi i/a = \frac{8}{3}\pi i/a$.

If the whole number of turns is N, we have $F = \frac{8}{3}\pi . Ni/R$, that is, *the force in the interior of the sphere is proportional directly to the current-strength i and to the number of turns N, and inversely to the radius R of the sphere.* We get the same result if we consider that this sphere behaves like a magnetized iron sphere whose intensity of magnetization is $J = i/a$.

In this way we can set up an almost constant magnetic field, which may be applied in the construction of instruments used to determine current-strength.

If the solenoid forms a *closed ring*, we obtain a system of currents which has many applications. Let AB (Fig. 93) be a circle whose centre is O and whose radius is r, and let R be the distance between the centre O and a straight line CD, which lies in the plane of the circle. If the circle rotates about the axis CD, it describes a circular ring. Suppose that on this ring there are N turns of wire, through which the current i flows. We replace the separate turns by magnetic shells, and determine the magnetic force in the interior of the ring. We reach this result most simply if we suppose a unit pole to move on a circle of radius R about the axis CD. The work done by the magnetic force $\mathfrak{H}$, which acts in the interior of the ring, when the unit pole has completed one revolution, is $2\pi R\mathfrak{H}$. This work is also equal to $4\pi Ni$ if the path of the pole is in the interior of the ring; it is equal to zero if the path is outside the ring.

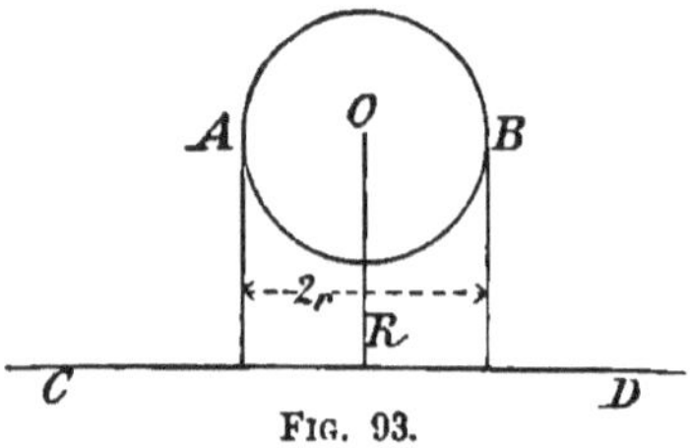

FIG. 93.

Hence we have, in the former case, $2\pi R\mathfrak{H} = 4\pi Ni$ and $\mathfrak{H} = 2Ni/R$; in the latter case, $\mathfrak{H} = 0$.

SECTION LXXX. THE FUNDAMENTAL EQUATIONS OF ELECTRO-MAGNETISM.

Up to this point we have considered the path of the electrical current as a geometrical line. In reality the current always occupies space, and is determined by its components along the coordinate axes. For example, if $dy . dz$ is a surface-element perpendicular to the x-axis, and if the quantity of electricity $u . dy . dz . dt$ passes through it in the positive direction in the time dt, u is the component of current in the direction of the x-axis. The components of current in the directions of the two other axes are represented by v and w. If Oy and Oz are drawn through the point O (Fig. 94), whose coordinates are x, y, z, parallel to the corresponding coordinate axes, and if the rectangle $OBDC$ is constructed with the sides dy and dz, the current $u.dy.dz$ flows through the element $OBDC$. If the components of the magnetic force are represented by α, β, γ, and if a unit pole moves about the rectangle in the direction $OBDCO$, the work done by the magnetic forces will be

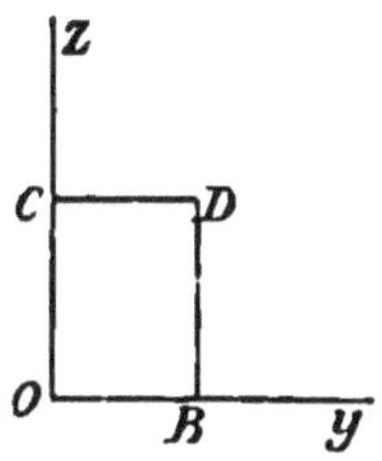

FIG. 94.

$$\beta . dy + (\gamma + \partial\gamma/\partial y . dy) . dz - (\beta + \partial\beta/\partial z . dz) . dy - \gamma . dz$$
$$= (\partial\gamma/\partial y - \partial\beta/\partial z) . dy . dz.$$

This is [LXXVIII. (d)] equal to $4\pi . u . dy . dz$. Hence, we obtain the equations

(a) $4\pi u = (\partial\gamma/\partial y - \partial\beta/\partial z)$, $4\pi v = (\partial\alpha/\partial z - \partial\gamma/\partial x)$, $4\pi w = (\partial\beta/\partial x - \partial\alpha/\partial y)$.

These equations express the current in terms of the magnetic force. In a region where there is no current we have $u = 0$, $v = 0$, $w = 0$, and therefore $\partial\gamma/\partial y = \partial\beta/\partial z$, $\partial\alpha/\partial z = \partial\gamma/\partial x$, $\partial\beta/\partial x = \partial\alpha/\partial y$,

or $\alpha . dx + \beta . dy + \gamma . dz = -dV$. *Therefore, in a region where there is no current the magnetic forces have a potential.* In this case the forces arise from magnets.

From equations (a) the magnetic force is not determined only by the components of current. If u, v, w are given and α, β, γ so determined that equations (a) are satisfied, these equations will also be satisfied if we replace α, β, γ by

$$\alpha' = \alpha + \partial V/\partial x, \quad \beta' = \beta + \partial V/\partial y, \quad \gamma' = \gamma + \partial V/\partial z,$$

where V is an arbitrary function. The potential due to the magnets present in the region is V.

We will now consider a few simple examples:

(*a*) Suppose the direction of the magnetic force to be parallel to the z-axis, and its magnitude to be a function of the distance r from this axis (Fig. 95). We then obtain from equations (a)

$$4\pi u = +d\gamma/dr . y/r, \quad 4\pi v = -d\gamma/dr . x/r, \quad w = 0.$$

The current is parallel to the xy-plane and perpendicular to r. The current-strength J is

$$J = u \cos(uJ) + v \cos(vJ), \quad J = -u . y/r + v . x/r = -1/4\pi . d\gamma/dr.$$

If γ is constant in the interior of a cylinder whose radius is $OA = r_1$ (Fig. 95), and equal to zero outside a cylinder of radius r_2, the current in the unit length of the cylinder is

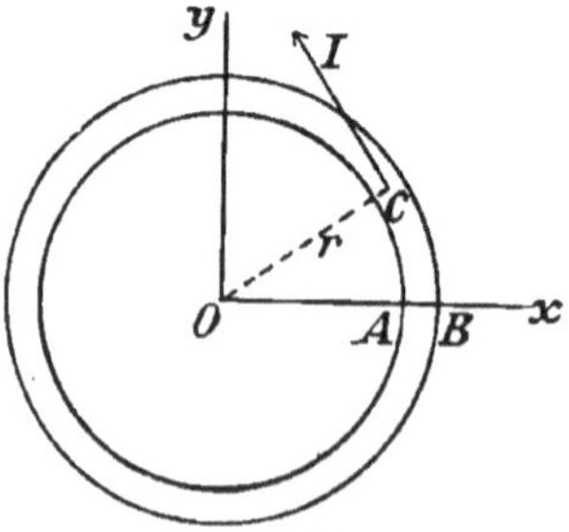

FIG. 95.

$$\int_{r_1}^{r_2} J \,.\, dr = -1/4\pi \,.\, \int_{r_1}^{r_2} d\gamma/dr \,.\, dr = \gamma/4\pi,$$

which agrees with LXXIX. (b).

(*b*) If the current-strength is given, we can find the magnetic force by integrating equations (a). Let u and v be zero, and w be a function of the distance r from the z-axis. We then have from (a)

$$0 = \partial\gamma/\partial y - \partial\beta/\partial z, \quad 0 = \partial\alpha/\partial z - \partial\gamma/\partial x, \quad 4\pi w = \partial\beta/\partial x - \partial\alpha/\partial y.$$

These equations will be satisfied if we assume that $\gamma = 0$ and that α and β are functions of x and y only. Suppose that the magnetic force is resolved into two components, one of which, R, acts in the direction of the prolongation of r, and the other, S, is perpendicular to r. We then obtain $\alpha = R \,.\, x/r - S \,.\, y/r$, $\beta = R \,.\, y/r + S \,.\, x/r$, and therefore $4\pi w = dS/dr + S/r = 1/r \,.\, d(Sr)/dr$.

If the conductor is a tube bounded by two coaxial cylinders whose radii are R_1 and R_2, and if w is constant in the conductor we have, if C_1, C_2, C_3 are constants, $S_1 r = C_1$, $2\pi w r^2 + C_2 = S_2 r$, $S_3 r = C_3$. The first of these equations holds for the interior, the second for the conductor, the third for the space outside the conductor. From the nature of the problem, S_1 must have a finite value in the axis; we therefore have $C_1 = 0$. Since the magnetic force changes continuously, we have $S_2 = 0$ when $r = R_1$, and therefore $C_2 = -2\pi w R_1^2$, $S_2 = 2\pi w r - 2\pi w R_1^2/r$. When $r = R_2$ we have $S_2 = S_3$, and therefore

$$C_3/R_2 = 2\pi w R_2 - 2\pi w R_1^2/R_2, \quad S_3 = 2\pi w(R_2^2 - R_1^2)/r.$$

Since $\pi w(R_2^2 - R_1^2)$ is equal to the current-strength i in the conductor, we have $S_3 = 2i/r$.

Therefore, *an infinitely long straight linear current exerts a magnetic force at a given point, which is inversely proportional to the distance of that point from the current.*

SECTION LXXXI. SYSTEMS OF CURRENTS IN GENERAL.

The components of current and the components of magnetic force are connected by the equations [LXXX. (a)]

(a) $4\pi u = \partial\gamma/\partial y - \partial\beta/\partial z, \quad 4\pi v = \partial\alpha/\partial z - \partial\gamma/\partial x, \quad 4\pi w = \partial\beta/\partial x - \partial\alpha/\partial y.$

From these equations it follows that (b) $\partial u/\partial x+\partial v/\partial y+\partial w/\partial z=0$. This equation corresponds with the *equation of continuity* in mechanics, and asserts *that the total quantity of electricity contained in a closed region is constant.* It thus appears *that the current, whose components are u, v, w, moves like an incompressible fluid.* There is never any accumulation of electricity, but only a displacement of it. This apparently contradicts experience; in order to be consistent with our method of treatment we assume with Faraday that an *electrical polarization* or an *electrical displacement* occurs. We represent the components of this displacement by f, g, h. If one of the components, say f, increases by the increment df in the time dt, $df/dt=\dot{f}$ represents the quantity of electricity which passes in unit time through a unit of area perpendicular to the x-axis, in consequence of the change of polarization. If p, q, r represent the components of the electrical current which is due to the flow of electricity through the body, we have

(c) $$u=p+df/dt,\quad v=q+dg/dt,\quad w=r+dh/dt.$$

These quantities, u, v, w, are the components of the actual current, which is made up of the current conducted by the body and the current arising from the change of polarization or the electrical displacement.

If the components of the current are finite, the components of magnetic force vary continuously when no magnets are present in the region. The components of force perpendicular to the surfaces of the magnets, if any are present, is in general discontinuous. We assume that currents of infinite strength do not occur in practice; however we sometimes consider the flow in a surface, in which case we must assume that the components of current in the surface are infinite. In this case the components of force parallel to the surface vary discontinuously on passage from one side of the surface to the other. If a_1 and a_2 represent these components of force, and J the quantity of electricity which flows through a unit of length perpendicular to the components, we have from LXXVIII. (d) $4\pi J=a_2-a_1$. We may obtain the same result from (a) as follows: We consider two surfaces whose equations are $z=c_1$ and $z=c_2$. We obtain from the first two equations (a)

$$4\pi\,.\int_{c_1}^{c_2}u\,.\,dz=\int_{c_1}^{c_2}\partial\gamma/\partial y\,.\,dz-\beta_2+\beta_1,\quad 4\pi\,.\int_{c_1}^{c_2}v\,.\,dz=a_2-a_1-\int_{c_1}^{c_2}\partial\gamma/\partial x\,.\,dz.$$

β_1 and β_2 are the components of the magnetic force in the direction of the y-axis on both sides of the plane surface; a_1 and a_2 have similar meanings. If $c_2-c_1=c$ is infinitely small, and u and v are infinitely great, the integrals $\int_{c_1}^{c_2}udz$ and $\int_{c_1}^{c_2}vdz$ are the quantities of electricity which flow in a surface. The integrals on the right side vanish simultaneously, and we obtain the value given in (d) for the difference between the components of magnetic force on the opposite sides of the surface.

SECTION LXXXII. THE ACTION OF ELECTRICAL CURRENTS ON EACH OTHER.

The work which must be done upon two conductors A and B (Fig. 96), carrying the currents i and i', in order to bring them nearer each other, can be determined in the following way: Let the position of the current A be fixed, while B is brought toward it from an infinite distance. B may be replaced by a magnetic shell whose surface-density is σ and whose thickness is e. Suppose CDE to be a straight line, normal to the shell, which cuts it on its negative face at C and on its positive face at D. A surface-element dS' at C carries with it the quantity of magnetism $-\sigma dS'$. If V represents the potential which A produces at C, the work done upon the element dS' at C is $-V\sigma.dS'$. Setting $CD=d\nu$, the work done upon the element dS' at D is $(V+\partial V/\partial\nu.d\nu).\sigma.dS'$. Hence the work arising from this portion of the shell is $\partial V/\partial\nu.d\nu.\sigma.dS'$. But $d\nu.\sigma=i'$, and hence the work done upon the whole shell is (a) $W=i'.\iint\partial V/\partial\nu.dS'$.

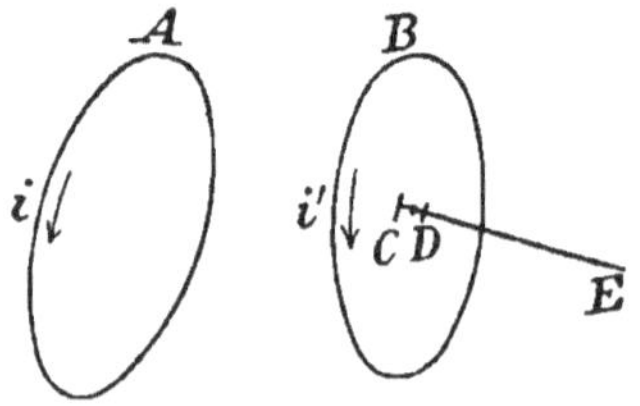

FIG. 96.

The force acting in the direction CE is $-\partial V/\partial\nu$. If CE makes angles with the axes whose cosines are l', m', n', we have

(b) $$W=-i'.\iint(l'\alpha+m'\beta+n'\gamma)dS'.$$

The quantities α, β, γ are determined in LXXVIII. (c), and may be put in the form

$$\alpha=i\int(\partial r^{-1}/\partial y.d\zeta/ds-\partial r^{-1}/\partial z.d\eta/ds)ds,$$

$$\beta=i\int(\partial r^{-1}/\partial z.d\xi/ds-\partial r^{-1}/\partial x.d\zeta/ds)ds,$$

$$\gamma=i\int(\partial r^{-1}/\partial x.d\eta/ds-\partial r^{-1}/\partial y.d\xi/ds)ds.$$

From the theorem of VI. (f) we have

$$\int(X.dx/ds'+Y.dy/ds'+Z.dz/ds')ds'$$
$$=\iint[l'(\partial Z/\partial y-\partial Y/\partial z)+m'(\partial X/\partial z-\partial Z/\partial x)+n'(\partial Y/\partial x-\partial X/\partial y)]dS'.$$

We now set

$$X=\int ii'/r\,.\,d\xi/ds\,.\,ds,\quad Y=\int ii'/r\,.\,d\eta/ds\,.\,ds,\quad Z=\int ii'/r\,.\,d\zeta/ds\,.\,ds,$$

and obtain

$$ii'\iint(d\xi/ds\,.\,dx/ds'+d\eta/ds\,.\,dy/ds'+d\zeta/ds\,.\,dz/ds')ds\,ds'/r$$
$$=\iiint ii'[l'(\partial r^{-1}/\partial y\,.\,d\zeta/ds-\partial r^{-1}/\partial z\,.\,d\eta/ds)$$
$$+m'(\partial r^{-1}/\partial z\,.\,d\xi/ds-\partial r^{-1}/\partial x\,.\,d\zeta/ds)$$
$$+n'(\partial r^{-1}/\partial x\,.\,d\eta/ds-\partial r^{-1}/\partial y\,.\,d\xi/ds)]ds\,dS',$$

or

$$ii'\iint(d\xi/ds\,.\,dx/ds'+d\eta/ds\,.\,dy/ds'+d\zeta/ds\,.\,dz/ds')ds\,ds'/r$$
$$=\iint i'(l'\alpha+m'\beta+n'\gamma)dS'=-W.$$

Hence from (b)

(c) $W=-ii'\iint(d\xi/ds\,.\,dx/ds'+d\eta/ds\,.\,dy/ds'+d\zeta/ds\,.\,dz/ds')ds\,ds'/r,$

where ds is an element of one conductor and ds' of the other.

If we represent the angle between two elements ds and ds' by ϵ, we obtain F. E. Neumann's expression for the *potential energy of two electrical currents,* (d) $W=-ii'\iint\cos\epsilon/r\,.\,ds\,ds'$.

If we represent the magnetic force which is perpendicular to an arbitrary surface containing the circuit B by $\mathfrak{H}$, we have from (a), (e) $W=-i'\,.\int\mathfrak{H}'\,.\,dS'=-i'N$, if N represents, in Faraday's nomenclature, the number of lines of force enclosed by the conductor. Therefore *the potential energy of a current equals the negative product of the current-strength and the number of lines of force enclosed by the conductor.* Hence it follows that a current always tends to move so that the number of lines of force enclosed by it shall become as great as possible. The positive direction of the lines of force is the direction in which a north pole moves under the action of the current [cf. LXXIV.] The above theorem has only been proved on the assumption that the magnetic force $\mathfrak{H}$ is due to another current. But because currents and magnets are equivalent, the law is true generally. Since the surface containing the circuit B may be arbitrarily chosen, the energy W depends only on the contour of the circuit.

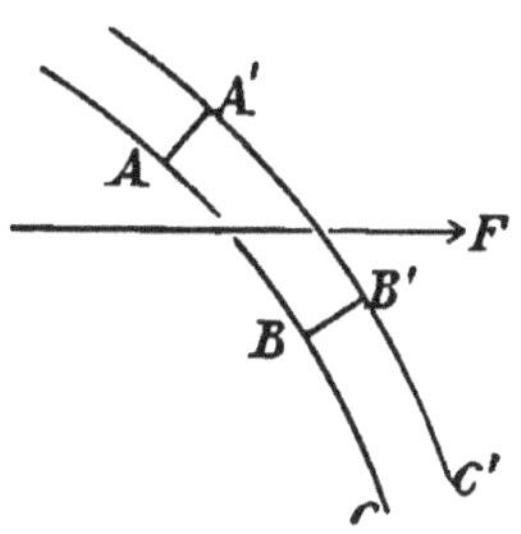

FIG. 97.

The force which acts on an element $AB=ds$ of the current ABC (Fig. 97) may be determined in the following way: Suppose that the conductor ABC moves so that AB is displaced to $A'B'$ in such a manner that AA' and BB' are perpendicular to AB. If $AA'=dp$, the area contained by the conductor increases by $ds.dp$. If the magnetic force at A is equal to $\mathfrak{H}$, and if its direction makes the angle α with the normal to the surface $ABB'A'$, the component K of the force normal to the surface is $K=\mathfrak{H}\cos\alpha$. Hence the increment of the potential energy of the conductor is, from (e), $dW = -iK.ds.dp$, if i is the current-strength. In order to cause the motion here described, a force X must act on ds in the direction AA', which is determined by $X.dp = -iK.ds.dp$, $X = -iK.ds$. Hence the force $-iK.ds$ acts on the current-element in the aforesaid direction, and if the current-element is free to move, the direction of its motion is perpendicular to the direction of the magnetic force as well as to its own direction. The direction of the motion is determined by laying the right hand on the current [cf. LXXVIII.].

It follows further that the force which acts on an element ds of the current i is perpendicular to the plane determined by the current and the direction of the magnetic force $\mathfrak{H}$. If we represent the angle between the direction of the force and the direction of the current by ϕ, this force will equal $\mathfrak{H}i.ds.\sin\phi$.

Section LXXXIII. The Measurement of Current-Strength or the Quantity of Electricity.

(a) *Constant Currents.*

To measure constant currents we generally use a galvanometer consisting of parallel circular conductors carrying the current whose strength is to be determined. A magnet whose dimensions are small in comparison with the radius of the coils, is suspended in the centre of the apparatus, which is so placed that the coils are parallel to the magnetic meridian. The current sets up a magnetic force whose value is Gi perpendicular to the direction of the earth's magnetic force, whose horizontal component is called H. G depends on the construction of the galvanometer. If G is constant in the region in which the magnet moves, the angle ϕ by which the magnet is turned from its position of rest by the current is determined by

(a) $$\operatorname{tg}\phi = Gi/H, \quad i = H/G.\operatorname{tg}\phi,$$

that is, *the current-strength is, in this case, proportional to the tangent of the angle of deflection.*

(b) *Variable Currents.*

It is very difficult to determine the strength of currents of short duration at any instant. We may, however, easily measure the total quantity of electricity Q which flows through the conductor. From LXXI. (d) the moment which tends to turn the magnet about a perpendicular axis is $-\mathfrak{M}a\sin\Theta$, if $\mathfrak{M}$ is the magnetic moment of the magnet, a the magnetic force, and Θ the angle between the directions of $\mathfrak{M}$ and a. Setting $a=Gi$, where G is the galvanometer constant, and assuming $\Theta=\frac{1}{2}\pi$, the *directive force* exerted by the current on the magnet is equal to $\mathfrak{M}Gi$. The total moment caused by the current is therefore $\int\mathfrak{M}Gi.dt=\mathfrak{M}G.Q$, if we write $Q=\int i.dt$. Q is the quantity of electricity which passes through the conductor during the discharge.

If J is the moment of inertia of the magnet, and ω its angular velocity, $J\omega$ will be its moment of momentum. We thus obtain the equation (b) $\mathfrak{M}GQ=J\omega$. If the period of oscillation of the magnet is called τ, we have, by LXXI. (f),

(c), (d) $\tau=\pi.\sqrt{J/\mathfrak{M}H}$ and therefore $Q=H\tau^2\omega/G\pi^2$.

The kinetic energy which the magnet receives from the impulse given to it by the current is $\frac{1}{2}J\omega^2$, in consequence of which it turns through the angle Θ. Its potential energy thereby increases from $-\mathfrak{M}H$ to $-\mathfrak{M}H\cos\Theta$; the work done on it is $\mathfrak{M}H(1-\cos\Theta)$. We therefore have $\frac{1}{2}J\omega^2=2\mathfrak{M}H\sin^2(\Theta/2)$, or, if Θ is very small, (e) $\Theta=\tau\omega/\pi$ and $Q=H\tau/\pi G.\Theta$, that is, *if there is no damping action on the magnet, and if its angular displacement is small, the quantity of electricity flowing through a section of the conductor is proportional to the angular displacement of the magnet.*

(c) *Damping Action.*

The oscillations of the magnet generally diminish rather rapidly in consequence of what is called damping or damping action. Damping arises from resistance of the air and the action of currents induced by the motion of the magnet in neighbouring conductors. If there is no damping, we have from LXXI. (e) and (f), when the oscillations are small, $d^2\Theta/dt^2=-\pi^2/\tau^2.\Theta$, τ is therefore the period of oscillation of the undamped magnet. We may assume that the damping action is proportional to the angular velocity $d\Theta/dt$. Taking the damping into account, we have, to determine the deflection Θ, the differential equation (f) $\ddot{\Theta}+2m\dot{\Theta}+\pi^2/\tau^2.\Theta=0$. The factor m depends on the size and character of the oscillating magnet, on the density of the air, and on the size, character, and position of the masses of metal in which currents are induced. If we set $\pi/\tau = n$ and $\Theta = e^{at}$, we have

$$a^2 + 2ma + n^2 = 0 \text{ and } a = -m \pm \sqrt{n^2 - m^2}\sqrt{-1}$$

in which it is assumed that $n > m$. Setting (g) $n^2 - m^2 = \pi^2/\tau_1^2$, we have $\Theta = \big(A \sin(\pi t/\tau_1) + B\cos(\pi t/\tau_1)\big) . e^{-mt}$. If $\Theta = 0$ at the time $t = 0$, we obtain $\Theta = A . e^{-mt} . \sin(\pi t/\tau_1)$. $d\Theta/dt = \omega$ at the time $t = 0$, and therefore $\Theta = \tau_1\omega/\pi . e^{-mt} . \sin(\pi t/\tau_1)$. To find the magnitude of the deflection we set $d\Theta/dt = 0$, and obtain (h) $\mathrm{tg}(\pi t/\tau_1) = \pi/m\tau_1$. If τ_0 is the smallest root of this equation, the successive roots are $\tau_0 + \tau_1$, $\tau_0 + 2\tau_1$, ... The oscillations are therefore isochronous. If we represent the deflections by Θ_1, Θ_2, Θ_3, ..., we have

$$\Theta_1 = \tau_1\omega/\pi . e^{-m\tau_0} . \sin(\pi\tau_0/\tau_1),$$
$$\Theta_2 = -\tau_1\omega/\pi . e^{-m(\tau_0+\tau_1)} . \sin(\pi\tau_0/\tau_1),$$
$$\Theta_3 = \tau_1\omega/\pi . e^{-m(\tau_0+2\tau_1)} . \sin(\pi\tau_0/\tau_1).$$

. . . .

If the position of equilibrium is designated by A_0, the first point of reversal by A_1, the second by A_2, etc., the ratio between the oscillations A_1A_2 and A_2A_3 is

$$(\Theta_1 - \Theta_2) : (\Theta_3 - \Theta_2) \text{ and } (\Theta_1 - \Theta_2)/(\Theta_3 - \Theta_2) = e^{m\tau_1}.$$

We set $m\tau_1 = \lambda$, and obtain (i) $\lambda = \log\text{nat}\,[(\Theta_1 - \Theta_2)/(\Theta_3 - \Theta_2)]$. λ is the *logarithmic decrement*, which can be very exactly determined from a series of oscillations. From (g) the *period of oscillation* τ_1, is (k) $\tau_1 = \tau . \sqrt{1 + \lambda^2/\pi^2}$. Therefore the period of oscillation is increased by the damping action. If we set $\tau = \tau_0$ in equation (h), we have

$$\mathrm{tg}(\pi\tau_0/\tau_1) = \pi/m\tau_1 \text{ and } \pi\tau_0/\tau_1 = \mathrm{arctg}(\pi/\lambda),\quad m\tau_0 = \lambda/\pi . \mathrm{arctg}(\pi/\lambda),$$
$$\sin(\pi\tau_0/\tau_1) = 1/\sqrt{1 + \lambda^2/\pi^2}.$$

Hence we have further $\Theta_1 = \tau_1\omega/\pi . e^{-\lambda/\pi . \mathrm{arctg}(\pi/\lambda)} . 1/\sqrt{1 + \lambda^2/\pi^2}$ and (l) $\omega = \pi\Theta_1/\tau_1 . \sqrt{1 + \lambda^2/\pi^2} . e^{\lambda/\pi . \mathrm{arctg}(\pi/\lambda)}$.

We obtain from (d) and (k) $Q = H\tau_1^2/G\pi^2 . \omega/(1 + \lambda^2/\pi^2)$, and using equation (l), (m) $Q = \Theta_1 . H\tau_1/G\pi . e^{\lambda/\pi . \mathrm{arctg}(\pi/\lambda)} . 1/\sqrt{1 + \lambda^2/\pi^2}$.

In order to determine the quantity of electricity sent through a conductor by an electrical current, whose duration is small in comparison with the period of oscillation of the magnet, we must determine the logarithmic decrement and the period of oscillation of the magnet. Q is then determined from these quantities, if we know in addition the intensity of the earth's magnetism and the constant of the galvanometer.

Setting $\operatorname{arctg} \pi/\lambda = \frac{1}{2}\pi - x$ we have $\operatorname{tg} x = \lambda/\pi$. If λ is very small, we have $x = \lambda/\pi$ and $\operatorname{arctg}(\pi/\lambda) = \frac{1}{2}\pi - \lambda/\pi$. If the damping action is insignificant, we can neglect higher powers in the series in which the exponential may be developed, and obtain

$$e^{\lambda/\pi \,.\, \operatorname{arctg}(\pi/\lambda)} = 1 + \lambda/2 \quad \text{and} \quad Q = \Theta_1 \,.\, H\tau_1/G\pi \,.\, (1 + \lambda/2).$$

SECTION LXXXIV. OHM'S LAW AND JOULE'S LAW.

We have up to this point assumed the existence of the electrical current and have not discussed the question of the way in which it is started and maintained. This mode of treatment in many respects lacks clearness. We will therefore state such facts as are well established by observation. The so-called galvanic elements can establish and maintain an almost constant current. In order to maintain a constant current in a conductor, an electromotive force must act in the direction of the current. If u is the quantity of electricity which flows in unit time through a unit of surface of the xy-plane in the direction of the x-axis, we can set $u = C \,.\, X$, if C is the *conductivity* and X the component of the electromotive force in the direction of the x-axis. C depends on the nature of the conductor, and may be supposed to have the same value in the conductor in all directions. If the components of current and of force in the other two directions are v, w, and Y, Z respectively, we have (a) $u = CX$, $v = CY$, $w = CZ$. Hence $\partial u/\partial x + \partial v/\partial y + \partial w/\partial z = C(\partial X/\partial x + \partial Y/\partial y + \partial Z/\partial z)$. If the *steady state* of the electrical current has been reached, the left side of the equation equals zero, and hence the right side is also equal to zero. If the electromotive forces have a potential V, we have (b) $\nabla^2 V = 0$. This equation states that *no free electricity is present within the conductor as soon as the current becomes steady*. The electromotive forces must therefore arise from the free electricity on the surface of the conductor.

Suppose ABC (Fig. 98) to be an electrical conductor. We will consider a portion of it which is bounded by the infinitely small cross-sections $AA' = S$ and $BB' = S$, separated from each other by the distance l, which is also infinitely small. If AB is parallel to the x-axis, the component of current u equals CX, and therefore the quantity of electricity $i = uS = CX \,.\, S$ flows through the cross-section S. If V and V' are the potentials at A and B respectively, we have $i = C \,.\, S \,.\, (V - V')/l$

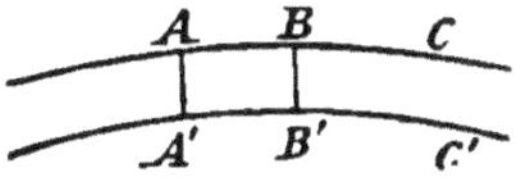

FIG. 98.

and further (c) $i=(V-V')/(l/CS)=(V-V')/R$. *The resistance R is directly proportional to the length of the conductor and inversely proportional to its cross-section, and to the conductivity of the substance constituting the conductor. The difference of potential* between A and B is $V-V'$. Equation (c) contains *Ohm's law*, according to which *the current-strength is directly proportional to the difference of potential and inversely proportional to the resistance.*

The quantity of electricity $i.dt$ flows through the cross-section AA' in the time dt and passes from A to B under the influence of the electromotive force X. The work done is therefore

$$i.dt.X.l=i.dt.(V-V').$$

The work done in this part of the conductor by the electromotive forces in unit time is (d) $A=i(V-V')=i^2R$. This work is transformed into heat in the conductor. Therefore *the quantity of heat developed in a conductor is proportional to the square of the current-strength and the resistance of the conductor.* This theorem was proved experimentally by Joule and deduced theoretically by Clausius.

CHAPTER X.

INDUCTION.

SECTION LXXXV. INDUCTION.

FARADAY was the first to demonstrate that a current is set up in a conductor if a magnet or a conductor carrying a current is moved in its neighbourhood. F. E. Neumann discovered the laws of these induced currents. Faraday himself afterwards described a method of determining the strength and direction of the induced current which possesses great advantages, because it makes it possible to visualise the process. Suppose ABC to be a closed conductor (Fig. 99), and DE, $D'E'$, etc., the lines of force enclosed by it. Let us designate an element of a surface bounded by the conductor by dS, the components of the magnetic force (cf. LXXI.) by α, β, γ, and the angles which the normal to dS makes with the axes by l, m, n. An electromotive force arises in the conductor if the integral

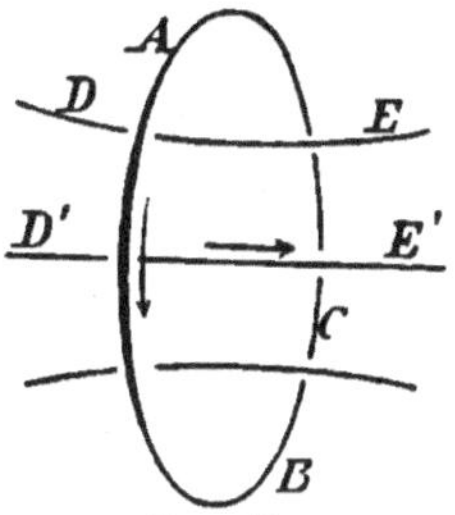

FIG. 99.

(a) $$N = \int(\alpha l + \beta m + \gamma n)dS$$

changes its value. If magnetizable bodies are enclosed by the circuit which have a greater permeability for lines of force than air, the components of force must be replaced by the components of magnetic induction. An electromotive force then arises in the conductor if the integral (cf. LXXIV.) $N = \int(al + bm + cn)dS$ changes its value. *An induced current arises if the number of lines of force enclosed by the conductor is changed.* If the change in the number of lines of force is an increase, the induced current tends to diminish the number of the enclosed lines of force, for the direction of the induced current is such that its own lines of force are opposite to the lines of force formerly existing. If the direction indicated in the figure by the arrow is taken as positive, the induced electromotive force acts in a negative direction.

According to Lenz's law, the current induced by the motion of a circuit tends, by its electrodynamic action, to oppose the motion by which it is induced.

In order to determine the magnitude of the induced electromotive force, we suppose that the current-strength at a given instant is equal to i. If we move the conductor in the magnetic field, we must, by LXXXII., do the work $-i.dN$, and, at the same time, the quantity of energy $Ri^2.dt$ is transformed into heat. We have at once $-i.dN = Ri^2.dt$, and therefore, because Ri equals the electromotive force e, (b) $e = -dN/dt$, that is, *the induced electromotive force is equal to the decrease in unit time of the number of lines of force enclosed by the circuit.*

The induced electromotive force depends on the value of the magnetic induction, whose components are a, b, c, not on the magnetic force, whose components are α, β, γ. If there is no magnet near, and if the coefficient of magnetization of the region is $k = 0$ (cf. LXXVI.), the induction and the magnetic force have the same value, and a, b, c may be replaced by α, β, γ.

For example, if the circuit ABC at the time t carries a current of strength i, the number N of lines of force passing through the circuit in the positive direction is $N = Li$. L is the number of lines of force if the current-strength is unity. It is called the *coefficient of self-induction.* If the current diminishes there arises an electromotive force (c) $e = -d(Li)/dt$. According to Ohm's law we have $e = Ri$, if we represent the resistance by R, and therefore

$$\text{(d)} \qquad Ri = -d(Li)/dt = -L.di/dt,$$

provided that the coefficient of self-induction L is constant. This coefficient depends on the permeability of the region and also on the form of the conductor.

If i_0 is the current-strength at the time $t = 0$, we have (e) $i = i_0.e^{-R/L.t}$. *The current-strength therefore diminishes the more rapidly the greater the resistance and the smaller the coefficient of self-induction.*

We obtain from (c)

$$\int_0^\infty e i dt = -L\int_0^\infty i.di = \tfrac{1}{2}Li_0^2.$$

From LXXXIV. (d), the left side of this equation is an expression for the work done, which appears as heat in the conductor. Hence we obtain for the electro-kinetic energy T of a conductor whose self-induction is L and which carries a current of strength i, (f) $T=\frac{1}{2}Li^2$. *The electro-kinetic energy of the circuit is therefore equal to half the product of the coefficient of self-induction L and the square of the current-strength i.*

If ABC and $A'B'C'$ (Fig. 100) are two conductors carrying currents whose respective strengths are i_1 and i_2, the current i_1 sets up a number L_1i_1 of lines of force which pass through the conductor ABC. The current i_2 also sets up lines of force, and the number of them which pass through ABC may be represented by $M_{21}i_2$. The total number of lines of force enclosed by ABC is therefore

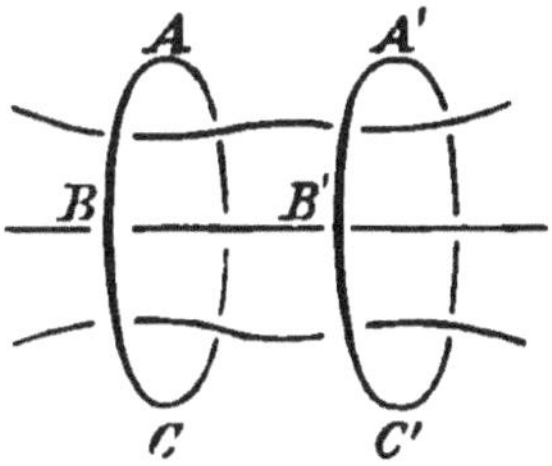

FIG. 100.

(g) $$N_1 = L_1i_1 + M_{21}i_2.$$

The number of lines of force enclosed by $A'B'C'$ is (h) $N_2 = L_2i_2 + M_{12}i_1$. From LXXXII. and the discussion at the beginning of this section, we have, in the usual notation,

$$\int(l_2a_1 + m_2b_1 + n_2c_1)dS_2 = i_1\int \cos\epsilon/r \,.\, ds_1ds_2,$$
$$\int(l_1a_2 + m_1b_2 + n_1c_2)dS_1 = i_2\int \cos\epsilon/r \,.\, ds_1ds_2,$$

The integral with respect to dS_2 equals i_1M_{12}, that with respect to dS_1 equals i_2M_{21}. From LXXXII. we have

$$M_{12} = M_{21} = \int \cos\epsilon/r \,.\, ds_1ds_2.$$

$M_{12}=M_{21}$ is *the coefficient of mutual induction of the two circuits.*

If R_1 and R_2 are the resistances of the conductors ABC and $A'B'C'$ respectively, we have

$$R_1i_1 = e_1 = -dN_1/dt = -L_1 \,.\, di_1/dt - M_{21} \,.\, di_2/dt,$$
$$R_2i_2 = e_2 = -dN_2/dt = -L_2 \,.\, di_2/dt - M_{12} \,.\, di_1/dt.$$

Hence we have, for the *electro-kinetic energy* T of a system of two conductors which carry the currents i_1 and i_2,

$$T = \int_0^\infty (e_1i_1 + e_2i_2)dt = \tfrac{1}{2}L_1i_1^2 + M_{12}i_1i_2 + \tfrac{1}{2}L_2i_2^2.$$

For the electro-kinetic energy T of any system of conductors, we find in the same way,

(i) $T = \frac{1}{2}(L_1i_1^2 + L_2i_2^2 + L_3i_3^2 + \ldots + 2M_{12}i_1i_2 + 2M_{13}i_1i_3 + 2M_{23}i_2i_3 + \ldots)$.

If N_1, N_2, $N_3\ldots$ denote the number of lines of force enclosed respectively by the conductors 1, 2, 3... so that, for example, $N_1 = L_1i_1 + M_{21}i_2 + M_{31}i_3 + \ldots$, the expression for the electro-kinetic energy T becomes (k) $T = \frac{1}{2}(N_1i_1 + N_2i_2 + N_3i_3 + \ldots) = \frac{1}{2}\Sigma Ni$, that is, *the electro-kinetic energy of a system of currents is equal to the sum of the products of the number of lines of force enclosed by each conductor and the strength of the current present in the conductor.*

Electrical currents arise not only if the neighbouring currents change in strength, but also if they change their position, so that M_{12}, M_{13} ... vary, and also if the conductor itself changes its form. In all cases the induced current is determined by the change in the number of lines of force enclosed by the conductor.

SECTION LXXXVI. COEFFICIENTS OF INDUCTION.

In the investigation of variable electrical currents flowing in wire coils, the coefficients of induction between different turns in any one coil and between separate coils are of great importance. The calculation of these coefficients is in most cases very difficult; we will consider only one simple case. We suppose two circular conductors whose radii are r_1 and r_2 (Fig. 101). They have a common axis, and are separated by the distance b. Suppose $r_2 > r_1$. We have to calculate the integral

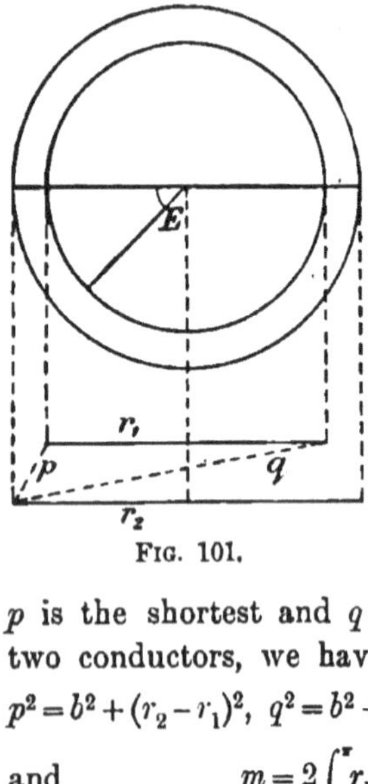

FIG. 101.

$$M_{12} = \int ds_2 \int ds_1 / r \,.\, \cos \epsilon.$$

We first evaluate the integral m,

$$m = \int ds_1 / r \,.\, \cos \epsilon.$$

We have $r^2 = b^2 + r_2^2 - 2r_1 r_2 \,.\, \cos\epsilon + r_1^2$. If p is the shortest and q the longest distance between points of the two conductors, we have

$$p^2 = b^2 + (r_2 - r_1)^2,\ q^2 = b^2 + (r_2 + r_1)^2,\ q^2 - p^2 = 4r_1 r_2,\ r^2 = p^2 + (q^2 - p^2)\sin^2\tfrac{1}{2}\epsilon$$

and
$$m = 2\int_0^\pi r_1 \,.\, \cos\epsilon \,.\, d\epsilon / \sqrt{p^2 + (q^2 - p^2)\sin^2\tfrac{1}{2}\epsilon}.$$

If α is a small angle, so chosen that $q\alpha$ is very great in comparison with p, we can set

$$m = 2\int_0^\alpha r_1 \,.\, d\epsilon / \sqrt{p^2 + q^2 \,.\, \epsilon^2/4} + 2\int_\alpha^\pi r_1 \,.\, d\epsilon \,.\, (1 - 2\sin^2\tfrac{1}{2}\epsilon)/q \sin\tfrac{1}{2}\epsilon.$$

$$m = 4r_1/q \,.\, \left[\int_0^\alpha d\epsilon/\sqrt{4p^2/q^2 + \epsilon^2} + \int_\alpha^\pi d\epsilon / 2\sin\tfrac{1}{2}\epsilon - \int_\alpha^\pi \sin\tfrac{1}{2}\epsilon \,.\, d\epsilon\right],$$

$$m = 4r_1/q \,.\, [\log(2r_1\alpha/p) - \log(\alpha/4) - 2],$$

$$m = 4r_1/q \,.\, [\log(8r_1/p) - 2] = 2[\log(8r_1/p) - 2].$$

With this value for m, we obtain $M_{12} = 4\pi r_2(\log(8r_1/p) - 2)$. On the assumptions which have been made we may set $r_1 = r_2 = R$, so that

(a) $$M_{12} = 4\pi R(\log(8R/p) - 2).$$

The coefficient of induction between two coils, the number of whose turns is n_1 and n_2 respectively, and for which the mean value of $\log p$ is expressed by $\log P$, is given by

(b) $$M_{12} = 4n_1 n_2 \pi R(\log(8R/P) - 2).$$

On the same assumptions the coefficient of self-induction L of a single coil, if n is the number of turns, is given by

(c) $$L = 4\pi n^2 R(\log(8R/P) - 2).$$

We will not go further into the calculation of coefficients of induction; in most cases they are determined experimentally by one of the following methods:

Methods of Determining the Coefficients of Induction.

(a) If the coefficient of mutual induction M of two coils L_1 and L_2 is known, we may determine in the following way the coefficient of mutual induction M' for two other coils L_1' and L_2'.

Let L_1 and L_2 (Fig. 102) be the coils whose coefficient is known, and L_1' and L_2' those which are to be investigated. A current is

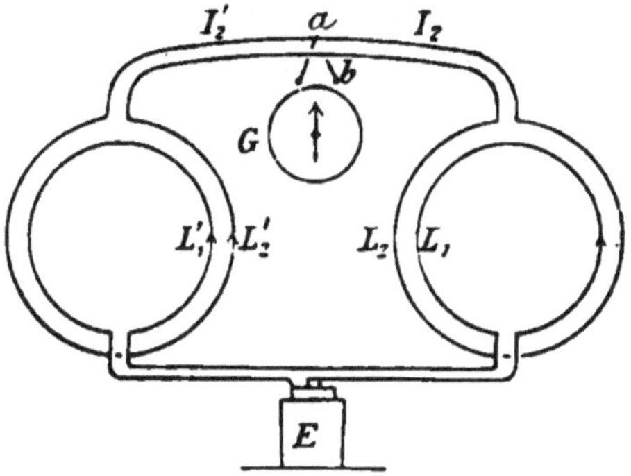

FIG. 102.

passed through the coils L_1' and L_2' from the voltaic cell E. The coils L_2 and L_2' are joined by conductors, and conductors are joined from the points a and b to the galvanometer G. If the current J which passes through L_1 and L_1' is suddenly broken, electromotive forces e and e' arise in L_2 and L_2'. If J_2 and J_2' are the strengths of the currents induced in L_2 and L_2', we have

$$e = -d(L_2J_2 + MJ)/dt, \quad e' = -d(L_2'J_2' + M'J)/dt,$$

where L_2 and L_2' represent the coefficients of self-induction. Applying Kirchhoff's laws to the circuit L_2G and $L_2'G$, it follows that

$$-d(L_2J_2 + MJ)/dt = R_2J_2 + G(J_2 - J_2'),$$
$$-d(L_2'J_2' + M'J)/dt = R_2'J_2' - G(J_2 - J_2'),$$

if the resistance of the galvanometer is designated by G, and the resistances of the coils L_2 and L_2' by R_2 and R_2' respectively. We multiply these equations by dt and integrate from $t=0$ to $t=\tau$, where τ is a very small time-interval. If the current J is broken at the instant $t=0$, a current is induced in the circuit $L_2'GL_2$ which, in the time τ, sets in motion in the circuit L_2 the quantity of electricity C_2, in L_2' the quantity C_2', and therefore in the galvanometer the quantity $C_2 - C_2'$. At the time $t=0$, $J=J$ and $J_2 = J_2' = C_2 = C_2' = 0$, and at the time $t=\tau$, $J=0$ and the induced current has also vanished, so that $J_2 = J_2' = 0$. Hence we have

$$MJ = R_2C_2 + G(C_2 - C_2'), \quad M'J = R_2'C_2' - G(C_2 - C_2').$$
$$C_2 - C_2' = J(M/R_2 - M'/R_2')/(1 + G/R_2 + G/R_2').$$

$C_2 - C_2'$ is the induced quantity of electricity which flows through the galvanometer, that is, the *total current*. The galvanometer shows no deflection if the resistances satisfy the equation $M'/M = R_2'/R_2$.

(*b*) The comparison between two coefficients of self-induction can be carried out in the following way: Let $ABCD$ (Fig. 103) be a

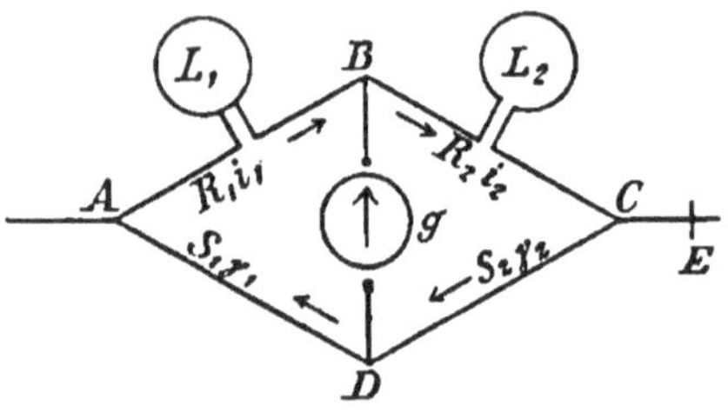

FIG. 103.

Wheatstone's bridge with a galvanometer inserted in the arm BD. L_1 and L_2 are two coils inserted in the arms AB and BC, whose coefficients of self-induction are to be compared. The current entering at A and passing out at C distributes itself in the conductors, and causes a deflection of the galvanometer needle. Let the resistances

R_1, R_2, S_1 and S_2 in AB, BC, AD and DC respectively, be so adjusted that no current passes through the galvanometer; we then have

$$R_1/R_2 = S_1/S_2.$$

When the circuit is broken at E, an electromotive force arises by induction in L_1 and L_2, in consequence of which a current of strength g flows through the galvanometer. Let i_1, i_2, γ_1, γ_2 respectively be the current-strengths in the conductors AB, BC, AD, DC. They are connected by the relations $i_1 = \gamma_1$, $i_2 = \gamma_2$. We then have

$$-d(L_1 i_1)/dt = (R_1 + S_1) i_1 + Gg, \quad -d(L_2 i_2)/dt = (R_2 + S_2) i_2 - Gg.$$

At the instant $t = 0$, when the circuit is broken at E, the same current i_0 was in AB and BC; whence

$$L_1 i_0 = (R_1 + S_1) \int_0^\tau i_1 \, . \, dt + G \int_0^\tau g \, . \, dt ; \quad L_2 i_0 = (R_2 + S_2) \int_0^\tau i_2 \, . \, dt - G \int_0^\tau g \, . \, dt,$$

where τ denotes a very short time. No deflection is caused in the galvanometer by the current g if $\int_0^\tau g \, . \, dt = 0$.

Since $i_1 = i_2 + g$ we have $i_1 = i_2$, if no current flows through the galvanometer, and hence in that case $L_1/L_2 = (R_1 + S_1)/(R_2 + S_2)$.

By the use of the relation $R_1 : R_2 = S_1 : S_2$, it follows that

$$L_1/L_2 = R_1/R_2,$$

that is, *the coefficients of self-induction of the coils are in the same ratio as the resistances of the two arms in which the coils are introduced.*

If therefore the needle of the galvanometer is not deflected either by a constant or a variable current passing through the circuits L_1 and L_2, we have a means of determining the ratio between the coefficients of self-induction L_1 and L_2.

SECTION LXXXVII. MEASUREMENT OF RESISTANCE.

The strength of the electrical current in a conductor is determined by its magnetic action; the electromotive force is determined by the change in the number of lines of force contained by the conductor. The resistance in the conductor may then be determined by the help of Ohm's law. Many methods of measuring resistance are in use. We will confine ourselves to the description of some of the simplest.

In one of these, used by W. Weber, the essential part of the apparatus is a wire coil which can rotate about a vertical axis. This coil is set perpendicular to the magnetic meridian and is then turned through 180°. The coil is in connection with a galvanometer; the total resistance of the circuit is R. If the plane of the coil makes an angle ϕ with the magnetic meridian at a particular instant, the number of lines of force passing through the coil is $SH\sin\phi$, if S represents the area enclosed by the coils and H the horizontal intensity of the earth's magnetism. If L denotes the coefficient of self-induction of the coil and galvanometer and i the current-strength, we have (a) $-d(SH\sin\phi)/dt - d(Li)/dt = Ri$.

Since ϕ changes, during the rotation of the coil, from $+\frac{1}{2}\pi$ to $-\frac{1}{2}\pi$, and since i is zero both at the beginning and at the end of the motion, we have (b) $2SH = RQ$, where Q denotes the total quantity of electricity which flows through the conductor. If Q is measured by the method described in LXXXIII., we have

$$R = 2SG\sqrt{\pi^2 + \lambda^2}/(\tau_1\Theta_1 e^{\lambda/\pi \,.\, \mathrm{arctg}\,\pi/\lambda}).$$

The absence of H from this expression shows that it is not necessary to know the intensity of the earth's magnetism in order to determine the resistance.

Sir William Thomson's (Lord Kelvin's) Method.

If the coil above described turns with a constant angular velocity ω, we have by (a) $-SH\omega\cos\phi - L\,.\,di/dt = Ri$. The integral of this equation is $i = i_0\,.\,e^{-Rt/L} - A\,.\cos(\phi - \alpha)$. If the rotation is continued for a considerable time, the exponential term vanishes and need be no longer considered. To determine A and α, we have

(c) $$A = SH\omega/(R\cos\alpha + L\omega\sin\alpha)\,; \quad \mathrm{tg}\,\alpha = L\omega/R,$$

and therefore $$A = SH\omega/R\sqrt{1 + L^2\omega^2/R^2} = SH\omega/R\,.\cos\alpha.$$

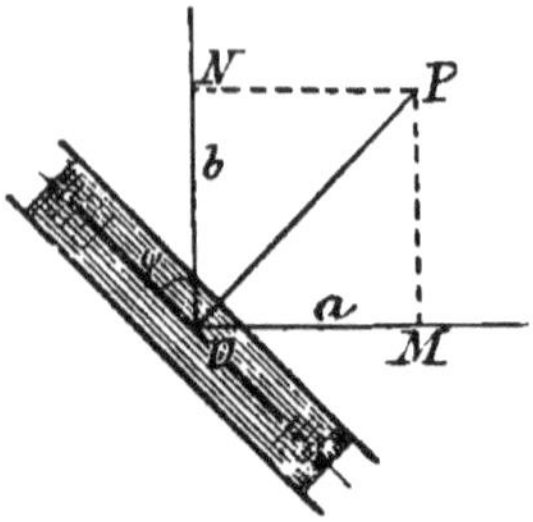

FIG. 104.

It thus appears that *the self-induction apparently increases the resistance.*

If ON (Fig. 104) is the magnetic meridian and if the line OM is perpendicular to it, the coil acts on a magnetic needle at its centre with the force Gi, whose direction is that of the line OP perpendicular to the plane of the coil. The components of this force are

$$OM = a = Gi\,.\cos\phi \quad \text{and} \quad ON = b = Gi\,.\sin\phi.$$

Let a_1 and b_1 denote the mean values of these forces. We then have

$$a_1 = 1/2\pi \,.\int_0^{2\pi} Gi \,.\, \cos\phi \,.\, d\phi\,;\quad b_1 = 1/2\pi \,.\int_0^{2\pi} Gi \,.\, \sin\phi \,.\, d\phi.$$

Now

$$\int_0^{2\pi}\cos(\phi - a)\,.\,\cos\phi\,.\,d\phi = \pi\,.\,\cos a,\quad \int_0^{2\pi}\cos(\phi - a)\,.\,\sin\phi\,.\,d\phi = \pi\,.\,\sin a,$$

and hence $\quad a_1 = -\tfrac{1}{2}GA\,.\,\cos a,\quad b_1 = -\tfrac{1}{2}GA\,.\,\sin a.$

The magnet at the centre of the coil turns from the meridian in the same sense as that in which the coil rotates. If its angular displacement is represented by Θ, we have

$$\operatorname{tg}\Theta = -a_1/(H + b_1) = GA\cos a/(2H - GA\sin a),$$

or, introducing the value of A, $\operatorname{tg}\Theta = GS\omega\cos^2 a/(2R - GS\omega\sin a\cos a)$. This equation, in connection with (c), serves to determine the resistance R. If a is very small we have $R = GS\omega/2\operatorname{tg}\Theta$.

L. Lorenz's Method.

Suppose that a metallic disk ABC (Fig. 105), whose radius is a, turns with constant velocity about an axis passing perpendicularly through its centre. Around the rim of the disk, and concentric with

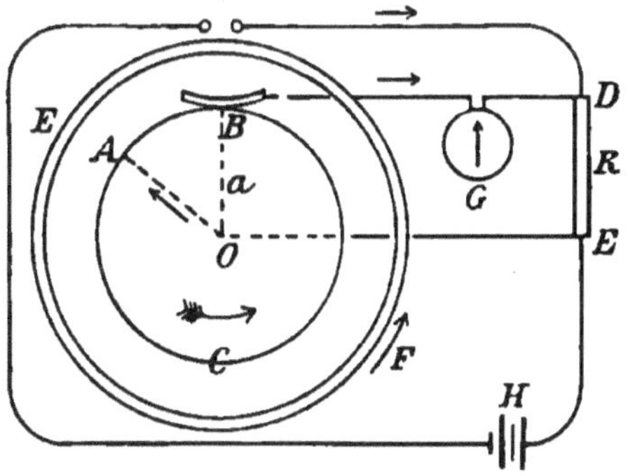

FIG. 105.

it, let there be placed a coil EF, through which flows an electrical current of strength i, arising from the voltaic battery H. This current sets up a magnetic force, whose component perpendicular to the plane of the disk may be set equal to mi, where m is a function of the distance from the centre of the disk O. If the disk turns from B to A, and if the current flows in the same direction, the electromotive force induced in the disk is directed from the centre to the periphery. A spring is placed at the point B, and is connected by the conductor $BGDEO$ with a rod which touches the disk at its centre. DE is the conductor whose resistance R is to be determined. If the current from the battery also flows through the conductor DE, we may so adjust the angular velocity of the disk that no current flows through the conductor which connects R with the disk. When this condition is attained, the galvanometer needle shows no deflection. If the electromotive force induced in the disk is represented by e, we then have $e=Ri$, where i denotes the current flowing through the resistance. To determine e, we consider the disk replaced by a ring BAC and a straight conductor OA. The circuit $OAGDEO$ is divided into the two circuits OAB and $BGDEOB$. The number of lines of force passing through the latter circuit is not changed during the motion. On the other hand, the number passing through the circuit OAB increases in unit time by

$$\int_0^a mir\omega \,.\, dr = i\omega\int_0^a mr\,.\,dr.$$

This change in the number of lines of force gives the induced electromotive force. We therefore have

$$Ri = i\omega\int_0^a mr\,.\,dr, \quad R=\omega\int_0^a mr\,.\,dr.$$

If n is the number of revolutions made by the disk in one second, we have $\omega=2\pi n$ and

$$R = n\int_0^a m\,.\,2\pi r\,.\,dr.$$

The integral gives the coefficient of mutual induction M between the coil EF and the disk ABC, or the number of lines of force which pass through the disk if a unit current is flowing in the coil. Hence we have $R=nM$. Therefore to measure the resistance we determine the number of revolutions per second which must be given to the plate in order that no current shall flow through G.

SECTION LXXXVIII. FUNDAMENTAL EQUATIONS OF INDUCTION.

We have hitherto determined the electromotive force induced in the conductor s by the change in the number N of lines of force which are enclosed by s. N represents the number of lines of force which pass through an arbitrary surface containing the conductor S; it is therefore determined by the conductor alone. Hence three quantities, F, G, H, can be so determined that the line integral

$$\int(F\,.\,dx/ds + G\,.\,dy/ds + H\,.\,dz/ds)ds$$

is equal to the surface integral $N = \iint(al+bm+cn)dS$. It is necessary for this, by VI. (f), that

(a) $$a = \partial H/\partial y - \partial G/\partial z,\ b = \partial F/\partial z - \partial H/\partial x,\ c = \partial G/\partial x - \partial F/\partial y.$$

We may also obtain these equations of condition by the assumption that, for example, dS is equal to the surface-element $dydz$, which is represented by $OBDC$ (Fig. 106). The line integral then becomes

$$Gdy + (H + \partial H/\partial y \,.\, dy)dz$$
$$- (G + \partial G/\partial z \,.\, dz)dy - Hdz$$
$$= (\partial H/\partial y - \partial G/\partial z)dydz.$$

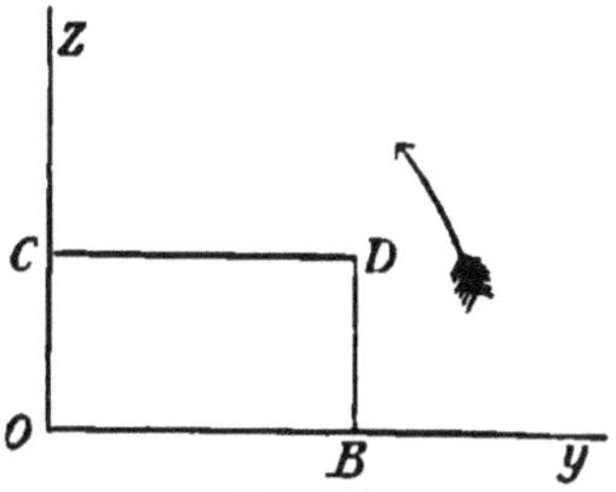

FIG. 106.

Since the surface integral in this case is $a \,.\, dydz$, we obtain the first of equations (a).

If these equations are supposed solved for F, G, H, we have

(b) $$N = \int(F \,.\, dx/ds + G \,.\, dy/ds + H \,.\, dz/ds)ds.$$

If the part of the region considered is at rest, the induced electromotive force e is determined by

(c) $$e = -dN/dt = -\int(dF/dt \,.\, dx/ds + dG/dt \,.\, dy/ds + dH/dt \,.\, dz/ds)ds.$$

If we set (d) $P = -dF/dt$, $Q = -dG/dt$, $R = -dH/dt$, we may consider P, Q, R, as the components of the electromotive force $\mathfrak{E}$, and if the integration is extended along the whole conductor, we obtain as an expression for the total induced electromotive force

(e) $$e = \int(P \,.\, dx + Q \,.\, dy + R \,.\, dz).$$

These components may also be determined directly by variation of the value of the magnetic induction, whose components are a, b, c. We thus obtain from (a) and (d)

(f) $$\begin{cases} -da/dt = \partial R/\partial y - \partial Q/\partial z, \\ -db/dt = \partial P/\partial z - \partial R/\partial x, \\ -dc/dt = \partial Q/\partial x - \partial P/\partial y. \end{cases}$$

Suppose that the electrical current, as has been remarked in LXXXI., is made up of two parts, namely, the current of conduction, which is proportional to the electromotive force, whose components are p, q, r, and the current due to the changes in the electrical polarization, whose components are f, g, h. Then for the components of the total electrical current, we obtain

$$u = p + df/dt,\ v = q + dg/dt,\ w = r + dh/dt.$$

If C represents the conductivity, we have (g) $p=CP$, $q=CQ$, $r=CR$ and (h) $f=kP/4\pi$, $g=kQ/4\pi$, $h=kR/4\pi$, where k is the dielectric constant measured in electromagnetic units. Hence we have

(i) $u=CP+k/4\pi \,.\, dP/dt$, $v=CQ+k/4\pi \,.\, dQ/dt$, $w=CR+k/4\pi \,.\, dR/dt$.

From LXXVI. the components of the magnetic induction $\mathfrak{B}$ and the components of the magnetic force $\mathfrak{H}$ are connected by the equations (k) $a=\mu\alpha$, $b=\mu\beta$, $c=\mu\gamma$, where μ is the magnetic permeability of the substances.

We have already found the following equations connecting the magnetic force and the components of current (cf. LXXX.):

(l) $4\pi u=\partial\gamma/\partial y-\partial\beta/\partial z$, $4\pi v=\partial\alpha/\partial z-\partial\gamma/\partial x$, $4\pi w=\partial\beta/\partial x-\partial\alpha/\partial y$.

SECTION LXXXIX. ELECTRO-KINETIC ENERGY.

By LXXXV. (k), the electro-kinetic energy of any system of conductors is expressed by $T=\frac{1}{2}\Sigma Ni$. By using LXXXVIII. (b) we obtain $T=\frac{1}{2}\Sigma\int(Fi\,.\,dx/ds+Gi\,.\,dy/ds+Hi\,.\,dz/ds)ds$. If u, v, w, are the components of current, a current i, in a conductor whose cross-section is A, may be expressed by $i=A\,.\,\sqrt{u^2+v^2+w^2}$, and we have also $i\,.\,dx/ds=uA$, etc. If we set $dxdydz=A\,.\,ds$, we obtain

(b) $$T=\tfrac{1}{2}\iiint(Fu+Gv+Hw)dxdydz.$$

If u, v, w are here expressed by the components of the magnetic force [LXXXVIII. (l)], we have

$$T=1/8\pi\,.\iiint[F(\partial\gamma/\partial y-\partial\beta/\partial z)+G(\partial\alpha/\partial z-\partial\gamma/\partial x)$$
$$+H(\partial\beta/\partial x-\partial\alpha/\partial y)]\,.\,dxdydz.$$

If the separate terms are integrated by parts, and the integration is extended over the whole infinite region, it follows, since at the boundary of this region α, β, γ are infinitely small of the third order, that

$$\iiint H\,.\,\partial\alpha/\partial y\,.\,dxdydz=-\iiint\alpha\,.\,\partial H/\partial y\,.\,dxdydz$$

and $$\iiint G\,.\,\partial\alpha/\partial z\,.\,dxdydz=-\iiint\alpha\,.\,\partial G/\partial z\,.\,dxdydz.$$

Analogous expressions hold for the other integrals.

By reference to LXXXVIII. we therefore obtain

(c) $$T=1/8\pi\,.\iiint(\alpha a+\beta b+\gamma c)dxdydz.$$

If no magnets or no bodies which can acquire an appreciable magnetization are present in the region, we have $a=\alpha$, $b=\beta$, $c=\gamma$,

and (d) $$T=1/8\pi\,.\iiint(\alpha^2+\beta^2+\gamma^2)dxdydz.$$

SECTION XC. ABSOLUTE UNITS.

In Physics we generally take the *centimetre*, *gram* and *second* as units of *length*, *mass* and *time* respectively. These are the units which are used in the theory of electricity. We will now proceed to express in terms of them the most important electrical and magnetic quantities. They are designated by the symbols L, M, T respectively (cf. Introduction).

(a) *The Electrostatic System of Units.*

In electrostatics the force $\mathfrak{F}$, with which two quantities of electricity e_1 and e_2 act on each other, is expressed by (cf. LIII.) $\mathfrak{F}=e_1e_2/r^2$, where r is the distance between the quantities. If $e_1=e_2=e$, we have $e=r\sqrt{\mathfrak{F}}$, and hence the dimensions of a quantity of electricity e are

$$[e]=[LL^{\frac{1}{2}}M^{\frac{1}{2}}T^{-1}]=[L^{\frac{3}{2}}M^{\frac{1}{2}}T^{-1}].$$

The electrical force F, which acts on a unit quantity of electricity, has the dimensions of the quantity e/r^2, and therefore

$$[F]=[L^{\frac{3}{2}}M^{\frac{1}{2}}T^{-1}/L^2]=[L^{-\frac{1}{2}}M^{\frac{1}{2}}T^{-1}].$$

The electrostatic potential Ψ (cf. LIV.) has the dimensions of the quantity e/r, and therefore $[\Psi]=[L^{\frac{3}{2}}M^{\frac{1}{2}}T^{-1}/L]=[L^{\frac{1}{2}}M^{\frac{1}{2}}T^{-1}]$.

The capacity C [cf. LV. (g)] has the dimensions of the quantity e/Ψ, and therefore $[C]=[L]$. The capacity, therefore, has the dimensions of a length.

The surface-density σ (cf. LIV.) is the quantity of electricity present on the unit of surface, and therefore $[\sigma]=[L^{\frac{3}{2}}M^{\frac{1}{2}}T^{-1}/L^2]=[L^{-\frac{1}{2}}M^{\frac{1}{2}}T^{-1}]$. The dimensions of surface-density are the same as those of electrical force [cf. LV. (e)]. Since the *electrical displacement* or *polarization* (cf. LXV.) is the quantity of electricity which passes through unit of surface in the dielectric, its dimensions are also $[L^{-\frac{1}{2}}M^{\frac{1}{2}}T^{-1}]$. The ratio between the electrical displacement $\mathfrak{D}$ and the electrical force is expressed by $K/4\pi$, where K is the dielectric constant. K is therefore a mere number.

The electrical energy W [cf. LXI. (a)] is measured by the product of the difference of potential and the quantity of electricity. Hence $[W]=[L^2MT^{-2}]$. These are also the dimensions of all other forms of energy.

(b) *The Electromagnetic System of Units.*

Two magnet poles which contain the quantities of magnetism μ_1 and μ_2, repel each other with the force F [cf. LXVIII. (a)], $F=\mu_1\mu_2/r^2$.

Hence quantity of magnetism has the same dimensions in the electromagnetic system as quantity of electricity in the electrostatic system. This relation holds throughout between the two systems for corresponding quantities in electrostatics and magnetism; we will therefore not consider each case separately.

The dimensions of the *strength* of the electrical current are determined from Biot and Savart's Law (cf. LXXVIII.). According to this law the force K, with which a current-element ds acts on a magnet pole containing the quantity of magnetism μ, is $K = \mu . ids . \sin\Theta/r^2$. Since K is a mechanical force, the dimensions of the current-strength i are

$$[i] = \left[\frac{LMT^{-2}L^2}{L^{\frac{3}{2}}M^{\frac{1}{2}}T^{-1} . L}\right] = [L^{\frac{1}{2}}M^{\frac{1}{2}}T^{-1}].$$

Since the *quantity of electricity* q may be considered as a product of current-strength and time, we have $[q] = [L^{\frac{1}{2}}M^{\frac{1}{2}}]$.

The electromotive force e which arises in a closed conductor has been defined [cf. LXXXV. (b)] by $e = -dN/dt$, where $N = \int\int(al + bm + cn)dS$.

Since magnetic induction, by definition, has the same dimensions as magnetic force, the dimensions of electromotive force are

$$[T^{-1} . L^{-\frac{1}{2}}M^{\frac{1}{2}}T^{-1} . L^2] = [L^{\frac{3}{2}}M^{\frac{1}{2}}T^{-2}].$$

In this system the electromotive force per unit of length of the conductor may be considered as the measure of the *electrical force*, whose components are P, Q, R. Hence the dimensions of electrical force are

$$[L^{\frac{3}{2}}M^{\frac{1}{2}}T^{-2}/L] = [L^{\frac{1}{2}}M^{\frac{1}{2}}T^{-2}].$$

According to Ohm's law the resistance is equal to the ratio between the electromotive force in the conductor and the current-strength; the dimensions of resistance are therefore

$$\left[\frac{L^{\frac{3}{2}}M^{\frac{1}{2}}T^{-2}}{L^{\frac{1}{2}}M^{\frac{1}{2}}T^{-1}}\right] = [LT^{-1}],$$

that is, *the dimensions of resistance are the same as those of velocity.*

Surface-density of electricity and *electrical polarization* have the dimensions of a quantity of electricity divided by an area, and are therefore

$$[L^{\frac{1}{2}}M^{\frac{1}{2}}/L^2] = [L^{-\frac{3}{2}}M^{\frac{1}{2}}].$$

By LXXXVIII. (b) the dielectric constant k in this system has he same dimensions as the ratio between the dielectric polarization and the electrical force, or $[L^{-\frac{3}{2}}M^{\frac{1}{2}}/L^{\frac{1}{2}}M^{\frac{1}{2}}T^{-2}] = [L^{-2}T^2]$.

(c) Comparison of the Two Systems.

If we measure a quantity of electricity electrostatically by Coulomb's torsion balance and electromagnetically by a galvanometer, we have two values for the same quantity. In the first system this quantity is expressed by $e.[L^{\frac{3}{2}}M^{\frac{1}{2}}T^{-1}]$, in the second system by $q.[L^{\frac{1}{2}}M^{\frac{1}{2}}]$; the ratio V between these expressions is $V=e/q.[LT^{-1}]$.

This ratio is therefore a velocity. It was first measured by Weber and Kohlrausch. Its value, as found by them, is $V=3,1.10^{10}$. which is very closely the velocity $3,0.10^{10}$ of light in air. Subsequent experiments have made it probable that V is actually the same as the velocity of light. It is thus shown that *an electromagnetic unit of electricity is equal to V electrostatic units.*

If a certain quantity of electricity flows through a portion AB of a conductor it produces heat in the conductor, which, considered as energy, must be independent of the system of measurement employed. The energy in electrostatic units is $e.\Psi_s$, in electromagnetic units $q\Psi_m$, where Ψ_s represents the difference of potential between A and B in electrostatic units, Ψ_m the same difference of potential in electromagnetic units. Hence we have $e\Psi_s=q\Psi_m$. We have shown that $e=Vq$, and therefore $\Psi_m=\Psi_s.V$; that is, *an electrostatic unit of potential is equal to V electromagnetic units of potential.*

If we designate electrical force in the electrostatic system by F_s, in the electromagnetic system by F_m, the difference of potential between two points, which are distant from each other by dx, is in the first system $\Psi_s=F_s.dx$, in the second system $\Psi_m=F_m.dx$. Since $\Psi_m=\Psi_s.V$ we have $F_m=V.F_s$. *Hence one unit of electrostatic force is equal to V units of electromagnetic force.*

The dielectric polarization $\mathfrak{D}$ is connected with the force F_s by the equation LXV. (d) $\mathfrak{D}=K/4\pi.F_s$, if the electrostatic system is used. In the electromagnetic system this equation takes the form $f=k/4\pi.F_m$. Since $\mathfrak{D}$ and f are quantities of electricity divided by areas, and since $e=qV$, we have $\mathfrak{D}=Vf$. Since F_m is an electrical force measured in electromagnetic units, we have

$$k=4\pi f/F_m=4\pi\mathfrak{D}/V^2F_s=K/V^2.$$

Hence in the electromagnetic system the equations connecting the components of dielectric polarization with the electrical force are

$$f=KP/4\pi V^2,\ \ g=KQ/4\pi V^2,\ \ h=KR/4\pi V^2.$$

We thus obtain ground for the assumption that *$V/\sqrt{K}$ is the velocity of light in a medium whose dielectric constant is K.*

(*d*) *Practical Units.*

In practical work these absolute units are often discarded in favour of others, called practical units. The unit of current-strength in this practical system is the *ampère*, equal to 10^{-1} electromagnetic units of current. The unit of resistance is the *ohm*, equal to 10^9 absolute units of resistance. The ohm is nearly equal to the resistance of a column of mercury, whose cross-section is 1 sq. mm. and whose length is 106,3 cm. The unit of electromotive force then follows from Ohm's law; it is called the *volt*, and is equal to 10^8 absolute units of electromotive force.

The unit of quantity of electricity is the quantity which flows in one second through any cross-section of a conductor in which the current-strength is an ampère. This unit is called a *coulomb*. The capacity of a condenser, one of whose coatings is charged with one coulomb when the difference of potential between its coatings is one volt, is called a *farad*; it is equal to $10^{-1}/10^8 = 10^{-9}$ absolute units of capacity. A body whose capacity is unity in the absolute electromagnetic system must be charged with unit quantity of electricity in order to reach unit potential. These quantities are the same as the quantity of electricity V and the potential $1/V$ in the electrostatic system. The electrostatic capacity of the body is therefore V^2. It follows from this that a farad is equal to $V^2/10^9$ electrostatic units of capacity. Since this capacity is very great, the millionth of a farad or a microfarad, is generally used as the practical unit of capacity.

CHAPTER XI.

ELECTRICAL OSCILLATIONS.

SECTION XCI. OSCILLATIONS IN A CONDUCTOR.

IF a conductor is traversed by alternating currents, that is, by currents which reverse their directions at regular intervals, we say that *electrical oscillations* exist in the conductor. Such alternating currents may be produced by induction, as in the well-known experiments of Feddersen. Let AB (Fig. 107) be a condenser whose plates are joined by conducting wires to two small metallic spheres C and D. If the condenser is so charged that A has the potential Ψ and B the potential zero, and if the distance CD is sufficiently diminished, a spark will pass from C to D. Careful investigation has proved that this spark consists of a series of sparks, which correspond to currents in opposite directions. Hence electrical oscillations will be set up by the discharge. But if the current-strength i varies in the conducting wires AC and DB, an electromotive force will be induced in them, which, from LXXXV., is equal to $-L\,.\,di/dt$, where L is the *coefficient of self-induction*.

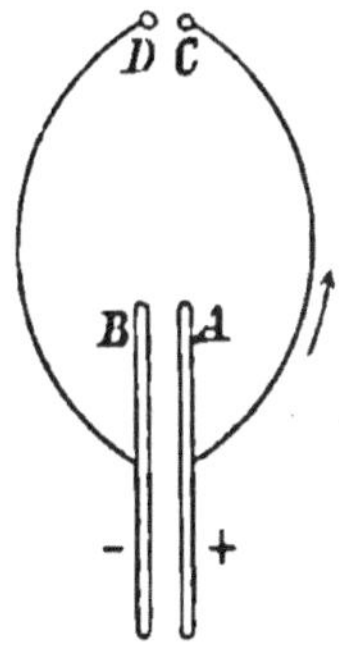

FIG. 107.

If we represent the electrical resistance in the conducting wires and in the spark-gap by r, the current-strength i is given by

$$\text{(a)} \qquad \Psi - L\,.\,di/dt = r\,.\,i.$$

If c is the capacity of the condenser in electromagnetic units, the charge q of the condenser at the time t is $c\Psi$; at the time $t+dt$, it is $c\Psi + c\,.\,d\Psi/dt\,.\,dt$. Hence we have

$$\text{(b)} \qquad c\Psi = c\Psi + c\,.\,d\Psi/dt\,.\,dt + idt\,; \quad i = -c\,.\,d\Psi/dt.$$

From (a) and (b) it follows that $Lc\,.\,d^2i/dt^2 + cr\,.\,di/dt + i = 0$. An integral of this equation is $i = A\,.\,e^{m_1t} + B\,.\,e^{m_2t}$, where m_1 and m_2 are the roots of the equation $Lcm^2 + crm + 1 = 0$. The roots of this equation are $m = -r/2L \pm \sqrt{r^2/4L^2 - 1/Lc}$. If r is sufficiently small, the roots m will be imaginary and we will have, neglecting $r^2/4L^2$,

$$m = -r/2L \pm \sqrt{-1}/\sqrt{Lc}.$$

Using the real part and the real coefficient of the imaginary part as particular solutions we can then express i by

$$i = e^{-rt/2L} . \left[A' . \sin(t/\sqrt{Lc}) + B' . \cos(t/\sqrt{Lc})\right].$$

Hence *the current-strength changes periodically and diminishes with the time.* This equation shows that the *period* T of the oscillation is given by $T/\sqrt{Lc} = 2\pi$, and hence $T = 2\pi\sqrt{Lc}$. Therefore *as the capacity of the condenser is diminished, the period T diminishes also.*

The amplitude of the oscillations is proportional to $e^{-rt/2L}$; it therefore diminishes continually as the time increases. The ratio between the amplitudes of two successive oscillations, the so-called *damping coefficient,* is $e^{-rt/2L} : e^{-r(t+T)/2L} = e^{rT/2L}$. The damping is therefore increased as the resistance of the connections is increased, and with a given period it is diminished as the self-induction L is increased. If we substitute for T its value $T = 2\pi\sqrt{Lc}$, the damping coefficient becomes $e^{\pi r\sqrt{c/L}}$. Hence the effect of damping on the oscillations is diminished when the capacity of the discharging conductor is diminished. Instead of making observations on the damping coefficient itself, we generally deal with its natural logarithm, the so-called *logarithmic decrement* δ. We have $\delta = rT/2L = \pi r . \sqrt{c/L}$. H. Hertz obtained very rapid oscillations by the use of the apparatus represented in Fig. 108. It consists of two large spheres of equal size at A and B, which are fastened on the ends of the copper rods AC and BD. The other ends of the rods AC and DB carry small spheres, separated by a distance of about 1 centimetre. The spheres are charged by the induction coil EF; the discharge occurs in the gap between the spheres C and D. If at a definite instant a current of strength i passes from A to B, and if the potentials of A and B have respectively the values Ψ_1 and Ψ_2, then $\Psi_1 - \Psi_2 - L . di/dt = ri$. If the capacity of each of the large spheres is represented by c, we have $c_1 . d\Psi_1/dt = -i$, $c_1 . d\Psi_2/dt = i$, or by using $c = \frac{1}{2}c_1$, $i = -c . d(\Psi_1 - \Psi_2)/dt$. The current-strength is therefore given by the same differential equation as before, and we obtain from it the same expression for the period T.

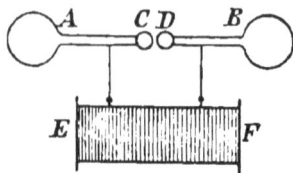

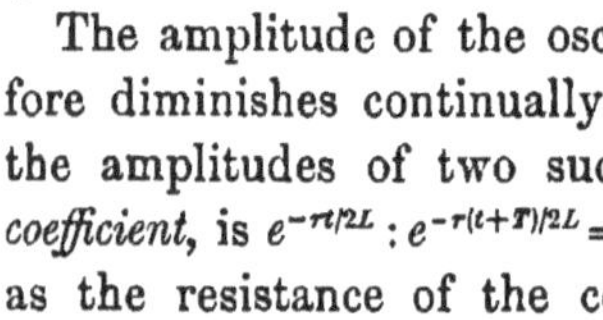

FIG. 108.

SECTION XCII. CALCULATION OF THE PERIOD.

In order to determine the period, we must first determine the *coefficient of self-induction.* The method of determining coefficients of self-induction for closed conducting circuits has already been given. In the present case we have to deal partly with actual currents in the cylindrical conductors, partly with polarization or displacement currents in the surrounding dielectric. The principal effect must be due to the induction in the conductor itself, since in it the distance between the inducing and the induced currents is least. The full treatment of the question would be very difficult, because the current-strengths in the different parts of the cross-section are not equal. We will therefore neglect these differences in the subsequent discussion, and calculate the coefficient of self-induction L in a cylinder on the assumption that the current-strength in all parts of the cross-section is the same. According to F. Neumann, the electromotive force E induced in a conductor s' by the action of a variable current i flowing through another conductor s, is determined by the variation of the integral Li, where $L = \iint \cos\epsilon/r \,.\, ds ds'$. In this integral, which is to be taken over all the elements of both conductors, ϵ denotes the angle between ds and ds', and r the distance between ds and ds.

Let AB and CD be two parallel lines (Fig. 109) which together with AC and BD form a rectangle. We set $AB = CD = l$, and $CF = s'$.

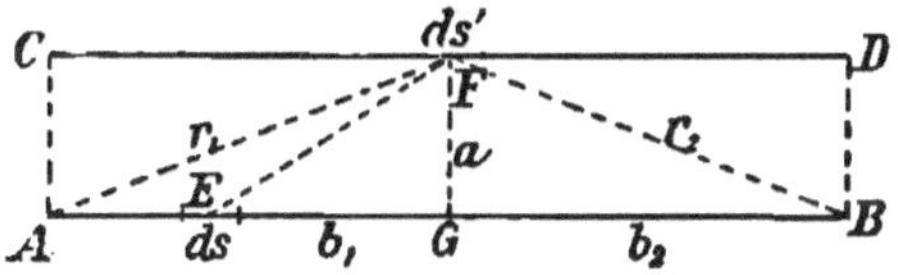

FIG. 109.

In the present case the integral which is to be evaluated becomes $\int_0^l \int_0^l ds \,.\, ds'/r$, because (Fig. 109) $\cos\epsilon = 1$. In order to find first the value of the integral $P = \int_0^l ds/r$ we draw FG perpendicular to CD and AB, and write $FG = a$. If $AG = b_1$, $BG = b_2$, $FA = r_1$, $FB = r_2$, we have $P = \int_0^{b_1} ds/r + \int_0^{b_2} ds/r$, where s is the distance of any point on AB from G. Now since

$$\int ds/r = \int ds/\sqrt{a^2+s^2} = \log\mathrm{nat}(s/a + \sqrt{1+s^2/a^2})$$

we obtain

$$P = \log\mathrm{nat}[(r_1+b_1)(r_2+b_2)/a^2]$$
$$= \log\mathrm{nat}[(AF+AG)(BF+BG)/a^2].$$

If a is very small in comparison with l, we may write

$$AF = AG = s' \text{ and } FB = GB = l - s',$$

and have $P = \log \text{nat}[4s'(l-s')/a^2]$. We then calculate the value of the integral $\int P\,.\,ds'$ and obtain

$$\int_0^l P\,.\,ds' = 2l\,.\,[\log \text{nat}(2l/a) - 1].$$

We will now calculate *the coefficient of self-induction of a wire with circular cross-section.* Let AB (Fig. 110) be the cross-section of a cylindrical conductor, of length l and radius R. Let the current-density u be constant throughout the cross-section. The current-strength i is then given by $i = r^2\pi u$. We will first consider the inductive action due to a filament D whose cross-section is ds; this filament is supposed to act on a line which is parallel to the axis of the cylinder, and passes through the point c. If $DC = a$, the inductive action is obtained by the variation of the integral

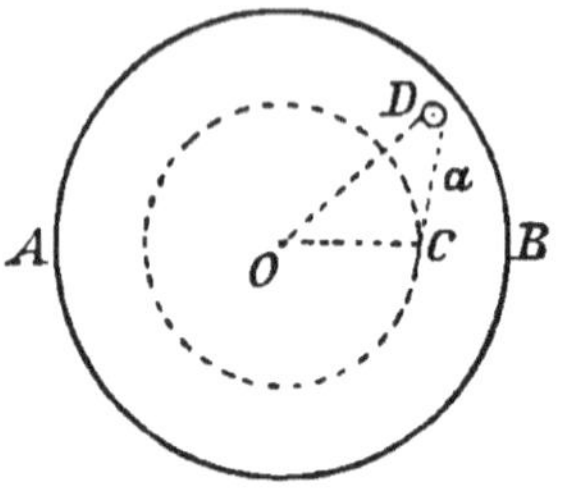

FIG. 110.

$$u\,.\,dS\,.\,2l\big(\log \text{nat}(2l/a) - 1\big) = u\,.\,dS\,.\,(M + N \log \text{nat}\, a),$$

where, for the sake of conciseness, we use M and N as symbols for the quantities $M = 2l(\log \text{nat}\, 2l - 1)$ and $N = -2l$. We must distinguish between two cases: $OD = r$ may be either greater or less than $OC = r_1$. First let $r > r_1$. The elements dS may be taken so as to form the surface of a ring whose area is $2\pi r\,.\,dr$. From the demonstration of XIII. $\log a$ equals the logarithm of half the sum of the greatest and least values which a can take. These values are respectively $r + r_1$ and $r - r_1$. The mean value required is therefore $\log r$. Hence we have the integral

$$u\int_{r_1}^{R} 2\pi r\,.\,dr\,.\,(M + N \log \text{nat}\, r) = \pi u\{M(R^2 - r_1^2) + N[R^2 \log \text{nat}\, R - r_1^2 \log \text{nat}\, r_1 - \tfrac{1}{2}(R^2 - r_1^2)]\}.$$

For that part of the cylinder whose distance from the axis is less than r_1 the mean value of the greatest and least values of a will equal r_1. Hence the integral for this part is

$$u\int_0^{r_1} 2\pi r\,.\,dr\,.\,(M + N \log \text{nat}\, r_1) = \pi u(M r_1^2 + N r_1^2\,.\,\log \text{nat}\, r_1).$$

The sum of both integrals is

$$\pi u\big\{MR^2 + N\big(R^2 \log \text{nat}\, R - \tfrac{1}{2}(R^2 - r_1^2)\big)\big\}.$$

In order to obtain the mean value of this quantity for all the filaments composing the cylinder, we need only find the mean value of r_1^2, since all the other quantities are constant. But since

$$1/\pi R^2 . \int_0^R r_1^2 . 2\pi r_1 . dr_1 = \tfrac{1}{2}R^2,$$

it follows that the mean value sought is $\pi u R^2(M + N(\log R - \frac{1}{4}))$. If we introduce into this equation the values

$$M = 2l(\log 2l - 1) \text{ and } N = -2l$$

and set $\pi u R^2 = i$, we obtain for the quantity, the variation of which gives the self-induction, $2li(\log(2l/R) - \frac{3}{4})$. We therefore obtain for the quantity L, $L = 2l(\log(2l/R) - \frac{3}{4})$. In Hertz's investigation, $l = 150$, $R = 0{,}25$ and therefore $L = 1902$, where all lengths are expressed in centimetres.

In order to calculate the period of oscillation, we will next determine the capacity of a sphere with a radius of 15 cm., such as Hertz used. If Q is its charge and Ψ its potential, the capacity C in electrostatic units is $C = Q/\Psi$. Representing the charge and the potential in electromagnetic units by Q' and Ψ' respectively, and using $V = 3 . 10^{10}$, the velocity of light in vacuo, we have $Q = VQ'$ and $\Psi = \Psi'/V$. The capacity c in electromagnetic units is therefore $c = Q'/\Psi' = C/V^2$. The period of oscillation is then given by $T = 2\pi\sqrt{LC}/V$.

Using the symbol introduced at the end of XCI., we have $c = \frac{1}{2}c_1$, where c_1 is the capacity in electromagnetic units of each of the two large spheres. Hence we must set $C = 15/2$, and obtain $T = 2{,}5/10^8$ seconds. The corresponding wave length in air is

$$2{,}5 . 10^{-8} . 3 . 10^{10} = 750 \text{ cm.}$$

SECTION XCIII. THE FUNDAMENTAL EQUATIONS FOR ELECTRICAL INSULATORS OR DIELECTRICS.

Maxwell shows that it follows, as a consequence of his theoretical views of the nature of electricity, that a change in the electrical polarization of the dielectric can set up electrical oscillations. The results which he obtained are so important that we will consider some of them here. For this purpose we will follow Hertz in substituting in the fundamental equations of LXXXVIII. electrostatic units for the electrical quantities, while the magnetic quantities shall be measured in electromagnetic units. The quantity of electricity which is displaced by the electrical force at a point in the dielectric through a surface-element which stands perpendicular to the direction of the force, is, according to LXV., equal to $K/4\pi$ multiplied by the magnitude of the force F. Representing the components of the electrical displacement by f, g, h, and the components of the electrical force by X, Y, Z, we have $f = KX/4\pi$, $g = KY/4\pi$, $h = KZ/4\pi$. If the component X increases by dx in the time dt, the quantity of electricity df flows through unit area in the direction of the x-axis; the component u of the current-strength in the dielectric is equal to df/dt, and we have

(a) $$u = K/4\pi \,.\, \partial X/\partial t, \quad v = K/4\pi \,.\, \partial Y/\partial t, \quad w = K/4\pi \,.\, \partial Z/\partial t.$$

The equations LXXX. (a) express the fact that the work done by the magnetic forces in consequence of the movement of a unit pole about the current is equal to the current-strength multiplied by 4π. If the current-strength is measured in electrostatic units, we have

(b) $$4\pi u/V = \partial\gamma/\partial y - \partial\beta/\partial z, \quad 4\pi v/V = \partial\alpha/\partial z - \partial\gamma/\partial x, \quad 4\pi w/V = \partial\beta/\partial x - \partial\alpha/\partial y,$$

since the electromagnetic unit of quantity of electricity equals V electrostatic units.

The electromotive force induced equals $-dN/dt$, if N represents the number of lines of force enclosed by the circuit. Since the electromotive force in electromagnetic units equals the electromotive force in electrostatic units multiplied by V, we have from LXXXVIII. (f) and (k)

(c) $$\begin{cases} -\mu/V \,.\, \partial\alpha/\partial t = \partial Z/\partial y - \partial Y/\partial z; \\ -\mu/V \,.\, \partial\beta/\partial t = \partial X/\partial z - \partial Z/\partial x; \\ -\mu/V \,.\, \partial\gamma/\partial t = \partial Y/\partial x - \partial X/\partial y. \end{cases}$$

From (a) and (b) we obtain

(d) $$\begin{cases} K/V \,.\, \partial X/\partial t = \partial\gamma/\partial y - \partial\beta/\partial z; \\ K/V \,.\, \partial Y/\partial t = \partial\alpha/\partial z - \partial\gamma/\partial x; \\ K/V \,.\, \partial Z/\partial t = \partial\beta/\partial x - \partial\alpha/\partial y. \end{cases}$$

If we now set $J = \partial X/\partial x + \partial Y/\partial y + \partial Z/\partial z$, we obtain from (c) and (d) $\mu K/V^2 \,.\, \partial^2 X/\partial t^2 = \nabla^2 X - \partial J/\partial x$. If we are dealing with a region in which K is constant, $K/4\pi \,.\, J = \partial f/\partial x + \partial g/\partial y + \partial h/\partial z$. If there is no electrical distribution in the region, we have, from LXVI. (d), $J = 0$; and hence

(e) $$\mu K/V^2 . \partial^2 X/\partial t^2 = \nabla^2 X; \quad \mu K/V^2 . \partial^2 Y/\partial t^2 = \nabla^2 Y; \quad \mu K/V^2 . \partial^2 Z/\partial t^2 = \nabla^2 Z.$$

These equations, in connection with equations (c) and (d), give

(f) $$\mu K/V^2 \,.\, \partial^2\alpha/\partial t^2 = \nabla^2\alpha; \quad \mu K/V^2 \,.\, \partial^2\beta/\partial t^2 = \nabla^2\beta; \quad \mu K/V^2 \,.\, \partial^2\gamma/\partial t^2 = \nabla^2\gamma.$$

at the same time $\partial\alpha/\partial x + \partial\beta/\partial y + \partial\gamma/\partial z = 0$ from LXXVI., if μ is constant.

From LXVII. (g), the electrical energy W is expressed by

(g) $$W = 1/8\pi \,.\, \iiint K(X^2 + Y^2 + Z^2)dxdydz.$$

The electrokinetic energy T, according to LXXXIX. (c), is

(h) $$T = 1/8\pi \,.\, \iiint \mu(\alpha^2 + \beta^2 + \gamma^2)dxdydz.$$

SECTION XCIV. PLANE WAVES IN THE DIELECTRIC.

We will now investigate the movement of plane waves in a dielectric. Let the plane waves be parallel to the yz-plane. The components of the electrical force are then functions of x only, and from the equations XCIII. (e), we have

$$\mu K/V^2 \,.\, \partial^2 X/\partial t^2 = \partial^2 X/\partial x^2 \,;\quad \mu K/V^2 \,.\, \partial^2 Y/\partial t^2 = \partial^2 Y/\partial x^2 \,;$$
$$\mu K/V^2 \,.\, \partial^2 Z/\partial t^2 = \partial^2 Z/\partial x^2.$$

At the same time also $\partial X/\partial x + \partial Y/\partial y + \partial Z/\partial z = 0$. Since Y and Z are independent of y and z, we have $\partial X/\partial x = 0$, and since, in this case, the only forces which occur are periodic, $X = 0$. *The direction of the electrical force is therefore parallel to the plane of the wave.* By a rotation of the coordinate axes we can make the y-axis coincide with the resultant of the components Y and Z. We therefore need to discuss only the equation $\mu K/V^2 \,.\, \partial^2 Y/\partial t^2 = \partial^2 Y/\partial x^2$. The integral of this equation is (a) $Y = b\sin[2\pi/T \,.\, (t - x/\omega)]$, where T is the *period of oscillation* and ω the *velocity of propagation.* The differential equation is satisfied if $\omega = V/\sqrt{\mu K}$. For vacuum $\mu = 1$, $K = 1$; hence V is the velocity of propagation of plane electrical oscillations in vacuo. For ordinary transparent bodies, $\mu = 1$. The velocity of propagation in such bodies is therefore $V/\sqrt{K}$. Maxwell assumed that electrical oscillations are identical with light waves. It has been shown by experiment that $\omega = V/N$, where N represents the index of refraction of the dielectric. The electromagnetic theory of light gives $\omega = V/\sqrt{K}$; hence we have $K = N^2$, that is, *the specific inductive capacity of a medium is equal to the square of its index of refraction.* The fact that this theorem holds for a large number of bodies is a strong confirmation of Maxwell's hypothesis. From this hypothesis almost all of the properties of light can be deduced.

According to XCIII. (c) we have, under the above conditions, $\alpha = 0$, $\beta = 0$, and

(b) $$\mu\gamma = Vb/\omega \,.\, \sin[2\pi/T \,.\, (t - x/\omega)] = Nb\sin[2\pi/T \,.\, (t - x/\omega)].$$

The direction of the magnetic force is therefore parallel to the plane of the wave and perpendicular to the direction of the electrical force.

We will supplement this discussion by the following examination of the relation between the electrical and the magnetic forces. Let an electrical force act in the yz-plane, parallel to the axis Oy (Fig. 111), and suppose it to increase uniformly from zero to Y_0 in one second. In consequence of this an electrical current v is set up in the same direction, and because $\partial Y/\partial t = Y_0$, we have from XCIII. (a),

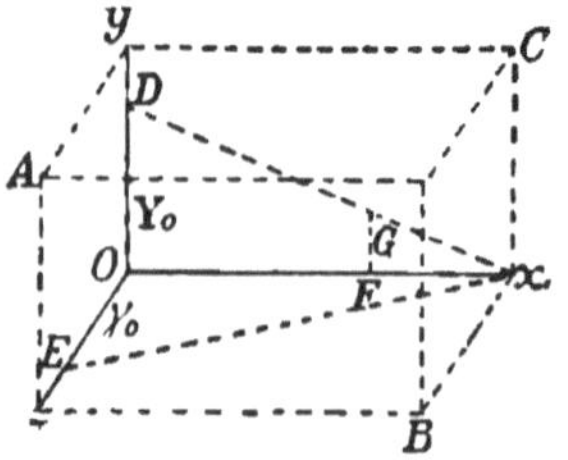

FIG. 111.

(c) $$v = K/4\pi \,.\, Y_0.$$

This electrical current will set up magnetic forces, which are parallel to the z-axis. We will assume that the magnetic force increases uniformly from zero to γ_0. In consequence of this an electromotive force will be induced in the surrounding region. We will assume that the electrical and magnetic actions advance in one second over the distance $Ox = \omega$ (Fig. 111).

The magnetic force decreases uniformly from $x = 0$ to $x = \omega$; the same statement holds for the electrical force $OD = Y_0$. The electrical current, on the other hand, has the same strength everywhere between O and x. This is explained by remarking that the electrical force at a point F, whose distance from x equals $1/n \,.\, Ox$, has acted only during $1/n$ seconds, and has, during this time interval, increased from 0 to $1/n \,.\, Y_0$; its increase in one second is therefore equal to Y_0.

Let a unit pole move in the rectangular path $OzBxO$ (Fig. 111). The magnetic force γ_0 acts only in the path Oz and acts in the direction of motion; hence the work done by the magnetic forces is equal to $\gamma_0 \,.\, Oz$. The quantity of current, measured in electromagnetic units, which the unit pole has encircled, is $v/V \,.\, Oz \,.\, Ox$. From LXXX. we have therefore $\gamma_0 \,.\, Oz = 4\pi v \,.\, Oz \,.\, Ox/V$, or because $Ox = \omega$, we obtain (d) $V\gamma_0 = 4\pi v\omega$.

The electromotive force, measured in electromagnetic units, which is induced by the motion about a closed path, is $e = -dN/dt$, if N represents the number of lines of force enclosed by the path. We have therefore $N = -\int e \,.\, dt$. The mean value of the electromotive force in the direction Oy is $\frac{1}{2} \,.\, Y_0 \,.\, V$, in electromagnetic units. The value of the before-mentioned integral, extended over the rectangular path $OyCx$, is $\frac{1}{2} Y_0 V \,.\, Oy$. The mean value of the magnetic force perpendicular to the surface $OyCz$ is $\frac{1}{2}\gamma_0$; hence the mean value of the magnetic induction is $\frac{1}{2} \,.\, \mu\gamma_0$. We therefore have the equation

$$\tfrac{1}{2} \,.\, Y_0 V \,.\, Oy = \tfrac{1}{2}\mu\gamma_0 \,.\, Oy \,.\, Ox,$$

and hence (e) $VY_0 = \mu\gamma_0\omega$, $NY_0 = \mu\gamma_0$, if the index of refraction N is substituted for V/ω. In this connection it must be noticed that the wave is propagated in the direction of the x-axis, that on our assumption the electrical force acts in the direction of the y-axis, and that then the magnetic force acts in the direction of the z-axis. *Hence if the right hand is held so as to point in the direction in which the wave is propagated, with the palm turned toward the direction of the electrical force, the thumb will point in the direction of the magnetic force.* If we represent the magnetic force by M and the electrical force by F, $NF = \mu M$. From (c) and (d) it follows that $V\gamma_0 = KY_0\omega$. Substituting in (e) the value of γ_0, we obtain $V^2 = \mu K\omega^2$. Hence the velocity of propagation is $\omega = V/\sqrt{\mu K}$. From (a) and (b) it follows that the relations (e) between the electrical and magnetic forces hold also in the case of plane waves. In vacuo both forces have the same numerical value.

SECTION XCV. THE HERTZIAN OSCILLATIONS.

H. Hertz succeeded in producing very rapid oscillations in a straight conductor, which also caused oscillations in the surrounding dielectric. We can form some idea of the nature of these oscillations in the following manner, due to Hertz:

Let the middle point of a conductor coincide with the origin of coordinates, and let the oscillations take place along the z-axis. The magnetic lines of force are then circles, whose centres lie on the z-axis. The electrical lines of force have a more complicated form. We start with the differential equation XCIII. (f) for the magnetic forces. For the sake of conciseness we set $V/K\mu = \omega$.

We first investigate an integral of the differential equation

$$\text{(a)} \qquad 1/\omega^2 \,.\, \partial^2 u/\partial t^2 = \nabla^2 u,$$

on the hypothesis that u is a function of t and of $r = \sqrt{x^2 + y^2 + z^2}$. We have then, from XV. (l), $\nabla^2 u = 1/r \,.\, \partial^2(ru)/\partial r^2$, and therefore $1/\omega^2 \,.\, \partial^2(ru)/\partial t^2 = \partial^2(ru)/\partial r^2$. If we set $k = 2\pi/T$ and $l = 2\pi/T\omega$, where T is a constant, then (b) $u = a/r \,.\, \sin(kt - lr)$ is a particular integral of the differential equation. The function u, as well as its differential coefficients taken with respect to x, y and z, therefore satisfies the differential equations XCIII. (f) for the components of magnetic force α, β, γ. In the case under consideration $\gamma = 0$, and hence we have $\partial\alpha/\partial x + \partial\beta/\partial y = 0$. As the simplest solution of the differential equation we obtain (c) $\alpha = -\partial^2 u/\partial t\partial y$, $\beta = \partial^2 u/\partial t\partial x$, where the differentiation with respect to t is introduced for use in the subsequent calculation. From (c) we obtain $\alpha = -\partial^2 u/\partial t\partial r \,.\, y/r$; $\beta = \partial^2 u/\partial t\partial r \,.\, x/r$. The resultant magnetic force is therefore

$$M = \partial^2 u/\partial t\partial r \,.\, \sqrt{x^2+y^2}/r = \partial^2 u/\partial t\partial r \,.\, \sin\Theta,$$

where Θ is the angle between the radius vector from the origin and the z-axis. The force M is perpendicular to the plane which contains the point considered and the z-axis.

If we set $kt - lr = \phi$, we have $M = ka(l \,.\, \sin\phi/r - \cos\phi/r^2) \,.\, \sin\Theta$. If r is very small in comparison with $1/l = \omega T/2\pi$, we have

$$M = -ka \,.\, \cos kt \,.\, \sin\Theta/r^2,$$

that is, *the force is determined by Biot and Savart's law, the oscillations in the conductor acting like a current element.*

For greater distances the magnetic force is

$$\text{(d)} \qquad M = 4\pi^2 a/T^2\omega r \,.\, \sin\left[2\pi/T \,.\, (t - r/\omega)\right] \,.\, \sin\Theta.$$

Hence the magnetic waves proceed forward in space with the velocity of light.

We will now calculate the electrical forces. From (c) and XCIII. (d) we obtain

$$KX/V = -\partial^2 u/\partial x\partial z\,;\quad KY/V = -\partial^2 u/\partial y\partial z\,;\quad KZ/V = \partial^2 u/\partial x^2 + \partial^2 u/\partial y^2.$$

Since u depends only on r and t, the carrying out of the differentiations gives

$$\text{(e)} \qquad \begin{cases} KX/V = (-\partial^2 u/\partial r^2 + 1/r \,.\, \partial u/\partial r) \,.\, xz/r^2\,; \\ KY/V = (-\partial^2 u/\partial r^2 + 1/r \,.\, \partial u/\partial r) \,.\, yz/r^2\,; \\ KZ/V = \partial^2 u/\partial r^2 \,.\, (r^2 - z^2)/r^2 + \partial u/\partial r \,.\, (r^2 + z^2)/r^3. \end{cases}$$

The electrical force R in the direction r is $(Xx + Yy + Zz)/r$, and hence from (e), $KR/V = 2/r \,.\, \partial u/\partial r \,.\, \cos\Theta = -2a(l\cos\phi/r^2 + \sin\phi/r^3)\cos\Theta$. If $R = 0$, the electrical force is tangent to a sphere whose radius is determined from the equation $\operatorname{tg}(kt - lr) = -lr$. Let the next spherical wave have the radius r'; then $\operatorname{tg}(kt - lr') = -lr'$. From this it follows that $\operatorname{tg} l(r' - r) = l(r' - r)/(1 + l^2 rr')$. If the radius of the waves is very large we may set $l(r' - r) = \pi$. But since $l = 2\pi/\lambda$, where $\lambda = T\omega$, we have $r' - r = \frac{1}{2}\lambda$. We thus obtain at last equidistant spherical waves.

SECTION XCVI. POYNTING'S THEOREM.

Let there be an electrical current i flowing from A to B through a long cylindrical conductor (Fig. 112) of circular cross-section. There is then a magnetic force M acting at every point in the region around the conductor, given by the equation $2\pi r . M = 4\pi i$, where $r = OC$, the distance of the point from the axis of the cylinder. We therefore have $M = 2i/r$. The equipotential surfaces of the electrostatic field of force within the conductor are planes perpendicular to the axis of the conductor. Outside the conductor they are likewise perpendicular to the axis, at least in the vicinity of the conductor. The equipotential surfaces of the magnetic field of force are planes which, like the plane OF, contain both the direction of the electrical force and the axis of the conductor. Let the electrical force in the surface of the conductor and in its vicinity be F'. If we designate by S the cross-section of the conductor, and by C its conductivity, we have from Ohm's law $i/S = CF'$. The heat produced in the conductor during one second is determined as follows: If the quantity of electricity i flows through the conductor, whose length we may call l, the electrical force does work equal to $F'il$. If J represents the mechanical equivalent of heat, the quantity of heat thus developed equals $F'il/J$. The work done by the electromotive force is therefore $F'il = \frac{1}{2} . MrF'l$.

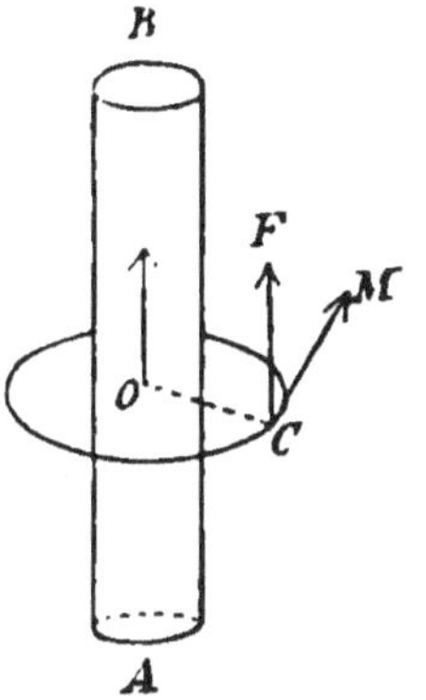

FIG. 112.

Now Poynting assumes that this quantity of energy enters the conductor through its surface. That portion of the surface which is to be considered is equal to $2\pi rl$. The quantity of energy which enters the conductor through unit area on its surface is therefore $1/4\pi . F'M$.

This quantity of energy moves in the direction CO, which is determined by the intersection of the electrical and magnetic equipotential surfaces. The relation of the direction in which the energy is propagated to the directions of the electrical and magnetic forces is determined in the same way as that given for the propagation of waves in XCIV.

The electrical force is here measured in electromagnetic units; if we express it in electrostatic units, and represent it by F, then $F' = VF$. The quantity of energy which enters in one second through a unit area of the surface, which is parallel to the directions both of the electrical and of the magnetic forces, equals $V/4\pi . F . M$.

We will now treat a more general case. Represent the magnetic force at a point in the region by M, the electrical force at the same point by F, and the angle between the two forces by (M, F). We assume that the quantity of energy passing in one second through unit area, which is parallel to the directions of M and F, is

$$V/4\pi \,.\, M \,.\, F \,.\, \sin(MF).$$

The direction in which the energy flows makes angles with the coordinate axes whose cosines are l, m, n. We have then

$$l = (\gamma Y - \beta Z)/MF \sin(MF)\,; \quad m = (\alpha Z - \gamma X)/MF \sin(MF)\,;$$
$$n = (\beta X - \alpha Y)/MF \sin(MF),$$

if α, β, γ, are the components of the magnetic force M, and X, Y, Z, the components of the electrical force F. From these equations it follows that $l\alpha + m\beta + n\gamma = 0$; $lX + mY + nZ = 0$; $l^2 + m^2 + n^2 = 1$. Hence the energy flows in a direction which is perpendicular to the directions both of the magnetic and of the electrical forces.

If we represent by E_x, E_y, E_z, the components of the flow of energy in the directions of the coordinate axes, we have

$$E_x = V/4\pi \,.\, MF \sin(MF) \,.\, l.$$

Hence we obtain the equations

$$E_x = V/4\pi \,.\, (\gamma Y - \beta Z)\,; \quad E_y = V/4\pi \,.\, (\alpha Z - \gamma X)\,; \quad E_z = V/4\pi \,.\, (\beta X - \alpha Y).$$

The energy present in a parallelepiped, whose edges are dx, dy, dz, increases, in the time dt, by an amount equal to

$$-(\partial E_x/\partial x + \partial E_y/\partial y + \partial E_z/\partial z)dxdydzdt.$$

Hence the increase of energy A which a unit of volume receives in the unit of time, is $A = -(\partial E_x/\partial x + \partial E_y/\partial y + \partial E_z/\partial z)$. If we substitute in these equations the values previously given for E_x, E_y, E_z, we have

$$A = V/4\pi \,.\, [X(\partial\gamma/\partial y - \partial\beta/\partial z) + Y(\partial\alpha/\partial z - \partial\gamma/\partial x)$$
$$+ Z(\partial\beta/\partial x - \partial\alpha/\partial y)] - V/4\pi \,.\, [\alpha(\partial Z/\partial y - \partial Y/\partial z)$$
$$+ \beta(\partial X/\partial z - \partial Z/\partial x) + \gamma(\partial Y/\partial x - \partial X/\partial y)].$$

By the help of equations (c) and (d) of XCIII., we obtain from this

$$A = K/8\pi \,.\, d(X^2 + Y^2 + Z^2)/dt + \mu/8\pi \,.\, d(\alpha^2 + \beta^2 + \gamma^2)/dt.$$

On comparing this expression with those given in (g) and (h) XCIII. for the electrostatic and electrokinetic energies, we see that A represents the total increase of energy which the unit of volume receives in the unit of time. Poynting's Theorem is thus proved, provided the dielectric is not in motion. The demonstration can easily be extended to the conductor if we use the developments of LXXXVIII.

and remember that a part of the energy absorbed by the conductor is transformed into heat.

We will now apply Poynting's theorem to a simple case. According to XCIV., we may, in the case of vibrations in a plane, express the electrical force F, there designated by Y, by $F = b \,.\, \sin[2\pi/T \,.\, (t - x/\omega)]$, and the magnetic force M, there designated by γ, by

$$M = Vb/\mu\omega \,.\, \sin[2\pi/T \,.\, (t - x/\omega)].$$

During any complete vibration there passes through unit area the quantity of energy

$$V^2b^2/4\pi\mu\omega \,.\, \int_0^T \sin^2[2\pi/T \,.\, (t - x/\omega)]dt = V^2b^2T/8\mu\pi\omega.$$

We may set $V = \omega$ and $\mu = 1$, and obtain for the quantity of energy which passes in one second through a unit area perpendicular to the plane of the wave, the quantity $Vb^2/8\pi$.

The quantity of heat which a square centimetre receives in one minute from the light of the sun is equal to about three gram-calories. This quantity of heat corresponds to the energy $3 \,.\, 4{,}2 \,.\, 10^7/60$ developed in one second. If we now set $V = 3 \,.\, 10^{10}$, we have $b = 0{,}04$. Since the unit of electrical force in the electrostatic system equals 300 volts, we obtain for the maximum electrical force of sunlight 12 volts per centimetre. The maximum magnetic force is 0,04, and amounts therefore to a fifth of the horizontal intensity of the earth's magnetism in middle latitudes.

The experimental basis for the mathematical treatment of electrostatics was given by Coulomb. Poisson handled a number of problems in electrostatics and gave the general method for their solution. Sir William Thomson (Lord Kelvin) also treated the same problems in part by a new and very ingenious method; his papers are specially recommended to the student (Reprint of Papers, 2nd Ed., 1884). Faraday (1837) developed new views of electrical polarization or displacement. On the foundation of these concepts, Maxwell constructed his development of the theory of electricity (*Treatise on Electricity and Magnetism*, 1873). Helmholtz treated electrostatics in a different way and solved new problems. His papers may be found in Wiedemann's *Annalen*.

The theory of magnetism advanced parallel with the theory of electrostatics. The same authors, and sometimes even the same works, deal with both subjects.

Ampère in his *Théorie Mathématique des Phénomènes Electrodynamiques*, Paris, 1825, discussed the theory of electrical currents. This work forms the foundation for all the recent development of that theory. The new concepts of the magnetic and inductive actions of electrical currents, developed by Faraday, were given a mathematical form by Maxwell in his work: *Treatise on Electricity and Magnetism*, 1873. We have followed Maxwell's methods in dealing with these topics. On the other hand Gauss, W. Weber, F. E. Neumann, Kirchhoff, and Lorenz, proceed from Ampère's theory.

We are indebted to William Thomson and G. Kirchhoff (Poggendorff's *Annalen*, 121) for the theory of electrical oscillations in conductors. Maxwell and Lorenz showed that electrical oscillations may also exist in the dielectric. By the investigations of H. Hertz, the theory of electrical oscillations has been given such extension and significance, that no one can predict the consequences to which it may lead.

CHAPTER XII.

REFRACTION OF LIGHT IN ISOTROPIC AND TRANSPARENT BODIES.

SECTION XCVII. INTRODUCTION.

As the number of facts discovered by the study of light increases, and as additional relations are found between light and other natural phenomena, it becomes increasingly difficult to construct a theory of light. According to the *emission theory*, which in its main features may be attributed to Newton, and which was handled mathematically by him, energy is transferred by minute bodies, called light corpuscles, which pass from the luminous to the illuminated body. It was supposed that these light corpuscles carried with them not only their kinetic energy but also another kind of energy, to which the luminous effects were due. In the last century the emission theory was sufficient to explain the phenomena then known. But its development could not keep pace with the advances of experimental knowledge; this became evident early in this century in connection with the great discoveries in optics which were made by Young, Fresnel and Malus. In opposition to this theory, Fresnel developed his first form of the *wave theory*, which originated with Huygens; in this form of the theory, the light waves were supposed to be longitudinal. According to the wave theory, the space between the luminous and illuminated bodies is filled with a material medium. By the action of the particles of this medium on each other, the energy emanating from the luminous body is propagated from particle to particle through this medium to the illuminated body. Hence energy resides in the medium during the transfer of light from one body to the other. The wave theory has many advantages over the emission theory. Notably the phenomena of interference are explained by it in a perfectly natural way. It succeeds also in explaining some of the phenomena of double refraction. But the explanation of the polarization of light by this theory offered difficulties which could be overcome only by the assumption that the direction of the light vibrations are perpendicular to the direction of the rays. Since Fresnel retained the idea that the medium in which the light vibrations are propagated, the *ether*, is a fluid, he encountered an obstinate resistance to his new form of the wave theory; Poisson rightly maintained that transverse vibrations can never be propagated in a fluid. Although the wave theory, in its original form, was somewhat open to criticism, and in many respects was insufficient, in that, among other matters, it could not explain the *dispersion of light*, yet it was a decided advance on the emission theory.

Since the phenomena of optics cannot be explained on the assumption that light is due to vibrations in an elastic medium, not even when this medium is supposed to be a solid, we must endeavour to explain them in another way. Among recent efforts in this direction, the *electromagnetic theory of light*, developed by Maxwell, has special advantages. In Maxwell's view, light is also a wave motion, but it consists of periodic electrical currents or displacements, which take the place of the etherial vibrations of Fresnel's theory. Maxwell determined on this assumption the velocity of light in vacuo and in transparent bodies, and reached conclusions which agree very well with the facts. Polarization and double refraction can also be readily explained by Maxwell's theory, and it has even been applied successfully to the study of dispersion.

Since Fresnel's formulas are of great importance for our subsequent study, we will develop them at the outset. We may here recall briefly the principal laws of light which hold for isotropic and perfectly transparent bodies. The knowledge of these laws is necessary for the deduction of Fresnel's formulas, but is not sufficient.

I. Light is propagated in any one medium with a velocity which depends on the wave length of the light, but not on its intensity. The velocity of light has different values in different media.

II. If a ray of light falls on a plane surface, separating two different media, both *refraction* and *reflection* occur at this surface. All three rays—that is, the incident, the refracted, and the reflected—lie in the same plane, which is perpendicular to the refracting surface. If α represents the angle of incidence, β the angle of refraction, and γ the angle of reflection, we have

$$\gamma = \alpha \text{ and } \sin \alpha / \sin \beta = N.$$

The *index of refraction* N is constant for homogeneous light.

III. If ω represents the velocity of light in the medium containing the reflected ray, and ω' its velocity in that containing the refracted ray, we have $N = \omega/\omega'$, and therefore $\sin\alpha/\sin\beta = \omega/\omega'$.

IV. Light can be considered a wave motion in a medium, called the ether. It is a matter of indifference whether we here consider the bodies themselves, or an unknown substance, or perhaps changes in the electrical or magnetic condition of the bodies. We wish only to indicate that the luminous motion may be expressed by one or more terms of the form $a\cos(2\pi t/T + \phi)$, where a is the *amplitude*, T the *period of vibration*, ϕ the *phase*, and t the *variable time*. The *intensity of light* is then expressed by a^2.

V. The motion of the ether is perpendicular to the direction of the ray of light; that is, the vibrations are *transverse*. Either the motion takes place always in the same direction, in which case the ray is *rectilinearly polarised*, or two or more simultaneous rectilinear motions may give the ether particles a motion in a curve, which is in general an *ellipse*. Rays of light of this sort are said to be *elliptically polarized*. If the path of the ether particle is a circle, the light is *circularly polarised*. Fresnel's conception of *natural light* was that its vibrations were also perpendicular to the direction of the ray and rectilinear, but that they changed their directions many times, and on no regular plan, in a very short interval of time.

SECTION XCVIII. FRESNEL'S FORMULAS.

Suppose the plane surface OP (Fig. 113) to be the surface of separation of two transparent isotropic media. We represent the velocity of light in the medium above the surface of separation OP by ω, and that in the medium below the surface by ω'. If N represents the index of refraction of the ray of light in its passage from the first to the second medium, we have $\omega = N\omega'$. We select the point O in the bounding plane as the origin of a system of rectangular co-ordinates, and draw the x-axis perpendicularly upward and the y-axis in the plane of incidence, that is, the plane passed through the normal to the surface at the point of incidence and the incident ray SO. The z-axis is therefore perpendicular to the plane of incidence. Further, let SO be the incident, OT the reflected, and OB the refracted ray. We designate the angle of incidence by α, the angle of refraction by β. The amplitude of the vibrations of the incident ray may be called u_1, and that of the vibrations of the refracted and reflected rays u_2 and u_3 respectively. The planes of vibration of these rays make angles with the plane of incidence, which are represented by ϕ_1, ϕ_2, ϕ_3 respectively. The components of motion along the coordinate axes are ξ_1, η_1, ζ_1, for the incident ray; ξ_2, η_2, ζ_2, for the refracted ray; ξ_3, η_3, ζ_3, for the reflected ray.

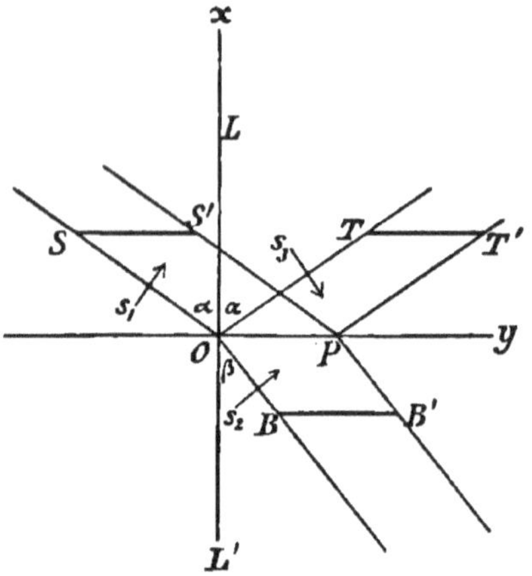

FIG. 113.

It is further advantageous to introduce symbols for the components of motion which lie in the plane of incidence and are perpendicular to the direction of the rays. These components of motion for the three rays are designated by s_1, s_2, s_3, respectively. We then obtain the following equations:

$$\text{(a)} \quad \begin{cases} \xi_1 = s_1 \sin \alpha, & \eta_1 = s_1 \cos \alpha, \quad u_1^2 = s_1^2 + \zeta_1^2, \quad \operatorname{tg}\phi_1 = \zeta_1/s_1\,; \\ \xi_2 = s_2 \sin \beta, & \eta_2 = s_2 \cos \beta, \quad u_2^2 = s_2^2 + \zeta_2^2, \quad \operatorname{tg}\phi_2 = \zeta_2/s_2\,; \\ \xi_3 = -s_3 \sin \alpha, & \eta_3 = s_3 \cos \alpha, \quad u_3^2 = s_3^2 + \zeta_3^2, \quad \operatorname{tg}\phi_3 = \zeta_3/s_3. \end{cases}$$

In order to express s_2, s_3, and ζ_2, ζ_3, in terms of s_1 and ζ_1, we must make certain assumptions with regard to the behaviour of the light at its passage from one medium to another.

I. Fresnel assumed first, that *no light is lost by reflection and refraction*, or, *the sum of the intensities of the reflected and refracted light is equal to that of the incident.* This law is only a statement of the *law of the conservation of energy* in a particular case, it being merely the assertion that the kinetic energy of the incident ray is equal to that of the reflected and refracted rays. Let *OPSS'* (Fig. 113) be a cylinder, the area of whose base *OP* is *A*, and whose slant height *SO* is equal to ω, the velocity of light. If we represent the density of the vibrating medium by ρ, the kinetic energy L_1 of the light contained in the cylinder considered is $L_1 = \frac{1}{2} . \rho . \omega \cos a . A . u_1^2$.

After the lapse of a second this kinetic energy is divided between the reflected and refracted rays. The kinetic energy L_3 in the reflected ray is $L_3 = \frac{1}{2} . \rho . \omega \cos a . A . u_3^2$, and the kinetic energy L_2 in the refracted ray is $L_2 = \frac{1}{2}\rho' . \omega' \cos \beta . A . u_2^2$, when ρ' represents the density of the vibrating medium below the bounding surface. Therefore, on the assumption made by Fresnel, we have

$$L_1 = L_2 + L_3 \text{ or } \rho\omega(u_1^2 - u_3^2)\cos a = \rho'\omega' . u_2^2 . \cos \beta.$$

Taking into account the relations $\omega = N . \omega'$ and $\sin a = N . \sin \beta$, we may give this equation the form

(b) $$\rho . (u_1^2 - u_3^2) . \sin a . \cos a = \rho' u_2^2 . \sin \beta . \cos \beta.$$

If the vibrations of the ray lie in the plane of incidence, $u_1 = s_1$, and we obtain

(c) $$\rho . (s_1^2 - s_3^2) . \sin a . \cos a = \rho' . s_2^2 . \sin \beta . \cos \beta.$$

But if the vibrations of the ray are perpendicular to the plane of incidence, we have $u_1 = \zeta_1$, and hence

(d) $$\rho(\zeta_1^2 - \zeta_3^2) \sin a . \cos a = \rho' . \zeta_2^2 . \sin \beta . \cos \beta.$$

II. Fresnel assumed, secondly, *that the components of the vibrations, which are parallel to the bounding surface, and on either side of it, are equal.* If the vibrations lie in the plane of incidence, we have, on this assumption, $\eta_1 + \eta_3 = \eta_2$, or

(e) $$(s_1 + s_3) \cos a = s_2 . \cos \beta.$$

But if the vibrations are perpendicular to the plane of incidence, we obtain (f) $\zeta_1 + \zeta_3 = \zeta_2$. It follows from (c) and (e) that

(g) $$\begin{cases} s_2 = s_1 . 2\rho . \sin a . \cos a / (\rho . \sin a . \cos \beta + \rho' . \cos a . \sin \beta), \\ s_3 = s_1(\rho . \sin a . \cos \beta - \rho' . \cos a . \sin \beta)/(\rho . \sin a . \cos \beta + \rho' . \cos a . \sin \beta), \end{cases}$$

and from (d) and (f) that

(h) $$\begin{cases} \zeta_2 = \zeta_1 . 2\rho . \sin a . \cos a / (\rho . \sin a . \cos a + \rho' . \sin \beta . \cos \beta), \\ \zeta_3 = \zeta_1 . (\rho . \sin a . \cos a - \rho' . \sin \beta . \cos \beta)/(\rho . \sin a . \cos a + \rho' . \sin \beta . \cos \beta). \end{cases}$$

III. Since the relation between ρ and ρ' is entirely unknown, Fresnel was compelled to make a third assumption, so he assumed *that the elasticity of the ether is everywhere the same, but that its density differs in different media.* On the other hand, F. E. Neumann assumed *that the density of the ether is the same in all media, but that its elasticity is different in different media.* Fresnel's assumption was natural, because he considered the ether as a gaseous body, but this is not justified, as has already been remarked. He further assumed that ω and ω'

can be expressed in the same way as in the theory of elasticity [cf. XXXV. (k)], and therefore set $\omega=\sqrt{\mu/\rho}$, $\omega'=\sqrt{\mu'/\rho'}$. Now, on Fresnel's assumption, $\mu=\mu'$, and hence, (i) $\rho'/\rho=\omega^2/\omega'^2=N^2$. By the use of the third assumption the equations (g) and (h) can be given the form

$$\text{(k)}\qquad \begin{cases} s_2=s_1\,.\,2\cos\alpha\,.\,\sin\beta/(\sin(\alpha+\beta)\cos(\alpha-\beta)), \\ \zeta_2=\zeta_1\,.\,2\cos\alpha\,.\,\sin\beta/\sin(\alpha+\beta), \\ s_3=-s_1\mathrm{tg}(\alpha-\beta)/\mathrm{tg}(\alpha+\beta), \\ \zeta_3=-\zeta_1\sin(\alpha-\beta)/\sin(\alpha+\beta). \end{cases}$$

These formulas are due to Fresnel.

Experiment alone can decide as to the value of these formulas. It follows, from the expression for s_3, that $s_3=0$, if $\alpha+\beta=\frac{1}{2}\pi$ or if $\mathrm{tg}\alpha=N$. Brewster showed that the light which is polarized perpendicularly to the plane of incidence, according to the definition of Malus, is not reflected when $\mathrm{tg}\alpha=N$. This value of the angle α is called the *angle of polarization. It is that at which the reflected and refracted rays are at right angles to each other.* This conclusion agrees with experiment, if we make the assumption that *the vibrations of polarized light are perpendicular to the plane of polarization.* On the whole, Fresnel's formulas agree very well with the results of experiments on the intensity of the reflected light.

In the notation already employed, the plane of vibration of the incident ray makes the angle ϕ_1 with the plane of incidence, the corresponding angle for the reflected ray is ϕ_3. Brewster found that $\mathrm{tg}\phi_3=\mathrm{tg}\phi_1\,.\cos(\alpha-\beta)/\cos(\alpha+\beta)$. This follows from Fresnel's formulas, since $\mathrm{tg}\phi_3=\zeta_3/s_3=\zeta_1\cos(\alpha-\beta)/s_1\cos(\alpha+\beta)=\mathrm{tg}\phi_1.\cos(\alpha-\beta)/\cos(\alpha+\beta)$. This agreement argues for the correctness of Fresnel's formulas.

Fresnel assumed that the elasticity of the vibrating medium is the same on both sides of the refracting surface. We have seen that this assumption is to some extent arbitrary. On the other hand F. E. Neumann assumed that $\rho=\rho'$. We obtain on this latter assumption from (g) and (h)

$$\text{(l)}\qquad \begin{cases} s_2=2\sin\alpha\,.\,\cos\alpha\,.\,s_1/\sin(\alpha+\beta), \\ \zeta_2=2\sin\alpha\,.\,\cos\alpha\,.\,\zeta_1/\sin(\alpha+\beta)\cos(\alpha-\beta), \\ s_3=\sin(\alpha-\beta)\,.\,s_1/\sin(\alpha+\beta),\ \ \zeta_3=\mathrm{tg}(\alpha-\beta)\,.\,\zeta_1/\mathrm{tg}(\alpha+\beta). \end{cases}$$

These equations agree with the results of experiment, on the assumption *that the vibrations occur in the plane of polarization,* as well as those of Fresnel.

We will now consider the components of motion along the normal during the passage of light from one medium to the other. This component above the bounding surface is $\xi_1 + \xi_3$, that below it is ξ_2. It follows from (a) and (g) that

$$\xi_1 + \xi_3 = 2\rho' . \sin \alpha . \cos \alpha . \sin \beta . s_1/(\rho . \sin \alpha . \cos \beta + \rho' . \cos \alpha . \sin \beta),$$
$$\xi_2 = 2\rho . \sin \alpha . \cos \alpha . \sin \beta . s_1/(\rho . \sin \alpha . \cos \beta + \rho' . \cos \alpha . \sin \beta).$$

From these equations it follows that $\xi_1 + \xi_3 = \zeta_2 . \rho'/\rho$. If we assume with Neumann that $\rho' = \rho$, we have $\xi_1 + \xi_3 = \xi_2$; that is, *the components of vibration perpendicular to the bounding surface are equal above and below it.* On the other hand we obtain from Fresnel's assumption (m) $\xi_1 + \xi_3 = N^2 . \xi_2$.

Fresnel's equations agree fully with experiment only when the index of refraction is about 1,5; it is in this case only that $s_3 = 0$ at the angle of polarization. In other cases s_3 is a minimum, but does not vanish. Several attempts have been made to explain this fact. Thus, for example, Lorenz assumed that the passage of the light from one medium to another occurs through an extremely thin intervening layer of varying density, and that therefore the change of density is not discontinuous.

SECTION XCIX. THE ELECTROMAGNETIC THEORY OF LIGHT.

In XCIV. it was proved that electrical vibrations are propagated in vacuo and in a great number of insulators with the velocity of light. This fact suggests the assumption that light consists of electrical vibrations. It was also shown that, when the wave is plane, the electrical and magnetic forces lie in the wave front. In the most simple case the electrical force F is perpendicular to the magnetic force M. If the permeability μ of the medium is equal to 1, which is approximately the case for most dielectrics, we have (XCIV.), (a) $M = NF$, where N is the index of refraction.

We will now develop the usual expressions for the reflected and transmitted light, considering first the boundary conditions. Since no free magnetism is present, and since the electrical current strength is everywhere finite, the magnetic force varies continuously during the passage from one medium to another. We take the refracting surface as the yz-plane, and the x-axis as the normal to it. Hence if α, β, γ are the components of the magnetic force on one side of the refracting surface and α', β', γ' those on the other side, we have

(b) $$\alpha = \alpha', \ \beta = \beta', \ \gamma = \gamma'.$$

The electrical force arises partly from induction and partly from the free electricity on the refracting surface. Let σ denote the density on this surface, X, Y, Z the components of the electrical force immediately above, and X', Y', Z' those immediately below this surface. We then have, from LXVI.,

$$4\pi\sigma = X - KX', \quad Y = Y', \quad Z = Z'.$$

The components of the electrical force which are parallel to the refracting surface vary continuously during the passage from one side of the surface to the other.

Let SO, OB, and OT (Fig. 114) be the directions of the incident, refracted, and reflected rays respectively. *Suppose the direction of the magnetic force to lie in the plane of incidence, and therefore the electrical force to be perpendicular to that plane.* The direction of the

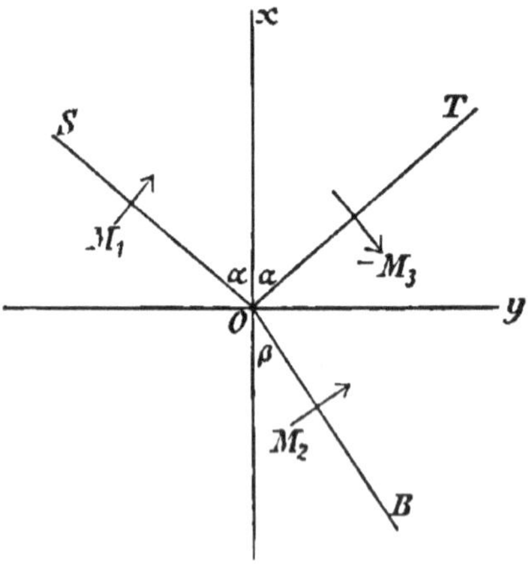

FIG. 114.

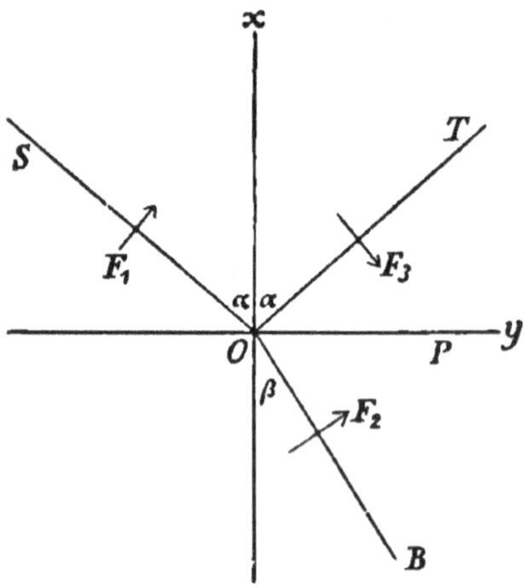

FIG. 115.

electrical force may be found by the rule given in XCIV. Since the magnetic force varies continuously, we obtain, by referring to the directions indicated in Fig. 114.

(d) $$M_1 \,.\, \cos\alpha - M_3 \,.\, \cos\alpha = M_2 \,.\, \cos\beta,$$

(e) $$M_1 \,.\, \sin\alpha + M_3 \,.\, \sin\alpha = M_2 \sin\beta.$$

Since the electrical force also varies continuously, we have

(f) $$F_1 + F_3 = F_2.$$

But from equation (a) $M_1 = F_1$, $M_3 = F_3$, and $M_2 = NF_2$, if $N = \sin\alpha/\sin\beta$ is the ratio between the velocities in the first and second media.

Hence equations (e) and (f) are identical. We obtain from (f) and (d)

$$\text{(g)} \qquad \begin{cases} F_2 = F_1 \,.\, 2\cos\alpha \,.\, \sin\beta/\sin(\alpha+\beta)\,; \\ F_3 = -F_1 \sin(\alpha-\beta)/\sin(\alpha+\beta). \end{cases}$$

Now let the electrical force be parallel to the plane of incidence. If the electrical forces are positive in the directions indicated in Fig. 115, the positive directions of the magnetic forces may be obtained by the rule given in XCIV.

The boundary conditions are

$$\text{(h)} \qquad F_1 \,.\, \cos\alpha + F_3 \,.\, \cos\alpha = F_2 \,.\, \cos\beta,$$

$$\text{(i)} \qquad M_1 - M_3 = M_2.$$

In this case also $M_1 = F_1$, $M_3 = F_3$, $M_2 = NF_2$, so that (k)

$$F_2 = F_1 . \cos\alpha . \sin\beta/\sin(\alpha+\beta) . \cos(\alpha-\beta) \text{ and } F_3 = -F_1 . \operatorname{tg}(\alpha-\beta)/\operatorname{tg}(\alpha+\beta).$$

Equations (g) and (k) correspond to Fresnel's equations XCVIII. (k). Hence, *the electromagnetic theory of light leads to the same results as those which are contained in Fresnel's formulas, provided that the electrical force is parallel to the direction of the vibrations assumed by Fresnel.*

We can further show, by the help of Poynting's theorem (XCVI.), that the energy which in a given time is transported to the refracting surface by the incident ray is equal to that which is carried away from it in the reflected and refracted rays. Since in this case the electrical and magnetic forces are perpendicular, the energy passing through unit area equals $VMF/4\pi$. The bounding surface S receives the quantity of energy $1/4\pi \,.\, VM_1F_1 \,.\, S\cos\alpha$ in unit time. We may write similar expressions for the energies of the reflected and transmitted rays, and have the relation

$$1/4\pi \,.\, VM_1F_1 \,.\, S \,.\, \cos\alpha = 1/4\pi \,.\, VM_2F_2 \,.\, S \,.\, \cos\beta + 1/4\pi \,.\, VM_3F_3 \,.\, S \,.\, \cos\alpha,$$

or $(M_1F_1 - M_3F_3) \,.\, \cos\alpha = M_2F_2 \,.\, \cos\beta$. By reference to the relations between the electrical and magnetic forces, we obtain

$$(F_1^2 - F_3^2)\cos\alpha = NF_2^2 \,.\, \cos\beta.$$

It appears from equations (g) and (k) that this equation is satisfied if the electrical force is either perpendicular or parallel to the plane of incidence.

SECTION C. EQUATIONS OF THE ELECTROMAGNETIC THEORY OF LIGHT.

If we at first consider only bodies in which there is no absorption of light, and in which the velocity of light is the same in all directions, we have, from XCIII. (e), the differential equations of the electrical force,

$$\text{(a)}\qquad \begin{cases} 1/\omega^2 . \partial^2 X/\partial t^2 = \nabla^2 X, & 1/\omega^2 . \partial^2 Y/\partial t^2 = \nabla^2 Y, \\ 1/\omega^2 . \partial^2 Z/\partial t^2 = \nabla^2 Z, & \partial X/\partial x + \partial Y/\partial y + \partial Z/\partial z = 0. \end{cases}$$

The boundary conditions are obtained by remarking that the components of the electrical and magnetic forces parallel to the bounding surface are equal on both sides of it. Therefore, if the x-axis is perpendicular to the refracting surface, we have

$$\text{(b)}\qquad Y = Y', \ Z = Z'; \ \beta = \beta', \ \gamma = \gamma'.$$

The last two conditions (b) may be put in the form [XCIII. (e)],

$$\text{(c)}\qquad \begin{cases} \partial X/\partial z - \partial Z/\partial x = \partial X'/\partial z - \partial Z'/\partial x; \\ \partial Y/\partial x - \partial X/\partial y = \partial Y'/\partial x - \partial X'/\partial y. \end{cases}$$

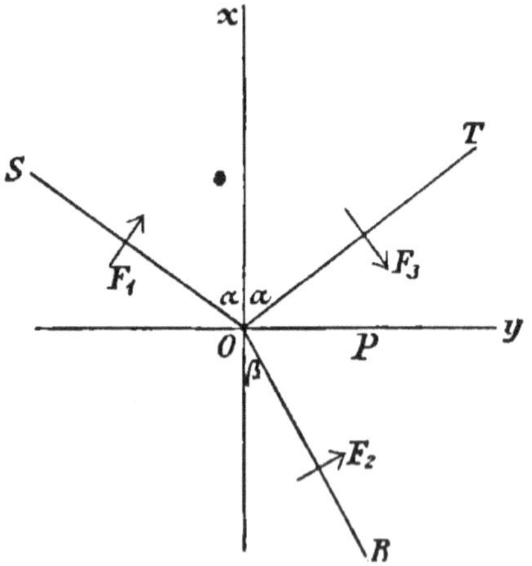

FIG. 116.

Let us suppose that a plane wave moves in a direction which makes angles with the axes whose cosines are l, m, n. Let the electrical force f at the origin be expressed by $f = F . \cos(2\pi t/T)$. The electrical force at a point whose coordinates are x, y, z is then

$$f = F . \cos\left[2\pi/T\left(t - (lx + my + nz)/\omega\right)\right].$$

If the direction of the electrical force makes angles with the axes whose cosines are λ, μ, ν, we have $X = \lambda f$, $Y = \mu f$, $Z = \nu f$. These expressions satisfy the equations (b); that they may also satisfy the last of equations (a) we must have $l\lambda + m\mu + n\nu = 0$, that is, *the direction of the electrical force is perpendicular to the direction of propagation.*

Let OP (Fig. 116) represent the refracting surface, and SO, OB and OT the incident, refracted, and reflected rays respectively. The system of coordinates is drawn in the same way as in XCVIII. For the incident wave, in which *the direction of the electrical force lies in the plane of incidence*, we can set

$$l = -\cos a,\quad m = \sin a,\quad n = 0;$$
$$\lambda = \sin a,\quad \mu = \cos a,\quad \nu = 0.$$

We have, for the reflected rays,

$$l = \cos a,\quad m = \sin a,\quad n = 0;$$
$$\lambda = -\sin a,\quad \mu = \cos a,\quad \nu = 0,$$

and for the refracted rays,

$$l = -\cos\beta,\quad m = \sin\beta,\quad n = 0;$$
$$\lambda = \sin\beta,\quad \mu = \cos\beta,\quad \nu = 0.$$

If the direction of the electrical force is perpendicular to the plane of incidence we have $\lambda = 0,\ \mu = 0,\ \nu = 1$.

If F_1, F_2, F_3 are the electrical forces in the incident, refracted, and reflected rays, when they lie in the plane of incidence, and if Z_1, Z_2, Z_3 are the electrical forces in the same rays when they are perpendicular to the plane of incidence, we have

$$\text{(c)}\quad \begin{cases} X = F_1 . \sin a . \cos V_1,\quad Y = F_1 . \cos a . \cos V_1,\quad Z = Z_1 . \cos V_1, \\ V_1 = \left[2\pi/T . \left(t - (-x\cos a + y\sin a)/\omega\right)\right]. \end{cases}$$

We obtain for the refracted ray, if ω' is the velocity of light in the second medium,

$$\text{(d)}\quad \begin{cases} X = F_2 . \sin\beta . \cos V_2,\quad Y = F_2 . \cos\beta . \cos V_2,\quad Z = Z_2 . \cos V_2, \\ V_2 = \left[2\pi/T . \left(t - (-x\cos\beta + y\sin\beta)/\omega'\right)\right]. \end{cases}$$

For the reflected ray we have

$$\text{(e)}\quad \begin{cases} X = -F_3 . \sin a . \cos V_3,\quad Y = F_3 . \cos a . \cos V_3,\quad Z = Z_3 . \cos V_3, \\ V_3 = \left[2\pi/T . \left(t - (x\cos a + y\sin a)/\omega\right)\right]. \end{cases}$$

To simplify the calculation we replace the trigonometrical form

$$\text{(f)}\qquad \cos k\left(t - (-x\cos a + y\sin a)/\omega\right)$$

by the expression (g) $e^{ki(t-(-x\cos a+y\sin a)/\omega)}$, where $i = \sqrt{-1}$ and $k = 2\pi/T$. In the final result we use only the real part of (g), namely (f). Both expressions satisfy the same differential equation, and therefore in calculations one of them may be replaced by the other.

If the refraction occurs at a plane surface, we may replace the expressions (c), (d) and (e) by the following:

$$\text{(h)}\qquad \begin{cases} X = F_1 . \sin a . e^{ki(t-(-x\cos a+y\sin a)/\omega)}, \\ Y = F_1 . \cos a . e^{ki(t-(-x\cos a+y\sin a)/\omega)}, \\ Z = Z_1 . e^{ki(t-(-x\cos a+y\sin a)/\omega)}; \end{cases}$$

(i)
$$\begin{cases} X = F_2 . \sin\beta . e^{ki(t-(-x\cos\beta+y\sin\beta)/\omega')}, \\ Y = F_2 . \cos\beta . e^{ki(t-(-x\cos\beta+y\sin\beta)/\omega')}, \\ Z = Z_2 . e^{ki(t-(-x\cos\beta+y\sin\beta)/\omega')} ; \end{cases}$$

(k)
$$\begin{cases} X = -F_3 . \sin\alpha . e^{ki(t-(x\cos\alpha+y\sin\alpha)/\omega)}, \\ Y = F_3 . \cos\alpha . e^{ki(t-(x\cos\alpha+y\sin\alpha)/\omega)}, \\ Z = Z_3 . e^{ki(t-(x\cos\alpha+y\sin\alpha)/\omega)}. \end{cases}$$

These equations express the components of the electrical force for the incident, refracted, and reflected rays.

In order to satisfy the conditions (b) and (c) it is necessary that $\sin\alpha/\omega = \sin\beta/\omega'$. Since the velocities of propagation ω and ω' are constant, we can set $N = \sin\alpha/\sin\beta$, where N is the *index of refraction.*

From equations (b) we have

(l) $$(F_1 + F_3)\cos\alpha = F_2 . \cos\beta ; \quad Z_1 + Z_3 = Z_2.$$

From (c) we obtain

(m) $$(F_1 - F_3)\sin\beta = F_2 . \sin\alpha ; \quad (Z_1 - Z_3)\cos\alpha . \sin\beta = Z_2 . \sin\alpha . \cos\beta.$$

From (l) and (m) we obtain Fresnel's equations [XCVIII. (k)] for the reflected and refracted waves. The problem is solved when β is not imaginary. β becomes imaginary when $\sin\beta > 1$, and therefore when $\sin\alpha > N$. In this case we must use the complete expressions (i) and (k).

If the electrical force is perpendicular to the plane of incidence, the reflected wave is determined by the real part of the expression

(n) $$-Z_1 . \sin(\alpha - \beta)/\sin(\alpha + \beta) . e^{ki(t-(x\cos\alpha+y\sin\alpha)/\omega)}.$$

We get this expression by the use of the last of equations XCVIII. (k). But since $\cos\beta = \sqrt{1 - \sin^2\alpha/N^2}$, and therefore

(o) $$Ni . \cos\beta = \sqrt{\sin^2\alpha - N^2}$$

we have

$$-\sin(\alpha - \beta)/\sin(\alpha + \beta) = (\cos\alpha + i\sqrt{\sin^2\alpha - N^2})/(\cos\alpha - i\sqrt{\sin^2\alpha - N^2}).$$

If we now set

(p) $$\cos\alpha = C . \cos\tfrac{1}{2}\gamma, \quad \sqrt{\sin^2\alpha - N^2} = C . \sin\tfrac{1}{2}\gamma,$$

we obtain

(q) $$\operatorname{tg}\tfrac{1}{2}\gamma = \sqrt{\sin^2\alpha - N^2}/\cos\alpha, \quad C^2 = 1 - N^2 ;$$

(r) $$-\sin(\alpha - \beta)/\sin(\alpha + \beta) = e^{i\gamma}.$$

Hence the real part of the expression (n) is

(s) $$Z_1 . \cos[k(t - (x\cos\alpha + y\sin\alpha)/\omega) + \gamma].$$

In this case *the reflection is total,* since the component Z_1 appears in the expression for the incident as well as in that for the reflected wave. But while, in the case of ordinary reflection, no difference of phase arises between the two waves, we have in this case a difference of phase γ, which may be determined from (q).

If the electrical forces for the incident wave are parallel to the plane of incidence, we determine the real part of the expression

(t) . $$-F_1 . \operatorname{tg}(a-\beta)/\operatorname{tg}(a+\beta) . e^{ki(t-(x\cos a+y\sin a)/\omega)}.$$

Using equation (o), we have

$$\operatorname{tg}(a-\beta)/\operatorname{tg}(a+\beta) = (N^2.\cos a + i\sqrt{\sin^2 a - N^2})/(N^2.\cos a - i\sqrt{\sin^2 a - N^2}).$$

If we set

(u) $$N^2.\cos a = D.\cos\tfrac{1}{2}\delta;\ \sqrt{\sin^2 a - N^2} = D.\sin\tfrac{1}{2}\delta,$$

so that (v) $\operatorname{tg}\tfrac{1}{2}\delta = \sqrt{\sin^2 a - N^2}/N^2\cos a$, $D^2 = N^4 . \cos^2 a - N^2 + \sin^2 a$, we obtain (x) $\operatorname{tg}(a-\beta)/\operatorname{tg}(a+\beta) = e^{i\delta}$. Hence the real part of the expression (t) is (y) $-F_1\cos[k(t-(x\cos a+y\sin a)/\omega)+\delta]$.

The reflection is therefore total. To determine the difference of phase δ between the reflected and incident waves, we may use equation (v). We obtain from (q) and (v) $\operatorname{tg}\tfrac{1}{2}(\delta-\gamma) = \sqrt{\sin^2 a - N^2}/\sin a \operatorname{tg} a$. If a_0 is the *limiting angle of total reflection* or *critical angle*, we have $N = \sin a_0$, and hence (z) $\operatorname{tg}\tfrac{1}{2}(\delta-\gamma) = \sqrt{\sin(a+a_0).\sin(a-a_0)}/\sin a \operatorname{tg} a$.

Since δ and γ are not equal, a linearly polarized ray of light, in which the vibrations make any angle with the plane of incidence, is elliptically polarized after reflection.

If the electrical force is perpendicular to the plane of incidence, the *transmitted light* is determined by the real part of the expression

(α) $$Z_1 . 2\cos a\sin\beta/\sin(a+\beta) . e^{ki(t-(-x\cos\beta+y\sin\beta)/\omega')}.$$

Referring to (p), we have

$$2\cos a\sin\beta/\sin(a+\beta) = 2\cos a/C . e^{1/2 i\gamma}$$

and $$(-x\cos\beta+y\sin\beta)/\omega' = (ix\sqrt{\sin^2 a - N^2}+y\sin a)/\omega.$$

Hence the real part of (α) is

(β) $$2\cos a/C . e^{kx\sqrt{\sin^2 a - N^2}/\omega} . Z_1 . \cos[k(t-y\sin a/\omega)+\tfrac{1}{2}\gamma].$$

Since $C^2 = 1 - N^2$, we obtain $4\cos^2 a/(1-N^2) . Z_1^2 . e^{4\pi x\sqrt{\sin^2 a - N^2}/\lambda}$, for the *intensity of the transmitted light*, where λ denotes the wave length.

The expression shows that, in this case also, a motion exists which corresponds to the refracted ray in the case of ordinary reflection; it is, however, appreciable only within a very small distance from the refracting surface.

Similar results are obtained in the investigation *of the refracted ray if the electrical force of the incident light is parallel to the plane of incidence.*

Remark: In order to obtain the real part of an expression of the form (n) we may use the following method. The expression (n) is thrown into the form

$$(A+Bi)\,.\,e^{i\Psi}=(A+Bi)(\cos\Psi+i\sin\Psi).$$

The real part of this is $R=A\,.\cos\Psi-B\,.\sin\Psi$. Now if we set $A=C\,.\cos\gamma$, $B=C\,.\sin\gamma$, we have (γ) $R=C\,.\cos(\Psi+\gamma)$, where C and γ are determined by

$$(\delta)\qquad \begin{cases} C^2=A^2+B^2=(A+Bi)(A-Bi);\\ \operatorname{tg}\gamma=-i[A+Bi-(A-Bi)]/[A+Bi+(A-Bi)]. \end{cases}$$

The expression (n) then takes the form

$$Z_1e^{i\Psi}\,.\,(\cos a+i\sqrt{\sin^2 a-N^2})/(\cos a-i\sqrt{\sin^2 a-N^2}).$$

In the case considered, therefore,

$$A+Bi=Z_1\,.\,(\cos a+i\sqrt{\sin^2 a-N^2})/(\cos a-i\sqrt{\sin^2 a-N^2}).$$

If $+i$ and $-i$ are interchanged, we obtain

$$A-Bi=Z_1(\cos a-i\sqrt{\sin^2 a-N^2})/(\cos a+i\sqrt{\sin^2 a-N^2}).$$

By multiplication of the two expressions we obtain $C^2=A^2+B^2=Z_1^{\,2}$. From (δ) it further follows that

$$\operatorname{tg}\gamma=2\cos a\sqrt{\sin^2 a-N^2}/(\cos^2 a-\sin^2 a+N^2).$$

This equation may be also obtained from (q).

Section CI. Refraction in a Plate.

We will consider the case of a plane wave of light falling on a plane parallel glass plate, whose thickness is a and whose index of refraction is N. We can determine the intensities of the reflected and transmitted light in the following way. We choose one surface A of the plate as the yz-plane, and draw the positive x-axis outward from this surface. Let a represent the angle of incidence, β the angle of refraction, ω and ω' the velocities of the light inside and outside the plate. A part of the refracted ray is reflected toward the surface A at a point E of the surface B. This part is again divided at the surface A, a part of it passing through that surface in the direction FG, while the other part is again reflected toward B. The light is thus reflected within the plate repeatedly. Since the plate is bounded on both sides by the same medium, the angle of exit is equal to the angle of incidence a. Now plane waves which move in the same direction may be compounded into a single plane wave.

Besides the incident wave we have to consider four others, namely, the wave reflected from A, the wave passing through B, and two waves in the plate itself.

I. We will first consider the case *in which the electrical force of the incident wave is perpendicular to the plane of incidence.* The component of the electrical force outside A is expressed as in the former paragraph by

$$Z = Z_1 . e^{ki(t-(-x\cos a+y\sin a)/\omega)} + Z_3 . e^{ki(t-(x\cos a+y\sin a)/\omega)}.$$

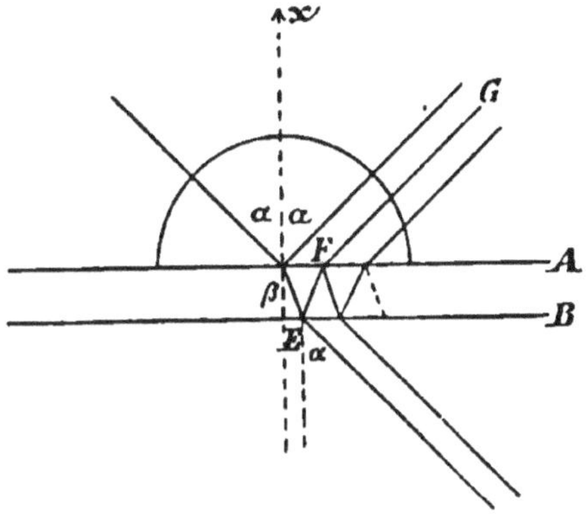

FIG. 117.

Similar expressions hold for the component Z' of the electrical force in the plate, which are obtained by replacing a by β, ω by ω', and introducing the new constants Z_2 and Z_4. Thus we obtain

$$Z' = Z_2 . e^{ki(t-(-x\cos\beta+y\sin\beta)/\omega')} + Z_4 . e^{ki(t-(x\cos\beta+y\sin\beta)/\omega')}.$$

We have for the component Z'' of the transmitted ray

$$Z'' = Z_5 . e^{ki(t-(-x\cos a+y\sin a)/\omega)}.$$

The boundary condition $Z = Z'$ when $x = 0$, gives (a) $Z_1 + Z_3 = Z_2 + Z_4$. Similarly $Z' = Z''$ when $x = -a$, or

$$Z_2 . e^{-kia.\cos\beta/\omega'} + Z_4 . e^{kia.\cos\beta/\omega'} = Z_5 . e^{-kia.\cos a/\omega}.$$

Now if we set $ka.\cos\beta/\omega' = u$, $ka.\cos a/\omega = v$, we can write the last condition in the form (b) $Z_2 . e^{-ui} + Z_4 . e^{ui} = Z_5 . e^{-vi}$.

We have, further, when $x = 0$, $\partial Z'/\partial x = \partial Z/\partial x$, and similarly, when $x = -a$, $\partial Z'/\partial x = \partial Z''/\partial x$. These equations of condition give

(c) $$(Z_1 - Z_3)\cos a . \sin\beta = (Z_2 - Z_4)\sin a . \cos\beta$$

and (d) $(Z_2 . e^{-ui} - Z_4 . e^{ui})\sin a . \cos\beta = Z_5 . e^{-vi} . \cos a . \sin\beta$. It follows from (b) and (d) that

$$Z_4/Z_2 = e^{-2ui} . \sin(a-\beta)/\sin(a+\beta),$$

or if $\sin(\alpha-\beta)/\sin(\alpha+\beta)=\epsilon$, $Z_4/Z_2=\epsilon\,.\,e^{-2ui}$. From (a) and (c) we have also $Z_3/Z_1=(-\epsilon Z_2+Z_4)/(Z_2-\epsilon Z_4)$, and therefore

(e) $$Z_3/Z_1=-(e^{ui}-e^{-ui})/(1/\epsilon\,.\,e^{ui}-\epsilon\,.\,e^{-ui}).$$

If N is greater than 1, u is always real, and we may set

$$Z_3/Z_1=-2i\epsilon\,.\,\sin u/[(1-\epsilon^2)\cos u+(1+\epsilon^2)i\,.\,\sin u].$$

Designating the intensity of the reflected light by C^2, we obtain, by the method indicated at the end of C.,

$$C^2=Z_1^2\,.\,4\epsilon^2\,.\,\sin^2 u/[(1-\epsilon^2)^2+4\epsilon^2\,.\,\sin^2 u].$$

But because $k=2\pi/T=2\pi\omega/\lambda$ and $u=2\pi\,.\,N\,.\cos\beta\,.\,a/\lambda$, it follows that

(f) $$\left\{\begin{array}{l} C^2=Z_1^2\,.\,4\epsilon^2\,.\,\sin^2(2\pi N\,.\cos\beta\,.\,a/\lambda)/[(1-\epsilon^2)^2 \\ \qquad +4\epsilon^2\,.\,\sin^2(2\pi N\,.\cos\beta\,.\,a/\lambda)]. \end{array}\right.$$

Hence no light is reflected if $2\pi N\,.\cos\beta\,.\,a/\lambda=p\pi$, where p is a whole number. This result is of special importance in the study of Newton's rings.

On the other hand, if $N<1$ and at the same time $\sin\alpha>N$, β will be imaginary. In this case we can no longer use equation (f). We then have [C. (o)] $Ni\,.\cos\beta=\sqrt{\sin^2\alpha-N^2}$, and hence

$$ui=2\pi a/\lambda\,.\,\sqrt{\sin^2\alpha-N^2}.$$

If we set $m=ui$ and $\epsilon=-e^{+i\gamma}$ [C. (r)], equation (e) takes the form

$$Z_3/Z_1=(e^m-e^{-m})/(e^{m-\gamma i}-e^{-m+\gamma i}).$$

Designating by C^2 the intensity of the reflected light, we obtain in the same way as before,

(g) $$C^2=Z_1^2\,.\,1/[1+4\sin^2\gamma/(e^m-e^{-m})^2],$$

where $\operatorname{tg}\frac{1}{2}\gamma=\sqrt{\sin^2\alpha-N^2}/\cos\alpha$ and $m=2\pi a/\lambda\,.\,\sqrt{\sin^2\alpha-N^2}$.

The relations which we have here considered occur in the case of two transparent bodies which are separated by a layer of air. If the thickness of the layer of air is very much greater than the wave length of the light, total reflection will occur. This is in accord with equation (g), which in this case gives $C^2=Z_1^2$. On the other hand, if a is small in comparison with the wave length, all the light passes through the layer of air. In consequence of this a black spot is seen if the hypotenuse of a right-angled glass prism is placed on the surface of a convex lens of long focus. If the angle of incidence α in the glass prism is less than the critical angle, a dark spot appears surrounded by coloured rings; but if the angle of incidence is greater than the critical angle, the rings disappear while the spot remains.

The spot is larger for red than for blue light. This result is contained also in the expression for the intensity of the reflected light. The transmitted light is complementary to the reflected light.

II. *If the direction of the electrical force of the incident light is parallel to the plane of incidence*, the disturbance outside the surface A is determined by

$$X = F_1 \,.\, \sin\alpha \,.\, e^{ki(t-(-x\cos\alpha+y\sin\alpha)/\omega)} - F_3 \,.\, \sin\alpha \,.\, e^{ki(t-(x\cos\alpha+y\sin\alpha)/\omega)},$$

$$Y = F_1 \,.\, \cos\alpha \,.\, e^{ki(t-(-x\cos\alpha+y\sin\alpha)/\omega)} + F_3 \,.\, \cos\alpha \,.\, e^{ki(t-(x\cos\alpha+y\sin\alpha)/\omega)}.$$

The disturbance inside the plate is given by

$$X' = F_2 \,.\, \sin\beta \,.\, e^{ki(t-(-x\cos\beta+y\sin\beta)/\omega')} - F_4 \,.\, \sin\beta \,.\, e^{ki(t-(x\cos\beta+y\sin\beta)/\omega')},$$

$$Y' = F_2 \,.\, \cos\beta \,.\, e^{ki(t-(-x\cos\beta+y\sin\beta)/\omega')} + F_4 \,.\, \cos\beta \,.\, e^{ki(t-(x\cos\beta+y\sin\beta)/\omega')};$$

and outside the surface B by

$$X'' = F_5 \,.\, \sin\alpha \,.\, e^{ki(t-(-x\cos\alpha+y\sin\alpha)/\omega)},$$

$$Y'' = F_5 \,.\, \cos\alpha \,.\, e^{ki(t-(-x\cos\alpha+y\sin\alpha)/\omega)}.$$

We must now determine the constants F_2, F_3, F_4, F_5. When $x=0$ the boundary conditions give $Y=Y'$, or

(h) $$(F_1+F_3)\cos\alpha = (F_2+F_4)\cos\beta.$$

Similarly, when $x=-a$,

$$F_2 \,.\, \cos\beta \,.\, e^{-kia\,.\,\cos\beta/\omega'} + F_4 \,.\, \cos\beta \,.\, e^{kia\,.\,\cos\beta/\omega'} = F_5 \,.\, \cos\alpha \,.\, e^{-kia\,.\,\cos\alpha/\omega}.$$

Using the same notation as before, we have

(i) $$F_2 \,.\, \cos\beta \,.\, e^{-ui} + F_4 \,.\, \cos\beta \,.\, e^{ui} = F_5 \,.\, \cos\alpha \,.\, e^{-vi}.$$

We have, further, when $x=0$,

$$\partial Y/\partial x - \partial X/\partial y = \partial Y'/\partial x - \partial X'/\partial y,$$

or (k) $$(F_1 - F_3)\sin\beta = (F_2 - F_4)\sin\alpha.$$

The same condition holds when $x=-a$, or

(l) $$F_2 \,.\, \sin\alpha \,.\, e^{-ui} - F_4 \,.\, \sin\alpha \,.\, e^{ui} = F_5 \,.\, \sin\beta \,.\, e^{-vi}.$$

We obtain from equations (i) and (l) $F_4/F_2 = e^{-2ui} \,.\, \mathrm{tg}(\alpha-\beta)/\mathrm{tg}(\alpha+\beta)$. But if we set $\mathrm{tg}(\alpha-\beta)/\mathrm{tg}(\alpha+\beta)=\epsilon'$, we will have $F_4/F_2 = \epsilon' \,.\, e^{-2ui}$. It follows from (h) and (k) that $F_3/F_1 = (-\epsilon' + \epsilon' \,.\, e^{-2ui})/(1-\epsilon'^2 \,.\, e^{-2ui})$, or (m) $F_3/F_1 = -(e^{ui}-e^{-ui})/(1/\epsilon' \,.\, e^{ui} - \epsilon' \,.\, e^{-ui})$. We thus obtain the intensity D^2 of the reflected light in the same way as we obtained the expression (f) from (e),

(n) $$D^2 = F_1^2 \,.\, \frac{4\epsilon'^2 \,.\, \sin^2(2\pi N\cos\beta a/\lambda)}{(1-\epsilon'^2)^2 + 4\epsilon'^2 \,.\, \sin^2(2\pi N\cos\beta a/\lambda)}.$$

If $\sin a > N$ and if β is therefore imaginary, we obtain the intensity of the reflected light in the following way: We have $\epsilon' = \mathrm{tg}(a-\beta)/\mathrm{tg}(a+\beta) = e^{\delta i}$, if, as in C. (u), we set

$$N^2 . \cos a = D . \cos \tfrac{1}{2}\delta, \quad \sqrt{\sin^2 a - N^2} = D . \sin \tfrac{1}{2}\delta.$$

If we further set $ui = m$, it follows that

$$F_3/F_1 = (e^m - e^{-m})/(e^{m-\delta i} - e^{-m+\delta i}).$$

The intensity D^2 of the reflected light is then

$$\text{(o)} \qquad D^2 = F_1^{\,2} . 1/\left(1 + 4\sin^2\delta/(e^m - e^{-m})^2\right),$$

in which expression

$$\mathrm{tg}\,\tfrac{1}{2}\delta = \sqrt{\sin^2 a - N^2}/N^2\cos a, \quad m = 2\pi a/\lambda . \sqrt{\sin^2 a - N^2}.$$

The expressions (n) and (o) for the intensity of the reflected light when the direction of the electrical force is parallel to the plane of incidence, lead to essentially the same results as equations (f) and (g), which hold when the direction of the electrical force is perpendicular to the plane of incidence. We only remark that, from equation (n), D^2 vanishes if $\epsilon' = 0$ or $(a+\beta) = \frac{1}{2}\pi$. In this case the angle of incidence is equal to the angle of polarization.

SECTION CII. DOUBLE REFRACTION.

Up to this point we have assumed that the value of the *dielectric constant* K is independent of the direction of the electrical force. Boltzmann, however, has shown that the dielectric constant of crystals has different values in different directions, and depends on the direction of the electrical force. Let K_1, K_2, K_3 represent the value of the dielectric constant in three perpendicular directions which are those of the coordinates x, y, z. Then in place of equations XCIII. (a), we use

$$\text{(a)} \qquad u = K_1/4\pi . \partial X/\partial t, \quad v = K_2/4\pi . \partial Y/\partial t, \quad w = K_3/4\pi . \partial Z/\partial t.$$

Equations XCIII. (d) and (c) become

$$K_1/V . \partial X/\partial t = \partial\gamma/\partial y - \partial\beta/\partial z, \quad K_2/V . \partial Y/\partial t = \partial a/\partial z - \partial\gamma/\partial x,$$
$$K_3/V . \partial Z/\partial t = \partial\beta/\partial x - \partial a/\partial y,$$

and if we set the magnetic permeability $\mu = 1$, we have

$$-1/V . \partial a/\partial t = \partial Z/\partial y - \partial Y/\partial z,$$
$$-1/V . \partial\beta/\partial t = \partial X/\partial z - \partial Z/\partial x,$$
$$-1/V . \partial\gamma/\partial t = \partial Y/\partial x - \partial X/\partial y.$$

Further, if we set

$$a^2 = V^2/K_1,\quad b^2 = V^2/K_2,\quad c^2 = V^2/K_3,\quad J = \partial X/\partial x + \partial Y/\partial y + \partial Z/\partial z,$$

we obtain

$$\text{(b)}\qquad \begin{cases} 1/a^2 . \partial^2 X/\partial t^2 = \nabla^2 X - \partial J/\partial x, \\ 1/b^2 . \partial^2 Y/\partial t^2 = \nabla^2 Y - \partial J/\partial y, \\ 1/c^2 . \partial^2 Z/\partial t^2 = \nabla^2 Z - \partial J/\partial z. \end{cases}$$

We will consider a plane wave moving through a body to which these equations apply. Its direction of propagation is determined by the angle whose cosines are l, m, n; the direction of the electrical force f is determined by the angle whose cosines are λ, μ, ν. We then have

(c) $X = \lambda f,\quad Y = \mu f,\quad Z = \nu f,\quad f = F . \cos\left[2\pi/T . \left(t - (lx + my + nz)/\omega\right)\right].$

F is constant, and the velocity of propagation ω depends only on the direction in which the wave is propagated. It follows from (c) that $\nabla^2 X = -4\pi^2\lambda f/T^2\omega^2$, and if $\cos\delta = l\lambda + m\mu + n\nu$, we obtain

(d) $J = 2\pi/T\omega . F . \cos\delta . \sin\left[2\pi/T . \left(t - (lx + my + nz)/\omega\right)\right].$

We obtain from the first of equations (b), (d′) $\lambda - l . \cos\delta = \omega^2 . \lambda/a^2$. This equation and the two similar to it take the forms

(e) $(a^2 - \omega^2)\lambda = a^2 l . \cos\delta,\quad (b^2 - \omega^2)\mu = b^2 m . \cos\delta,\quad (c^2 - \omega^2)\nu = c^2 n . \cos\delta.$

We use these equations to obtain the physical meaning of the magnitudes a, b, c. If $\omega = a$ we have either $l = 0$ or $\cos\delta = 0$. In the latter case $\mu = \nu = 0$ and $\lambda = \pm 1$ and therefore also $l = 0$. Hence a plane wave parallel to the x-axis is propagated with the velocity a when the electrical force is parallel to the same axis. The meaning of the magnitudes b and c is obtained in a similar way. By the *optical axes of elasticity* we mean the three directions in a body which have the property that a plane wave, in which the electrical force or the direction of vibration is parallel to one of the axes, for example a, is propagated with the velocity a in all directions perpendicular to the axis a.

We can find the velocity of propagation and the direction of the force from equations (e) and (d), in connection with the relation $\lambda^2 + \mu^2 + \nu^2 = 1$. If equations (e) are multiplied by l, m, n, respectively, and added, it follows from (d) that

$$a^2 l^2/(a^2 - \omega^2) + b^2 m^2/(b^2 - \omega^2) + c^2 n^2/(c^2 - \omega^2) = 1.$$

For brevity we will write for this equation $\Sigma a^2 l^2/(a^2 - \omega^2) = 1$. But because $a^2 = a^2 - \omega^2 + \omega^2$, we also have

$$\Sigma a^2 l^2/(a^2 - \omega^2) = \Sigma l^2 + \Sigma \omega^2 l^2/(a^2 - \omega^2) = 1.$$

Since $\Sigma l^2 = 1$, it follows that

(f) $$l^2/(a^2-\omega^2)+m^2/(b^2-\omega^2)+n^2/(c^2-\omega^2)=0.$$

This equation is of the fourth degree in ω. Since two of its roots are numerically equal to the other two but of opposite sign, *the electrical wave has two velocities of propagation,* ω_1 *and* ω_2.

We may give equation (f) the form

(g) $$\omega^4-\big(l^2(b^2+c^2)+m^2(a^2+c^2)+n^2(a^2+b^2)\big)\omega^2+l^2b^2c^2+m^2a^2c^2+n^2a^2b^2=0.$$

If $l=0$, that is, if the plane wave is parallel to the x-axis, we have

$$\omega^4-(a^2+m^2c^2+n^2b^2)\omega^2+a^2(m^2c^2+n^2b^2)=0.$$

The roots of this equation are $\omega_1=a$, $\omega_2=\sqrt{m^2c^2+n^2b^2}$.

This result can be represented by drawing lines in the yz-plane from the point O (Fig. 118), which are proportional to the velocities of propagation. The ends of these lines then lie on two curves, one of which is given by $\omega_1=a$, and is a *circle*; the other is given by ω_2, and is an *oval*. If $a>b>c$, the minor semi-axis c of the curve given by ω_2 lies in the y-axis, and its major semi-axis b in the z-axis. The relations of the plane waves which are parallel to the y- and z-axes respectively, are given in Figs. 119 and 120. The relation in the xz-plane is especially peculiar (Fig. 119). In that case, we have, for $m=0$, $\omega_1=b$, $\omega_2=\sqrt{l^2c^2+n^2a^2}$

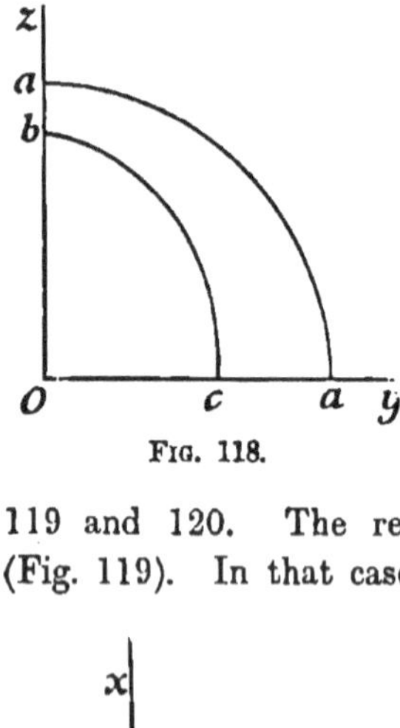

FIG. 118.

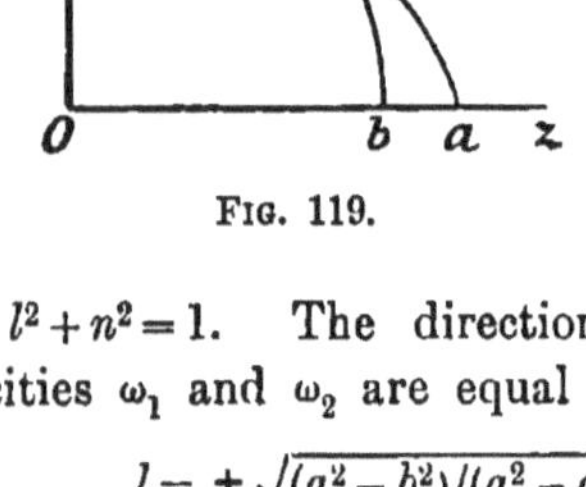

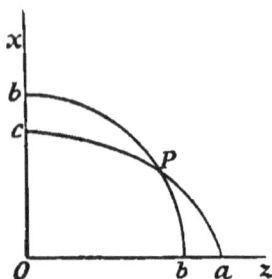

FIG. 119.

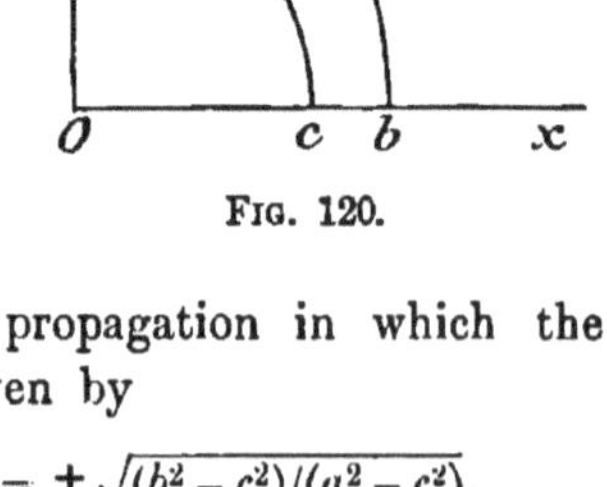

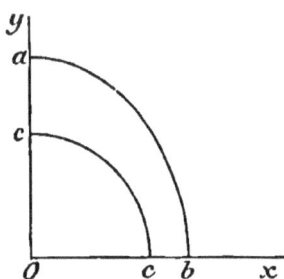

FIG. 120.

and $l^2+n^2=1$. The direction of propagation in which the two velocities ω_1 and ω_2 are equal is given by

$$l=\pm\sqrt{(a^2-b^2)/(a^2-c^2)}\,;\quad n=\pm\sqrt{(b^2-c^2)/(a^2-c^2)}.$$

SECTION CIII. DISCUSSION OF THE VELOCITIES OF PROPAGATION.

If ω_1 and ω_2 represent the two velocities of propagation of the same plane wave, we have [CII. (g)]

$$\text{(a)}\qquad \begin{cases} \omega_1^2 + \omega_2^2 = l^2(b^2+c^2) + m^2(a^2+c^2) + n^2(a^2+b^2), \\ \omega_1^2 \,.\, \omega_2^2 = l^2b^2c^2 + m^2a^2c^2 + n^2a^2b^2, \end{cases}$$

and
$$(\omega_1^2 - \omega_2^2)^2 = \left(l^2(b^2+c^2) + m^2(a^2+c^2) + n^2(a^2+b^2)\right)^2 - 4(l^2b^2c^2 + m^2a^2c^2 + n^2a^2b^2).$$

If we multiply the last term on the right side of this equation by $l^2+m^2+n^2=1$ we obtain

$$(\omega_1^2 - \omega_2^2)^2 = l^4(b^2-c^2)^2 + m^4(c^2-a^2)^2 + n^4(a^2-b^2)^2 + 2m^2n^2(a^2-b^2)(a^2-c^2) + 2l^2n^2(b^2-c^2)(b^2-a^2) + 2l^2m^2(c^2-a^2)(c^2-b^2).$$

If $a > b > c$, it follows that

$$(\omega_1^2 - \omega_2^2)^2 = l^4(b^2-c^2)^2 + m^4(a^2-c^2)^2 + n^4(a^2-b^2)^2 + 2m^2n^2(a^2-b^2)(a^2-c^2) - 2l^2n^2(b^2-c^2)(a^2-b^2) + 2l^2m^2(a^2-c^2)(b^2-c^2),$$

or

$$\text{(b)}\qquad \begin{cases} (\omega_1^2 - \omega_2^2)^2 = \left(l^2(b^2-c^2) + m^2(a^2-c^2) + n^2(a^2-b^2)\right)^2 - 4l^2n^2(a^2-b^2)(b^2-c^2), \\ (\omega_1^2 - \omega_2^2)^2 = \left[m^2(a^2-c^2) + (l\sqrt{b^2-c^2} + n\sqrt{a^2-b^2})^2\right].\left[m^2(a^2-c^2) + (l\sqrt{b^2-c^2} - n\sqrt{a^2-b^2})^2\right]. \end{cases}$$

Hence the two velocities ω_1 and ω_2 are equal for certain directions of the wave normals. This equality exists when $m=0$ and either

$$l\sqrt{b^2-c^2} + n\sqrt{a^2-b^2} = 0, \text{ or } l\sqrt{b^2-c^2} - n\sqrt{a^2-b^2} = 0.$$

These conditions are satisfied by $m=0$ and $l/n = \pm\sqrt{(a^2-b^2)/(b^2-c^2)}$. These equations represent four directions, which are parallel to the xz-plane and perpendicular to the axis of mean elasticity b. If we represent the cosines of the angles made by these directions with the coordinate axes by l_0, m_0, n_0, we have

$$\text{(c)}\qquad m_0 = 0,\ l_0 = \pm\sqrt{(a^2-b^2)/(a^2-c^2)},\ n_0 = \pm\sqrt{(b^2-c^2)/(a^2-c^2)}.$$

We call the directions in the crystal, defined by equations (c), the *optic axes.* There are two such axes, since each of these equations represents two opposite directions.

If Oa and Oc (Fig. 121) represent the axes of greatest and least elasticity a and c, and if OA_1 is one of the directions in which ω_1 and ω_2 are equal, they are equal not only in the opposite direction OB but also in the directions OA_2 and OB_2, if OA_2 makes the same angle with Oa as that made by OA_1.

We will now express *the velocity of propagation in any direction* in terms of the angles made by this direction with the optic axes OA_1 and OA_2. The cosines of the angles which the direction of propagation of the plane wave makes with the axes are l, m, n. We then have

$$\text{(c}'\text{)} \quad \begin{cases} \cos E_1 = l \cdot \sqrt{(a^2-b^2)/(a^2-c^2)} + n \cdot \sqrt{(b^2-c^2)/(a^2-c^2)}, \\ \cos E_2 = l \cdot \sqrt{(a^2-b^2)/(a^2-c^2)} - n \cdot \sqrt{(b^2-c^2)/(a^2-c^2)}. \end{cases}$$

From this it follows that

$$\text{(d)} \quad \begin{cases} 2l = (\cos E_1 + \cos E_2) \cdot \sqrt{(a^2-c^2)/(a^2-b^2)} \quad \text{and} \\ 2n = (\cos E_1 - \cos E_2) \cdot \sqrt{(a^2-c^2)/(b^2-c^2)}. \end{cases}$$

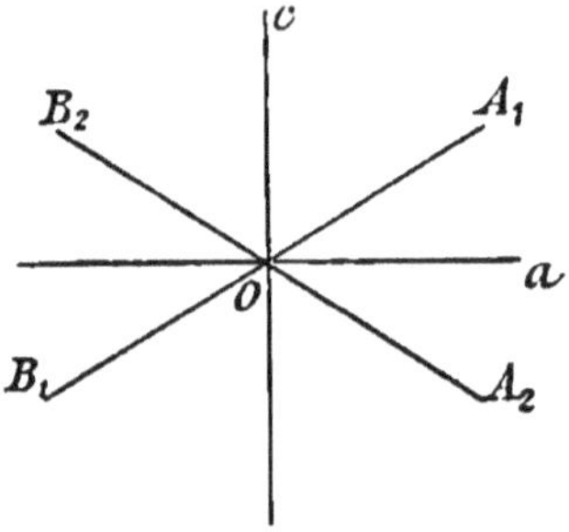

FIG. 121.

If we eliminate m from equation (a) by the help of the equation $l^2+m^2+n^2=1$, we obtain $\omega_1^2+\omega_2^2=a^2+c^2-l^2(a^2-b^2)+n^2(b^2-c^2)$, from which we obtain by use of equation (d),

$$\text{(e)} \quad \omega_1^2+\omega_2^2 = a^2+c^2-(a^2-c^2) \cdot \cos E_1 \cdot \cos E_2.$$

From equations (c′) we obtain

$$(a^2-c^2) \cdot \sin^2 E_1 = a^2-c^2-l^2(a^2-b^2)-n^2(b^2-c^2)-2ln \cdot \sqrt{a^2-b^2} \cdot \sqrt{b^2-c^2},$$
$$(a^2-c^2) \cdot \sin^2 E_2 = a^2-c^2-l^2(a^2-b^2)-n^2(b^2-c^2)+2ln \cdot \sqrt{a^2-b^2} \cdot \sqrt{b^2-c^2}.$$

Further, since $l^2+m^2+n^2=1$, we also have

$$l^2(b^2-c^2)+m^2(a^2-c^2)+n^2(a^2-b^2) = a^2-c^2-l^2(a^2-b^2)-n^2(b^2-c^2).$$

By help of these relations we obtain from the first of equations (b) (f) $\omega_1^2-\omega_2^2 = \pm(a^2-c^2) \cdot \sin E_1 \cdot \sin E_2$, and from (e) and (f),

$$\text{(g)} \quad \begin{cases} 2\omega_1^2 = a^2+c^2-(a^2-c^2) \cdot \cos(E_1-E_2), \\ 2\omega_2^2 = a^2+c^2-(a^2-c^2) \cdot \cos(E_1+E_2). \end{cases}$$

The greatest value of the velocity of propagation is a and the least c. This follows if we set $E_1=E_2$ and $E_1+E_2=\pi$. If the normal to the waves is parallel with one of the optical axes, for example OA_1, we have $E_1=0$ and $\cos\frac{1}{2}E_2=l_0$, and hence $\omega_1=\omega_2=b$. The velocity of propagation is then equal to the axis of mean elasticity.

Section CIV. The Wave Surface.

Suppose a plane wave to start from the origin of the system of coordinates, in the direction in which its normal makes angles with the axes whose cosines are l, m, n. After the lapse of a unit of time, the distance of the wave from the origin is ω. If about each of the points of the plane wave we construct a wave surface as it would appear after the lapse of unit time, the plane wave thus propagated is the envelope of all the wave surfaces. If x, y, z are the coordinates of a point of the plane wave in its new position, we have

(a) $$lx + my + nz = \omega,$$

and ω is determined by

(b) $$l^2/(a^2 - \omega^2) + m^2/(b^2 - \omega^2) + n^2/(c^2 - \omega^2) = 0.$$

We have, further, (c) $l^2 + m^2 + n^2 = 1$. If l, m, n, and ω vary, the planes (a) envelope a surface, which is called the *wave surface*. *Hence if we consider all possible plane waves passed through a point, and if we determine the position of the same waves after unit time, the wave surface is the envelope of all the plane waves thus determined.* We will now investigate the equation of this wave surface. We obtain from (a), (c), and (b),

(d) $$x\,.\,dl + y\,.\,dm + z\,.\,dn = d\omega,$$

(e) $$l\,.\,dl + m\,.\,dm + n\,.\,dn = 0,$$

(f) $$l\,.\,dl/(a^2 - \omega^2) + m\,.\,dm/(b^2 - \omega^2) + n\,.\,dn/(c^2 - \omega^2) + F\omega\,.\,d\omega = 0,$$

where (g) $$F = l^2/(a^2 - \omega^2)^2 + m^2/(b^2 - \omega^2)^2 + n^2/(c^2 - \omega^2)^2.$$

If we eliminate $d\omega$ by means of (d) from equation (f), we have

$$[F\omega x + l/(a^2 - \omega^2)]dl + [F\omega y + m/(b^2 - \omega^2)]dm + [F\omega z + n/(c^2 - \omega^2)]dn = 0.$$

We add, to the left side of this equation, equation (e) multiplied by a factor A. Since dl, dm, dn may be considered as arbitrary quantities, we have

(h) $$\begin{cases} l/(a^2 - \omega^2) + F\omega x + Al = 0, \quad m/(b^2 - \omega^2) + F\omega y + Am = 0, \\ n/(c^2 - \omega^2) + F\omega z + An = 0. \end{cases}$$

If these equations are multiplied in order by l, m, n respectively and added, we obtain, by reference to (a) and (b), $A = -F\omega^2$. Therefore

(i) $$\begin{cases} l/(a^2 - \omega^2) = F\omega(l\omega - x), \quad m/(b^2 - \omega^2) = F\omega(m\omega - y), \\ n/(c^2 - \omega^2) = F\omega(n\omega - z). \end{cases}$$

If we square both sides of these equations, add them, and use equation (g), we have $1 = F\omega^2\big(\omega^2 - 2\omega(lx + my + nz) + r^2\big)$, in which $r^2 = x^2 + y^2 + z^2$.

Further, by reference to (a) we obtain (k) $F\omega^2(r^2-\omega^2)=1$. F may now be eliminated from equations (i) by means of k, and we have

(l)
$$\left\{\begin{array}{c} x(a^2-\omega^2)=l\omega(a^2-r^2),\quad y(b^2-\omega^2)=m\omega(b^2-r^2),\\ z(c^2-\omega^2)=n\omega(c^2-r^2).\end{array}\right.$$

These equations enable us to determine the point of contact between the wave surface and the plane wave, and therefore *the direction of propagation* of the ray. The plane wave moves in the direction determined by l, m, n.

If we multiply both sides of equations (l) by x, y, z respectively, and add, we have

$$x^2(a^2-\omega^2)/(a^2-r^2)+y^2(b^2-\omega^2)/(b^2-r^2)+z^2(c^2-\omega^2)/(c^2-r^2)=\omega^2,$$

since by (a) $lx+my+nz=\omega$. This equation may be written in the abbreviated form $\Sigma x^2(a^2-\omega^2)/(a^2-r^2)=\omega^2$, or in the form

$$\begin{aligned}\Sigma x^2(a^2-\omega^2)/(a^2-r^2)&=\Sigma x^2(a^2-r^2+r^2-\omega^2)/(a^2-r^2)\\ &=\Sigma x^2+(r^2-\omega^2)\Sigma x^2/(a^2-r^2).\end{aligned}$$

But since in this notation $\Sigma x^2=x^2+y^2+z^2=r^2$, we have finally

$$(r^2-\omega^2)\big(1+\Sigma x^2/(a^2-r^2)\big)=0.$$

The equation of the wave surface is therefore

$$x^2/(a^2-r^2)+y^2/(b^2-r^2)+z^2/(c^2-r^2)+1=0.$$

But because

$$\Sigma x^2/r^2=1 \quad\text{and}\quad \Sigma\big(x^2/(a^2-r^2)+x^2/r^2\big)=\Sigma a^2x^2/(a^2-r^2)=0$$

we may also write the equation of the wave surface in the form

(m)
$$a^2x^2/(a^2-r^2)+b^2y^2/(b^2-r^2)+c^2z^2/(c^2-r^2)=0.$$

We can easily transform this equation into

(n)
$$(a^2x^2+b^2y^2+c^2z^2)r^2-(b^2+c^2)a^2x^2-(a^2+c^2)b^2y^2-(a^2+b^2)c^2z^2+a^2b^2c^2=0.$$

The equation of the wave surface is therefore of the fourth degree. In order to investigate this equation we set $x=fr$, $y=gr$, $z=hr$. By substitution of these values in the equation of the wave surface, it becomes

$$2r^2(a^2f^2+b^2g^2+c^2h^2)-[(b^2+c^2)a^2f^2+(a^2+c^2)b^2g^2+(a^2+b^2)c^2h^2]=\pm R,$$
$$R^2=[(b^2+c^2)a^2f^2+(a^2+c^2)b^2g^2+(a^2+b^2)c^2h^2]^2-4a^2b^2c^2(a^2f^2+b^2g^2+c^2h^2).$$

From which we get

$$\begin{aligned}R^2=&[(a^2-c^2)b^2g^2+(af\sqrt{b^2-c^2}+ch\sqrt{a^2-b^2})^2]\\ &\times[(a^2-c^2)b^2g^2+(af\sqrt{b^2-c^2}-ch\sqrt{a^2-b^2})^2].\end{aligned}$$

Hence a straight line drawn from the origin of coordinates cuts the surface in two points, which coincide when $R=0$ or when

(o)
$$g=0 \text{ and } f/h=\pm c/a\,.\sqrt{(a^2-b^2)/(b^2-c^2)}.$$

In this case

$$f = \pm c/b \,.\, \sqrt{(a^2-b^2)/(a^2-c^2)}, \quad h = \pm a/b \,.\, \sqrt{(b^2-c^2)/(a^2-c^2)}.$$

There are therefore four such points in the wave surface, all of which lie in the xz-plane. Hence the wave surface is a surface of the fourth degree with two nappes. The four points which the two nappes have in common are called *umbilical points*.

To exhibit the form of this surface we will determine the curves formed by the intersection of the wave surface, and the coordinate planes yz, xz, yx. If, for this purpose, we set in equation (n) $x=0$, $y=0$, $z=0$ successively, we obtain

$$(y^2+z^2-a^2)(b^2y^2+c^2z^2-b^2c^2)=0,$$
$$(z^2+x^2-b^2)(c^2z^2+a^2x^2-a^2c^2)=0,$$
$$(x^2+y^2-c^2)(a^2x^2+b^2y^2-a^2b^2)=0,$$

Hence the curves formed by the intersection of the wave surface with the coordinate axes are circles and ellipses, as represented in figures 122, 123, and 124. The curves in the xz-plane are of

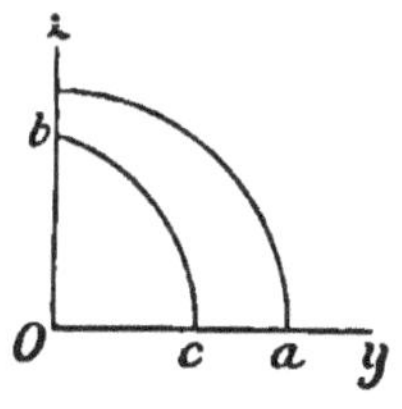

FIG. 122.

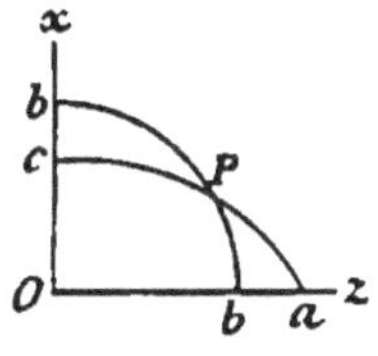

FIG. 123.

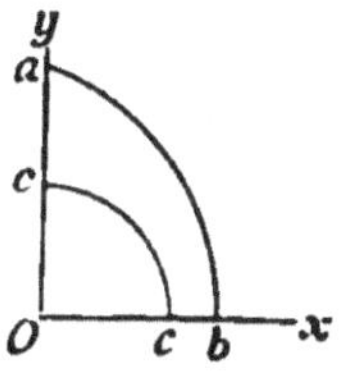

FIG. 124.

special interest. The equation $z^2+x^2=b^2$ represents a circle of radius b. The equation $c^2z^2+a^2x^2-a^2c^2=0$ represents an ellipse whose semi-axes are a and c. On the assumption that $a>b>c$, the circle and the ellipse intersect at a point P, and this point is one of the umbilical points.

Equations (l) and (h) serve to determine the coordinates of the point of contact between the wave surface and a plane wave which moves in a direction determined by l, m, n.

The case *in which the wave is propagated in the direction of one of the optic axes* is of special interest. In this case [CIII.], the velocity equals b, and the direction of propagation is given by the equations

$$m=0, \quad l=\sqrt{(a^2-b^2)/(a^2-c^2)}, \quad n=\sqrt{(b^2-c^2)/(a^2-c^2)},$$

since we here consider only that optic axis which lies between the positive directions of the z- and x-axes. Equations (l) then become

$$x(a^2-b^2)=lb(a^2-r^2), \quad z(b^2-c^2)=nb(r^2-c^2).$$

If we introduce in these equations the values for l and n given above, we have

(p) $x\sqrt{(a^2-b^2)(a^2-c^2)}=b(a^2-r^2),\quad z\sqrt{(b^2-c^2)(a^2-c^2)}=b(r^2-c^2).$

These equations represent two spheres, in whose lines of intersection lie the points of contact of the wave plane and the wave surface, *therefore in this case the plane of the wave touches the wave surface in a circle.*

We may also obtain this result in the following way. By the use of equations CIII. (c), we give (p) the form

(q) $x=b(a^2-x^2-y^2-z^2)/l_0(a^2-c^2),\quad z=b(x^2+y^2+z^2-c^2)/n_0(a^2-c^2).$

The curve represented by these two equations is a *plane* curve because (r) $xl_0+zn_0=b$.

We now introduce a new system of coordinates with the same origin; suppose the η-axis to coincide with the y-axis, while the ζ-axis coincides with the optic axis. To effect this, we set

(s) $$x=\xi n_0+\zeta l_0,\quad y=\eta,\quad z=-\xi l_0+\zeta n_0.$$

The equation (r), which represents a plane, then becomes (t) $\zeta=b$, that is, *the plane of the curve of intersection is perpendicular to the direction of the optic axis and passes through its end point.* The first of equations (q), by the use of (s) and (t), takes the form

(u) $$\xi^2+\xi\,.\,n_0l_0(a^2-c^2)/b+\eta^2=0.$$

This represents a *circle,* which passes through the point $\xi=0$, $\eta=0$, and $\zeta=b$, or through the end point of the optic axis. The radius r of the circle is $r=\sqrt{(b^2-c^2)(a^2-b^2)}/2b$, and the coordinates of its centre are $\xi=-r$, $\eta=0$, $\zeta=b$. Thus the circle is determined in which the plane perpendicular to one of the optic axes at its end point touches the wave surface.

SECTION CV. THE WAVE SURFACE (*continued*).

Let ON (Fig. 125) be the normal to a plane wave; the direction of the normal is determined by the cosines l, m, n. Let OP_1 and OP_2 be the two velocities of propagation of the wave considered. Let Q_1 and Q_2 represent the points of contact between the plane wave and the wave surface. We then have $OQ_1=r_1$ and $OQ_2=r_2$. We represent the coordinates of the points Q_1 and Q_2 by x_1, y_1, z_1 and x_2, y_2, z_2 respectively. If $Q_1P_1=p_1$ and $Q_2P_2=p_2$ are the perpendiculars let fall from the points of contact on the directions of propagation, we have $p_1{}^2=r_1{}^2-\omega_1{}^2$ and $p_2{}^2=r_2{}^2-\omega_2{}^2$.

The connection between the direction of the normal and the points of contact is given by equations (l) CIV. We will investigate more particularly the directions of the lines p_1 and p_2. The projection of PQ on the x-axis is $\omega l - x$. If we represent the cosines of the angles which p makes with the axes by λ', μ', ν', we will have

$$\lambda' = (\omega l - x)/p, \quad \mu' = (\omega m - y)/p,$$
$$\nu' = (\omega n - z)/p.$$

Introducing in this equation the values of x, y, z, given in CIV. (l), we obtain

(a) $$\lambda' = l\omega p/(a^2 - \omega^2), \quad \mu' = m\omega p/(b^2 - \omega^2),$$
$$\nu' = n\omega p/(c^2 - \omega^2).$$

In order to find the angle between P_1Q_1 and P_2Q_2, we determine its cosine

$$\cos(P_1Q_1 \wedge P_2Q_2) = \lambda_1'\lambda_2' + \mu_1'\mu_2' + \nu_1'\nu_2'$$

or

$$\cos(P_1Q_1 \wedge P_2Q_2) = p_1p_2\omega_1\omega_2\Sigma l^2/(a^2 - \omega_1^2)(a^2 - \omega_2^2).$$

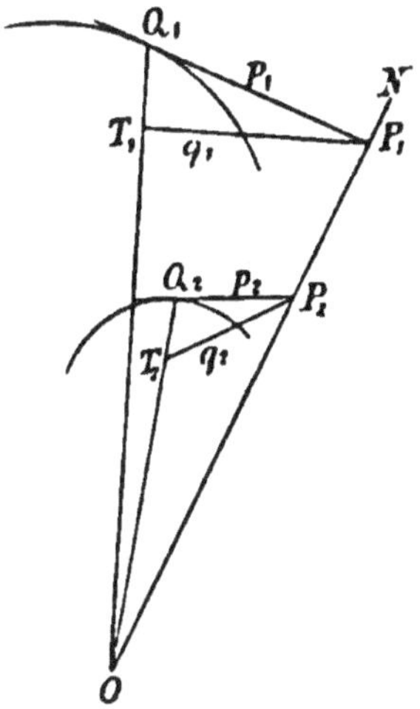

FIG. 125.

But because [CII. (f)]

$$\Sigma l^2/(a^2 - \omega_1^2) = 0 \text{ and } \Sigma l^2/(a^2 - \omega_2^2) = 0,$$

we also have $$(\omega_1^2 - \omega_2^2) . \Sigma l^2/(a^2 - \omega_1^2)(a^2 - \omega_2^2) = 0.$$

Hence, if the values of ω_1 and ω_2 are different, we have $\cos(P_1Q_1 \wedge P_2Q_2)$ equal to zero, and the angle between P_1Q_1 and P_2Q_2 a right angle. But if ω_1 equals ω_2, the points P_1 and P_2 coincide, as we saw in CIV. In this case, there is an infinite number of points of contact which lie on a circle passing through the wave normals.

If the lines P_1T_1 and P_2T_2 are drawn from P_1 and P_2 perpendicular to OQ_1 and OQ_2 respectively, and if we set $P_1T_1 = q_1$ and $P_2T_2 = q_2$, we have $q : p = \omega : r$, and therefore $q = p\omega/r$. Further, $OT = \omega^2/r$. If λ, μ, ν are the cosines of the angles which q makes with the coordinate axes, we have $\lambda = (\omega l - OT . x/r)/q$, etc., and hence [CIV. (l)],

(b) $$\lambda(a^2 - \omega^2) = la^2p/r, \quad \mu(b^2 - \omega^2) = mb^2p/r, \quad \nu(c^2 - \omega^2) = nc^2p/r.$$

If we compare this result with the expressions in CII. (e), which determine the direction of the electrical force F, whose components are X, Y, Z, we see that the electrical force is parallel to q. If we introduce in the equation CII. (d) the values for λ, μ, ν given above, and notice that $lx + my + nz = \omega$, we have $\cos\delta = p/r$. Since there are two directions of q, namely q_1 and q_2, there are two directions, q_1 and q_2, of the force in any plane wave. These lie in two planes perpendicular to the plane wave. There are two values for δ, namely, $\angle OQ_1P_1$ and $\angle OQ_2P_2$; these angles are equal to $\angle T_1P_1O$ and $\angle T_2P_2O$ respectively.

The electrical forces X, Y, Z cause an electrical polarization, whose variation may be looked on as an electrical current. The components of current [CII. (a) and (c)] are

$$u = K_1/4\pi \,.\, \partial X/\partial t = K_1\lambda/4\pi \,.\, \partial U/\partial t, \text{ etc.}$$

If λ_0, μ_0, ν_0 are the cosines of the angles made by the axes with the direction of the current, we have $\lambda_0 : \mu_0 : \nu_0 = \lambda/a^2 : \mu/b^2 : \nu/c^2$. But we obtain, by the help of CII. (e),

$$\lambda_0 : \mu_0 : \nu_0 = l/(a^2 - \omega^2) : m/(b^2 - \omega^2) : n/(c^2 - \omega^2).$$

Hence the current has two directions, corresponding to the two values of ω. From equation (a) the same ratio holds between the cosines of the angles which p makes with the axes as between the cosines determining the directions of the current. Hence the two directions of the current are parallel to p_1 and p_2 respectively.

In order to determine the direction of the electrical force and the current, we proceed in the following way. If a plane wave moves in the direction determined by the normal ON, we construct two planes which touch the wave surface and are parallel to the plane wave. These planes are those constructed at Q_1 and Q_2. We then draw Q_1P_1 and Q_2P_2 perpendicular to the wave normal. The electrical currents, which are in the wave planes, are parallel to Q_1P_1 and Q_2P_2. The corresponding velocities of propagation are OP_1 and OP_2. There are two directions of current in every plane wave, which are perpendicular to each other. The electrical forces, which are connected with these directions of current, are parallel to P_1T_1 and P_2T_2.

SECTION CVI. THE DIRECTION OF THE RAYS.

When a plane wave is propagated in an isotropic medium, the direction of the normal to the wave coincides with the direction of the ray. In doubly refracting media, the direction of the ray is in general different from the direction of the wave-normals. We will now determine the direction of the ray. Let MN (Fig. 126) be the surface of a doubly refracting body on which the cylinder of rays $KOPL$ falls perpendicularly. By Huygen's principle, the separate points in the bounding surface OP may be considered as centres of luminous disturbance. The luminous disturbance is propagated through the body in such a way that, after unit time, it reaches the wave surfaces which are constructed about the separate points of the bounding surface OP. Therefore, if the wave surfaces RA, SC, etc., are constructed about O, P, and the intervening points, we obtain a plane RS which touches every wave surface and is congruent to and similarly situated with OP. The direction OR or PS is then the direction of the rays. If from the point O we let fall a perpendicular OB on the plane RS tangent to the wave surface RA, $OB=\omega$ is the velocity of propagation of the wave. If l, m, n represent the direction cosines of the normal to the wave surface, ω is determined by equation CII. (f)

(a) $$l^2/(a^2-\omega^2)+m^2/(b^2-\omega^2)+n^2/(c^2-\omega^2)=0.$$

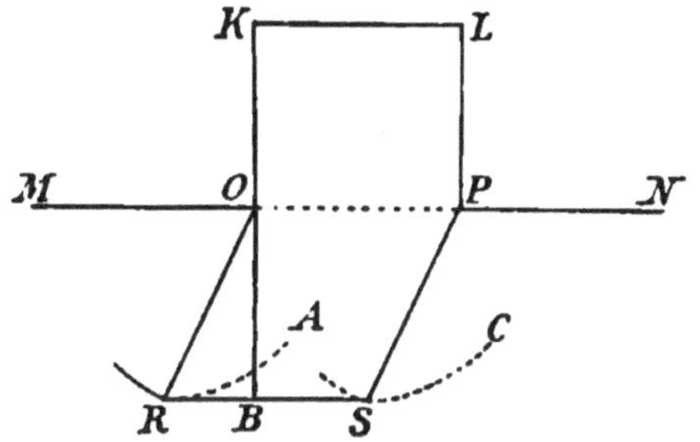

FIG. 126.

The position of the point of contact of the plane RS and the wave surface RA is given [CIV. (l)] from the equations

(b) $$x(a^2-\omega^2)=l\omega(a^2-r^2),\quad y(b^2-\omega^2)=m\omega(b^2-r^2),\quad z(c^2-\omega^2)=n\omega(c^2-r^2),$$

where x, y, z are the coordinates of the point desired, and $OR=r$ is its distance from the origin of coordinates. OB represents the *velocity of propagation of the wave, OR the velocity of propagation of the ray.*

Instead of the wave surface itself we may sometimes use to advantage another surface, called the *reciprocal wave surface.* Let O be the centre of the wave surface, AR a part of the surface itself, and BR a plane which touches the wave surface at the point R. From the point O we let fall the perpendicular OB on the tangent plane. The point B', in the perpendicular OB produced, is determined in such a way that (c) $OB'=r'=s^2/\omega$, where $\omega=OB$, and s is constant. The reciprocal wave surface is then the locus of the points determined by (c). This surface, like the wave surface, is a surface of two nappes. Its equation is obtained in the following way. If l, m, n

represent the direction cosines of $OB=\omega$, and x', y', z' the coordinates of the point B', we have (d) $x'=lr'$, $y'=mr'$, $z'=nr'$. But because

$$l^2/(a^2-\omega^2)+m^2/(b^2-\omega^2)+n^2/(c^2-\omega^2)=0,$$

it follows by (c) and (d) that

$$x'^2/(a^2r'^2-s^4)+y'^2/(b^2r'^2-s^4)+z'^2/(c^2r'^2-s^4)=0.$$

We set (e) $a'=s^2/a$, $b'=s^2/b$, $c'=s^2/c$, and obtain the *equation of the reciprocal wave surface* in the form

(f) $$a'^2x'^2/(a'^2-r'^2)+b'^2y'^2/(b'^2-r'^2)+c'^2z'^2/(c'^2-r'^2)=0.$$

This surface differs from the ordinary wave surface [cf. CIV. (m)] only in that its constants a', b', c' are the reciprocals of the constants a, b, c of the wave surface.

If we draw through the point B' (Fig. 127) a plane tangent to the reciprocal wave surface $B'A'$, we can show that the plane

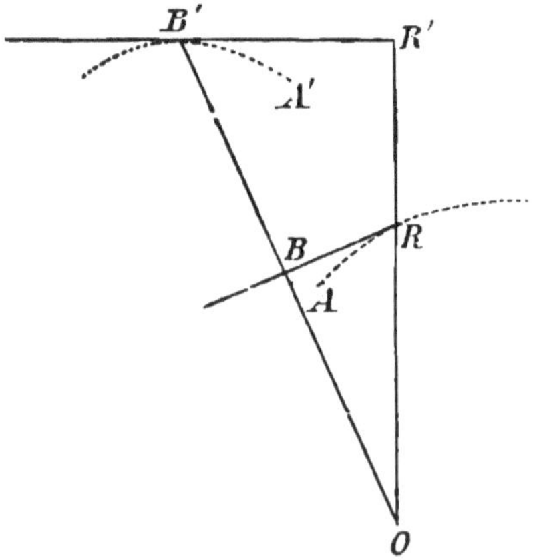

FIG. 127.

$B'R'$ is perpendicular to the prolongation of OR, and that therefore OR' is perpendicular to $B'R'$. Further, if $OR'=\omega'$, $OR=r$, we have (g) $\omega'=s^2/r$. This follows by the same method by which we have passed from one surface to the other. We can also prove it directly. If the direction OR' is determined by the cosines l', m', n', we have [CIV. (l)] (h) $x'(a'^2-\omega'^2)=l'\omega'(a'^2-r'^2)$ etc.

The equations (h) determine ω', l', m', n'. Setting $\omega'=s^2/r$ and

$$l'=x/r,\quad m'=y/r,\quad n'=z/r,$$

and using equations (c), (d), (e), (g), equation (h) takes the form $x(a^2-\omega^2)=l\omega(a^2-r^2)$, etc. Since these equations are identical with those in CIV. (l), it follows that the point of intersection of OR'

and the wave surface is the point at which the tangent plane touches the wave surface. It follows further from (e) and (g) that (i) $r'\omega = r\omega'$ or $OB.OB' = OR.OR'$. *In order to determine the direction of a ray from the reciprocal wave surface, we produce the wave normal until it cuts that surface. The direction of the ray is then perpendicular to the tangent plane at the point of intersection.*

SECTION CVII. UNIAXIAL CRYSTALS.

If two of the constants a, b, c are equal, for example if $b=c$, the equations become much simplified. The bodies for which this relation holds are called *uniaxial crystals*. In order to find the velocities of propagation ω_1 and ω_2, we apply equation CII. (g) which is transformed into (a) $\omega^4 - [b^2 + l^2b^2 + (1 - l^2)a^2]\omega^2 + b^2[l^2b^2 + (1 - l^2)a^2] = 0$. From this equation we obtain (b) $\omega_1{}^2 = b^2$, $\omega_2{}^2 = l^2b^2 + (1 - l^2)a^2$. Hence the velocity ω_1 is constant; the velocity ω_2 depends on the direction of the wave normal, or on the angle which the wave normal makes with the axis of elasticity a. This axis is called the *optic axis*; it coincides with the principal axis of the crystal. In the direction of this optic axis there is only one wave velocity, and therefore also only one ray velocity. If we designate the angle between the wave normal and the optic axis by ϵ, we have (c) $\omega_2{}^2 = a^2\sin^2\epsilon + b^2\cos^2\epsilon$.

Hence, a plane wave, on its passage from an isotropic to an uniaxial medium, is divided into two waves, one of which is propagated with a velocity ω_1, which is independent of the direction of the wave normal. This wave is called the *ordinary wave*. The velocity of the other or *extraordinary* wave changes with the direction of the wave normal.

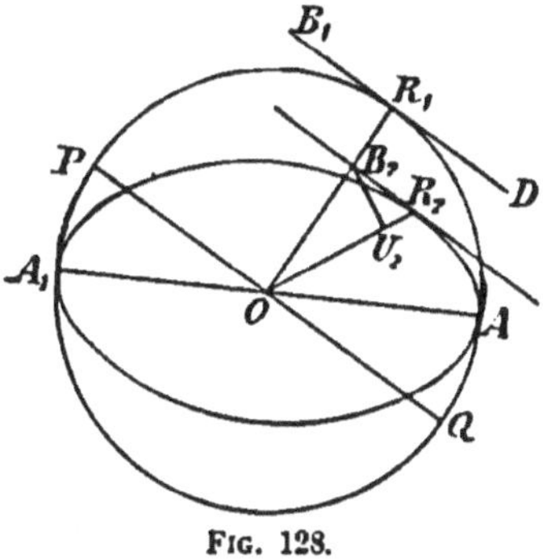

FIG. 128.

We obtain the equation of the wave surface for uniaxial crystals from CIV. (n), by setting $b=c$. We thus obtain

(d) $(r^2 - b^2)\left(a^2x^2 + b^2(y^2+z^2) - a^2b^2\right) = 0.$

Hence, the wave surface consists of a sphere whose radius is b, and an ellipsoid of revolution whose polar and equatorial axes are $2b$ and $2a$ respectively; the sphere and the ellipsoid touch at the extremities of the polar axis. In Fig. 128, AA_1 represents the polar or optic axis, AR_1A_1 a plane section through the sphere and AR_2A_1 a plane section through the ellipsoid. Let OB_2R_1 be the normal to the plane wave POQ, B_1D and B_2R_2, two planes tangent to the wave surface, which are both perpendicular to the wave normal. OR_1 and OB_2 are the velocities of propagation in the direction of the wave normal. We call such a plane section, which contains the optic axis as well as the wave normal, the *principal section.* The direction of the rays of the extraordinary wave is represented by OR_2, if the plane B_2R_2 touches the ellipsoid at the point R_2. The direction of the electrical force is given by B_2U_2, which is perpendicular to OR_2. The direction of the rays and the wave normal of the ordinary wave coincide, and the direction of the electrical force is perpendicular to the plane of the figure.

The polar axis $AA_1 = 2b$ (Fig. 128) is greater than the equatorial axis $2a$; the crystals for which this occurs are called *positive crystals.* If $a > b$, the crystal is called *negative.* The sphere can enclose the ellipsoid or inversely; *crystals of the first kind are called positive, those of the second kind negative.* Iceland spar is a negative crystal, quartz is a positive crystal.

If we set $\omega = b$ in CII. (e), we obtain (e) $\lambda_1 = 0$ and $\delta_1 = \frac{1}{2}\pi$, that is, *the direction of the electrical force in the ordinary wave is perpendicular to the optic axis as well as to the wave normal; it is therefore perpendicular to the principal section.*

In order to obtain the direction of the electrical force in the extraordinary waves from CII. (e), we introduce in it the value for ω_2 given in (b), and notice that $l = \cos\epsilon$, $m = \sin\epsilon$, $n = 0$. We then obtain

$$\lambda_2 = \frac{a^2 \cos\delta_2}{(a^2 - b^2)\cos\epsilon}, \quad \mu_2 = -\frac{b^2 \cos\delta_2}{(a^2 - b^2)\sin\epsilon}, \quad \nu_2 = 0.$$

Hence the direction of the electrical force in the extraordinary wave is parallel to the principal section. It follows from the last equation that $1/\cos^2\delta_2 = (a^4 \sin^2\epsilon + b^4 \cos^2\epsilon)/(a^2 - b^2)^2 \sin^2\epsilon \cos^2\epsilon$, and hence

$$\text{(f)} \qquad \operatorname{tg}\delta_2 = \pm(a^2 \sin^2\epsilon + b^2 \cos^2\epsilon)/(a^2 - b^2)\sin\epsilon\cos\epsilon$$

We thus obtain the equations

$$\text{(g)} \qquad \begin{cases} \lambda_2 = \pm a^2 \sin\epsilon / \sqrt{a^4 \sin^2\epsilon + b^4 \cos^2\epsilon}, \\ \mu_2 = \mp b^2 \cos\epsilon / \sqrt{a^4 \sin^2\epsilon + b^4 \cos^2\epsilon}, \\ \nu_2 = 0. \end{cases}$$

SECTION CVIII. DOUBLE REFRACTION AT THE SURFACE OF A CRYSTAL.

When a ray of polarized light falls on the plane surface of a doubly refracting medium both reflection and refraction occur. Let the x-axis of the system of coordinates be parallel to the normal to the surface drawn outwards, and the z-axis perpendicular to the plane of incidence; the y-axis is then parallel to the line of intersection between the plane of incidence and the refracting surface. For the components of the electrical force of the incident ray we have, as in C.,

(a) $X_i = \lambda_i f_i,\ Y_i = \mu_i f_i,\ Z_i = \nu_i f_i,\ f_i = F_i \cos\left[2\pi/T.\left(t - (-x\cos\alpha + y\sin\alpha)/\Omega\right)\right]$,

or, using only the real part, (b) $f_i = F_i e^{ki[t-(-x\cos\alpha + y\sin\alpha)/\Omega]}$. In these equations α is the angle of incidence and Ω the velocity of the light outside the crystal. In addition to these equations we have the condition that the electrical force is perpendicular to the direction of the ray. Hence, since the direction of the incident ray makes the angles $\pi - \alpha$, $\frac{1}{2}\pi - \alpha$ and $\frac{1}{2}\pi$ with the axes, we have

(c) $$-\lambda_i \cos\alpha + \mu_i \sin\alpha = 0.$$

In the corresponding notation we have for the reflected ray

(d) $$X_r = \lambda_r f_r,\quad Y_r = \mu_r f_r,\quad Z_r = \nu_r f_r,\quad f_r = F_r \,.\, e^{ki[t-(l_r x + m_r y + n_r z)/\Omega]}.$$

That the electrical force shall be perpendicular to the direction of the ray, we must have (e) $\lambda_r l_r + \mu_r m_r + \nu_r n_r = 0$. Finally, for the refracted ray, we have

(f) $X_b = \lambda_b f_b,\quad Y_b = \mu_b f_b,\quad Z_b = \nu_b f_b,\qquad$ (g) $f_b = F_b e^{ki[t-(l_b x + m_b y + n_b z)/\omega]}$.

ω depends on l_b, m_b, n_b, or on the direction of the propagation of the refracted wave. The boundary conditions are the same as those of isotropic bodies. We have everywhere in the bounding surface, for which $x = 0$, $Y_i + Y_r = Y_b$, or

$$\mu_i F_i e^{ki(t - y\sin\alpha/\Omega)} + \mu_r F_r e^{ki(t-(m_r y + n_r z)/\Omega)} = \mu_b F_b e^{ki(t-(m_b y + n_b z)/\omega)}.$$

Since this equation must hold for all values of y and z, we have (h) $\sin\alpha/\Omega = m_r/\Omega = m_b/\omega$ and (i) $0 = n_r/\Omega = n_b/\omega$. By the last equation $0 = n_r = n_b$, that is, *the wave normals of the reflected and refracted waves lie in the plane of incidence.* It follows from (h) that $m_r = \sin\alpha$, that is, *the angle of reflection is equal to the angle of incidence.* Therefore the direction of the reflected ray is determined in the same way as in the case of reflection by an isotropic body.

If β represents the angle of refraction, we have

$$l_b = -\cos\beta,\quad m_b = \sin\beta,\quad n_b = 0,$$

therefore from (h) $\sin\alpha/\Omega = \sin\beta/\omega$. If we determine the direction of the wave normal by the cosines of the angles which it makes with the axes of elasticity, we have, to determine ω, the equation

$$\text{(l)} \qquad l^2/(a^2-\omega^2) + m^2/(b^2-\omega^2) + n^2/(c^2-\omega^2) = 0.$$

If (xa), (ya), etc., denote the angles between the axes of elasticity and the coordinate axes, we have $l = l_b\cos(xa) + m_b\cos(ya) + n_b\cos(za)$. Introducing here the values for l_b, m_b, etc., given above, we obtain

$$\text{(m)} \qquad \begin{cases} l = -\cos\beta\,.\cos(xa) + \sin\beta\,.\cos(ya) \\ m = -\cos\beta\,.\cos(xb) + \sin\beta\,.\cos(yb) \\ n = -\cos\beta\,.\cos(xc) + \sin\beta\,.\cos(yc). \end{cases}$$

By the help of equations (m) and (l), ω can be expressed in terms of β. The equation thus obtained in connection with (k) determines the angle of refraction. In general we obtain two values for β, one or both of which may be imaginary; if this is the case the reflection is total.

We can find the direction of the wave normal and that of the ray by a construction given by Huygens. About the point O (Fig. 129) as centre construct the sphere PD, whose radius is $OD = \Omega$, where Ω denotes the velocity of light in air. If the incident ray is produced, it meets the sphere at the point D. The plane which the sphere touches at D cuts the refracting surface in a straight line, whose projection on the plane of the figure is Q. The plane QR containing this line is drawn tangent to the wave surface FR, whose centre is at O.

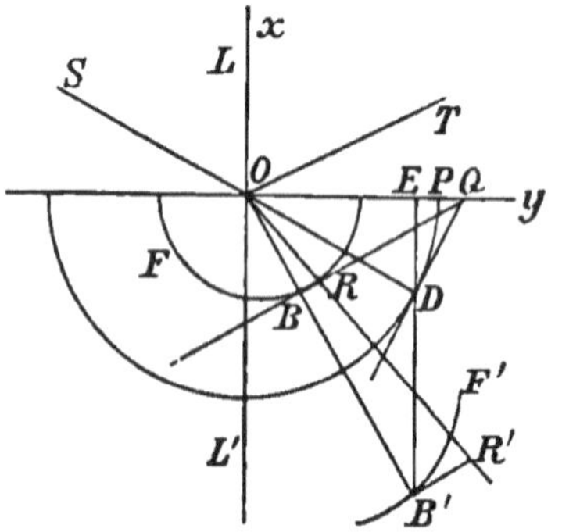

FIG. 129.

The perpendicular $OB = \omega$ is let fall from O on the tangent plane QR. The normal to the refracted wave is then OB and $L'OB = \beta$, if LOL' is the normal to the surface. Now $OB = \omega$ is the velocity of propagation in the direction OB, and also $OQ = OD/\sin\alpha = OB/\sin\beta$, or $\Omega/\sin\alpha = \omega/\sin\beta$, so that equation (k) is also satisfied. OB is the direction of the wave normal of the refracted wave, and OR the direction of the corresponding ray. Since the wave surface in general has two nappes, two planes tangent to the wave surface can be drawn through Q. The construction therefore determines two wave normals and two ray directions.

This construction really serves only as a representation of the refraction; it cannot be used for the determination of the direction of propagation so long as the construction is confined to the plane, because the point of contact R does not lie in the plane of incidence; we can, however, obtain the direction of the wave normal by a construction in the plane of incidence given by MacCullagh.

If we draw through D (Fig. 129) the line DE perpendicular to the refracting surface, the point of intersection B' of DE and the wave normal OB is so situated that $OB \,.\, OB' = OD^2$, for we have $OB = OQ \,.\, \sin\beta$, $OB' = OE/\sin\beta$, and further, as may easily be seen from Fig. 129, $OQ \,.\, OE = OD^2$. From this follows the relation

$$OB \,.\, OB' = OD^2.$$

But we have $OB = \omega$, $OD = \Omega$, and if we set $OB' = r'$, it follows that (n) $r' = \Omega^2/\omega$. Therefore the point B' lies on the reciprocal wave surface whose equation is [CVI. (f)]

$$a'^2x^2/(a'^2 - r^2) + b'^2y^2/(b'^2 - r^2) + c'^2z^2/(c'^2 - r^2) = 0$$

if the coordinate axes are parallel to the axes of elasticity. In this equation $a' = \Omega^2/a$, $b' = \Omega^2/b$, $c' = \Omega^2/c$. If we set

$$N_1 = \Omega/a, \quad N_2 = \Omega/b, \quad N_3 = \Omega/c,$$

and choose as the unit of length the velocity of light Ω in the surrounding medium, it follows that

$$\text{(o)} \qquad N_1^2x^2/(N_1^2 - r^2) + N_2^2y^2/(N_2^2 - r^2) + N_3^2z^2/(N_3^2 - r^2) = 0.$$

This is the equation of the reciprocal wave surface. It follows further from the discussion of CVI. that the direction of the rays OR is perpendicular to the plane tangent to the reciprocal wave surface at the point B'.

We can therefore construct the wave normal in the following way. About the point O as centre with unit radius we construct the circle PD; about the same point we draw the curve of intersection between the plane of incidence and the surface (o). This curve is represented in (Fig. 29) by $B'F'$. We then produce the incident ray to the point D lying on the circle, and draw the straight line DB' perpendicular to the refracting surface and cutting $F'B'$ at B'. OB is then the direction of the wave normal, while the direction of the ray OR' is perpendicular to the plane tangent to the surface $F'B'$ at the point B'. We can easily derive the condition for total reflection from this construction. It can also be applied to the reflection of light within the crystal itself.

SECTION CIX. DOUBLE REFRACTION IN UNIAXIAL CRYSTALS.

Using the same notation as in CVIII., we have, to determine the angle of refraction of the wave normal, the equation

(a) $$\sin \alpha/\Omega = \sin \beta/\omega.$$

In the case of uniaxial crystals ω has the values ω_1 and ω_2 which are [CVII. (b), (c)] $\omega_1^2 = b^2$, $\omega_2^2 = a^2\sin^2\epsilon + b^2\cos^2\epsilon$. In the first case the angle of refraction β_1 is obtained from the equation

$$\sin \alpha = N_0 \sin \beta_1 \text{ where } N_0 = \Omega/\omega_1,$$

the so-called ordinary index of refraction. The corresponding direction of the electrical force is perpendicular to the principal section. If the second wave normal makes the angle β_2 with the normal to the surface, we have $\sin \alpha/\Omega = \sin \beta_2/\omega_2$. If we represent the angles made by the axis of the crystal with the coordinate axes by (xa), (ya), (za) and notice that $\pi - \beta_2$, $\frac{1}{2}\pi - \beta_2$, $\frac{1}{2}\pi$, are the angles made by the refracted ray with the coordinate axes, we have

$$\cos \epsilon = -\cos(xa)\cos \beta_2 + \cos(ya)\sin \beta_2.$$

Hence, for the calculation of β_2, we have the equation

$$\Omega^2\sin^2\beta_2/\sin^2\alpha = a^2 - (a^2 - b^2)\big(\cos(xa)\cos \beta_2 - \cos(ya)\sin \beta_2\big)^2.$$

The corresponding direction of the electrical force is parallel to the principal section. If the optic axis lies in the plane of incidence we set $\cos(xa) = \cos \psi$, $\cos(ya) = \sin \psi$, and then obtain

$$\Omega^2\sin^2\beta_2/\sin^2\alpha = a^2 - (a^2 - b^2) \,.\, \cos^2(\psi + \beta_2).$$

If $A = a^2\sin^2\psi + b^2\cos^2\psi$, $B = a^2\cos^2\psi + b^2\sin^2\psi$, $C = (a^2 - b^2)\sin \psi \cos \psi$, we have $AB - C^2 = a^2b^2$ and $\Omega^2/\sin^2\alpha = A \operatorname{cotg}^2\beta_2 + 2C \operatorname{cotg} \beta_2 + B$. From this follows

(b) $$A \operatorname{cotg} \beta_2 = -C + \sqrt{A\Omega^2/\sin^2\alpha - a^2b^2}.$$

If the axis of the crystal is perpendicular to the plane of incidence, we have $(xa) = (ya) = \frac{1}{2}\pi$, from which $\sin \alpha = N_e \sin \beta_2$, where $N_e = \Omega/a$, is the extraordinary index of refraction. If a and b are expressed in terms of N_e and N_0, we have from (b)

(c) $$\begin{cases} (N_0^2\sin^2\psi + N_e^2\cos^2\psi)\operatorname{cotg} \beta_2 = -(N_0^2 - N_e^2)\sin \psi \cos \psi \\ \qquad + N_0N_e\sqrt{\sin^{-2}\alpha(N_0^2\sin^2\psi + N_e^2\cos^2\psi) - 1}. \end{cases}$$

In order to obtain the equation of the reciprocal wave surface, we set $\Omega = 1$, and substitute N_e for a, N_0 for b, in the equation for the wave surface. Thus we obtain [CVII. (d)],

(d) $$(r^2 - N_0^2)[N_e^2x^2 + N_0^2(y^2 + z^2) - N_0^2N_e^2] = 0,$$

as the equation for the reciprocal wave surface referred to the axes of elasticity as coordinate axes. We obtain the same result from CVIII. (o), if we set $N_1 = N_e$ and $N_2 = N_3 = N_0$.

In Fig. 130, OP is the refracting surface, and OA the optic axis, supposed to lie in the plane of incidence; AM_1 and AM_2 are the curves in which the plane of incidence cuts the reciprocal wave surface. AM_1 is a circle with radius N_0, AM_2 an ellipse whose semi-major axis OA equals N_0, and whose semi-minor axis OM_2 equals N_e. We draw a circle of radius $OD = 1$, which cuts at D the prolongation of the incident ray. The line ED, perpendicular to the refracting surface, cuts the reciprocal wave surface at the points B_1 and B_2. The normals to the refracted waves are then OB_1 and OB_2. For the ordinary wave the direction of the ray coincides with the wave normal OB_1; for the extraordinary wave it is perpendicular to the plane tangent to the ellipsoid at the point B_2.

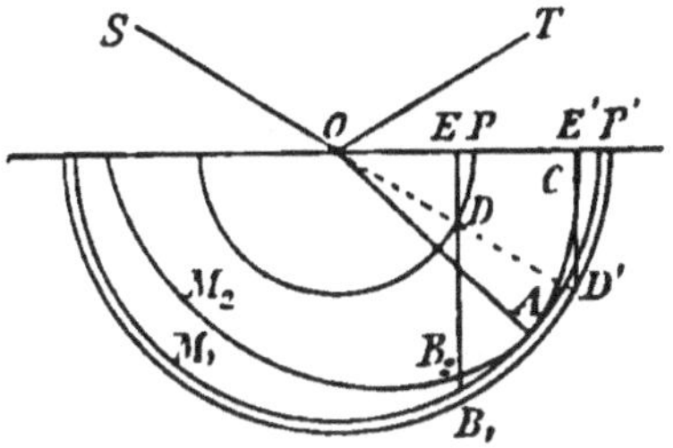

FIG. 130.

If the crystal is immersed in a fluid whose index of refraction is greater than that of the crystal, the circle PD is replaced by another circle of greater radius, for example $P'D'$. If this circle cuts the prolongation of the incident ray at D', the directions of the wave normals are determined by the point of intersection between the reciprocal wave surface and the line $D'E'$, perpendicular to the refracting surface. In this case total reflection can occur. If $D'E'$ does not cut the reciprocal wave surface there will be no refraction; if $D'E'$ cuts only one curve, there is only one refracted ray. If, as in Fig. 130, $D'E'$ touches the ellipse at a point C, refraction will occur; the direction of the ray is parallel to the bounding surface OP.

Our presentation of optics is based on Maxwell's conception of light as an electrical vibration. A more extended discussion on this same basis has been given by H. A. Lorenz. Glazebrook published a discussion of the most important optical theories in the Report for 1885 of the British Association for the Advancement of Science. von Helmholtz has lately given a theory of the dispersion of light in which he employs the electromagnetic theory of light.

CHAPTER XIII.

THERMODYNAMICS.

SECTION CX. THE STATE OF A BODY.

IF the particles of a system are in motion and exert force on one another, the system possesses a certain *energy* U. The energy of a system of discrete particles is made up of their *kinetic* and *potential energies.* The former depends on the velocities of the particles at any instant, the latter on their distances apart, or on the configuration of the system; together they determine *the state* of the body. Thus the energy at any instant depends only on the state of the system at that instant, and is independent of its previous states. The principle of energy has been proved only for a system of discrete particles; we make the assumption in the mechanical theory of heat, that the same principle or a corresponding one holds for all systems of particles.

A certain amount of energy is inherent in every body. This we call its *internal energy,* since we take no account of that part of its energy which arises from its mutual actions with other bodies. By the possession of this internal energy the body is in a condition to do work; thus variations occur in its form, volume, temperature etc. The energy is determined solely by the state of the body; if the body in a certain state possesses the energy U, and if it is subjected to any variations of form, magnitude, etc., and finally returns to its original state, the internal energy will be again equal to U.

To determine the internal energy of a body it is necessary to know the quantities which determine its state. From Boyle's and Gay-Lussac's laws the state of an ideal gas is completely determined by its pressure and volume. The temperature is given if these two quantities are known. Boyle's and Gay-Lussac's laws furnish an equation which expresses the relations between pressure, temperature, and volume; we call it *the equation of state of a gas*, because it enables us to determine the state of an ideal gas under any conditions, if it is known under definite conditions, for instance, at 0° C. and 760 mm. pressure. The behaviour of real gases cannot be accurately represented by an equation embodying Boyle's and Gay-Lussac's laws, but conforms to other equations which include these laws as a limiting case. The state of a fluid is in general determined by the same quantities; it depends to some extent on the form of the surface and the nature of the bodies in contact with it. The actions of electrical and magnetic forces may come into play in both gases and fluids. As a rule the knowledge of a great number of quantities is required to express the state of a solid, especially if it is subjected to the action of forces. The equation which unites all quantities which determine the state of a body is called *the equation of state.*

Since the state of a gas only depends on the pressure p and the volume v, it may be represented by a point in a plane with the coordinates p and v; a series of such points, or a curve, represents a series of successive states. The v-axis of this system (Fig. 131) is drawn horizontal; and the p-axis vertical. We represent the volume and pressure of the gas in its original state by v_0 and p_0; its state is then given by the point A. If the gas expands under constant pressure, its state is represented by a horizontal line AB, parallel to the v-axis. This is called the *curve of constant pressure.* The *curves of constant volume* are vertical straight lines. If heat

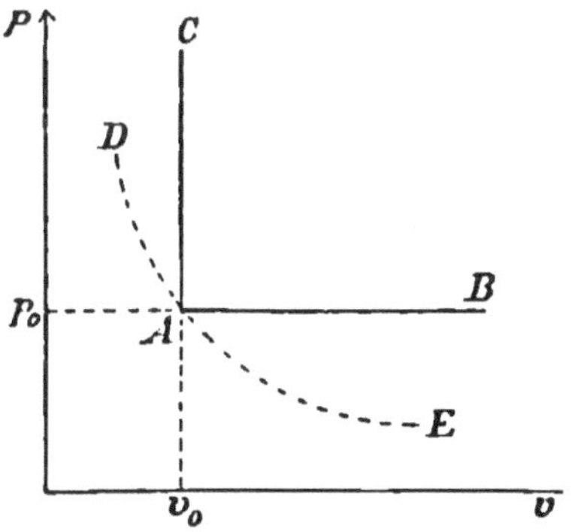

FIG. 131.

is communicated to the gas whose original state is given by A, at constant volume v_0, the variation of the state of the gas is represented by the straight line AC, and its pressure increases. If the temperature of a gas remains constant during its successive states, we have, from Boyle's law, $v.p = \text{const.}$ Hence the *curves of constant temperature* or the *isothermal lines* are rectangular hyperbolas whose asymptotes are the coordinate axes. In order to change the state of a gas in such a way that its temperature remains constant, there is required either *compression with abstraction of heat* or *expansion with communication of heat.* If a gas whose original state is represented by A is subjected to compression with abstraction of heat, or to expansion with communication of heat, in such a way that its temperature remains constant, its successive states will be represented by the hyperbola DAE. We may suppose the gas enclosed in a receptacle put in connection with an infinitely great source of heat, whose temperature is equal to that of the gas at the point A. If we change the volume of the gas, the source of heat sometimes takes up heat and sometimes gives it out, but the gas retains the temperature of the source. If the gas is enclosed in an envelope through which heat cannot pass, it is heated by compression so that its temperature rises, or cooled by expansion so that its temperature falls. In this case the changes of state are called *adiabatic* and the curve which represents them is called an *adiabatic* or *isentropic curve*.

The state of a solid cannot in general be represented in a plane, since it depends on more than two variables.

A series of changes by which the state of a body is altered in any manner, and which is such that the body finally returns to its original state, is called a *cyclic process*. If a body goes through a cyclic process the energy which it receives from surrounding bodies is equal to that which it gives up to them. The steam-engine is a system of bodies which periodically returns to the same state. It appears from the action of the steam-engine, that *heat and work are similar or equivalent quantities*, which can be transformed into one another, and are both, therefore, forms of energy. This conclusion has been established by accurate experiment. The quantity of energy produced in the one form is always proportional to that applied in the other form. This law of the *equivalence of heat and energy* was first formulated by R. Mayer (1842). The later observations of Joule and others have shown that the quantity of work which is equivalent to a unit of heat, or to the quantity of heat which will raise the temperature of a gram of water by 1° C. is equal to $4,2 \, . \, 10^7$ absolute units of work (C.G.S.). This result is called *the first law of thermodynamics*. It may be thus stated: Heat and work are equivalent; work can be obtained from heat and heat from work. The *work equivalent* or the *mechanical equivalent* of the unit of heat is designated by J.

If the quantity of heat dQ is communicated to a body it receives the energy $J \, . \, dQ$. This goes partly to increase the internal energy U of the body, partly to do the work dW. We then have

(a) $$J \, . \, dQ = dU + dW.$$

This equation is called the first fundamental equation. We will apply it to the case in which the work dW is done by expansion against external pressure.

We consider the body ABC (Fig. 132), which is subjected to the hydrostatic pressure p at every point on its surface. When the body expands its volume becomes $A'B'C'$. The normals AA', BB' are drawn from the surface-element $AB = dS$ to the new surface. We set $AA' = \nu$ and obtain for the work done by the body,

$$\int \nu p \, . \, dS = p \int \nu \, . \, dS = p \, . \, dv,$$

where dv denotes the total increase in volume of the body. Equation (a) then becomes (b) $J \, . \, dQ = dU + p \, . \, dv$. If the state of a body is determined by the independent variables p and v, the definite values p_1

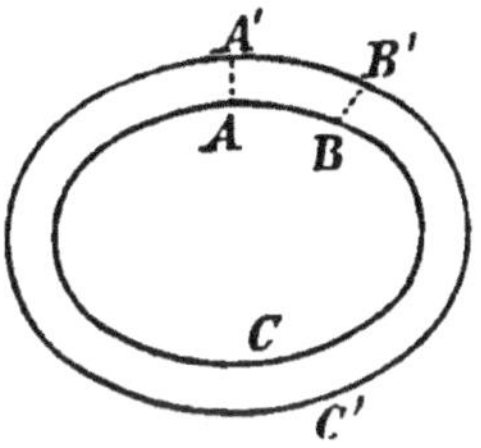

FIG. 132.

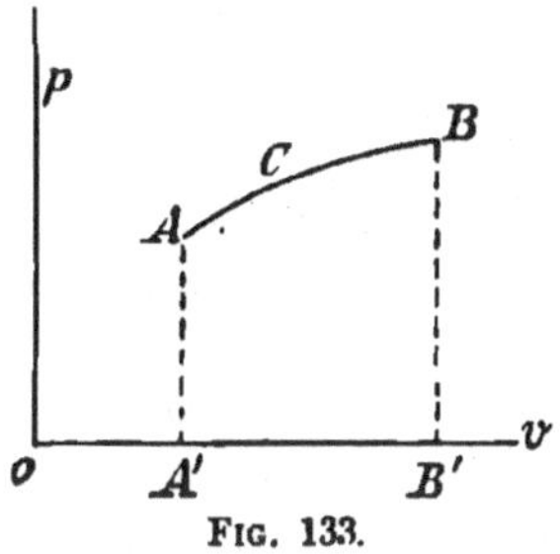

FIG. 133.

and v_1 correspond to a point A (Fig. 133). Suppose the body to pass through a series of states represented by the curve ACB; the values p_2 and v_2 correspond to the point B. We then have from (b)

$$\text{(c)} \qquad JQ = U_2 - U_1 + \int_1^2 p \, . \, dv.$$

Q is the quantity of heat introduced during the change of state, $U_2 - U_1$ the increase of the internal energy, and $\int_1^2 p \, . \, dv$ the work done. The increase $U_2 - U_1$ is determined by the initial and final values of p and v, or by the position of the points A and B. The external work is measured by the area of the figure $A'ABB'A$; *this work therefore depends on the process by which the change from one state to the other is effected.* This holds also for Q. Since U is a function of p and v, we obtain (d) $J \, . \, dQ = \partial U / \partial p \, . \, dp + (\partial U / \partial v + p) dv$. If the function U is known, it is possible to find the quantity of heat necessary to produce any change in the state of the body. U is determined from equation (c), by measuring the quantity of heat received by the body and the quantity of work done by it. Our knowledge of the quantity U is still very limited.

SECTION CXI. IDEAL GASES.

Clément and Desormes and subsequently Joule showed that the temperature of a gas which expands without overcoming resistance, that is, without doing work, remains unchanged.* The initial and final states of a gas which expands without doing work lie on the same isothermal, that is, *the internal energy of a gas is a function of its temperature only*, and is therefore independent of its volume if the temperature remains constant. If we take the temperature θ and the volume v of the gas as independent variables, we have

$$J\,.\,dQ=\partial U/\partial\theta\,.\,d\theta+\partial U/\partial v\,.\,dv+p\,.\,dv.$$

Now $\partial U/\partial v=0$ and therefore $J\,.\,dQ=\partial U/\partial\theta\,.\,d\theta+p\,.\,dv$. If the mass of gas contained in the volume v is equal to unity then $\partial U/\partial\theta=Jc_v$, where c_v denotes the *specific heat* of the gas at constant volume, that is, the quantity of heat which must be communicated to its unit of mass in order to raise its temperature one degree in such a way that, while its pressure changes, its volume remains constant. *If the specific heat of the gas at constant volume is constant, its internal energy must be a linear function of its temperature.*

For ideal gases the equation giving the relation between pressure, volume and temperature is $pv=R\theta$, where R is a constant. If θ and v are the *independent variables of the gas*, we have (a) $J.dQ=Jc_v.d\theta+p.dv$. From the observations of Regnault, c_v is independent of the pressure and temperature of the gas. If θ and p are chosen as *the independent variables*, v must be considered as a function of them, so that

$$dv=\partial v/\partial\theta\,.\,d\theta+\partial v/\partial p\,.\,dp,$$

and substituting this in equation (a) we have

$$J\,.\,dQ=(Jc_v+p\,.\,\partial v/\partial\theta)\partial\theta+p\,.\,\partial v/\partial p\,.\,dp.$$

From the equation $pv=R\theta$, it follows that

$$p\,.\,\partial v/\partial\theta=R \text{ and } p\,.\,\partial v/\partial p=-v, \text{ and } J\,.\,dQ=(Jc_v+R)d\theta-v\,.\,dp.$$

In order to obtain the *specific heat c_p at constant pressure*, that is, the quantity of heat which must be communicated to the unit of mass of the gas to raise its temperature one degree, in such a way that, while its volume changes, its pressure remains constant, we set $dp=0$ and obtain $c_p=c_v+R/J$, (b) $J\,.\,dQ=Jc_p\,.\,d\theta-v\,.\,dp$. *If p and v are*

* More exact measurements show that the gas, in these circumstances, is slightly cooled. From this it follows that there are attractive forces between its separate particles.

chosen as the independent variables, we have $d\theta = \partial\theta/\partial p \,.\, dp + \partial\theta/\partial v \,.\, dv$. It follows from $pv = R\theta$ that

$$R \,.\, \partial\theta/\partial p = v, \quad R \,.\, \partial\theta/\partial v = p, \quad R \,.\, d\theta = v \,.\, dp + p \,.\, dv,$$

and from (b) that (c) $R \,.\, dQ = c_v v \,.\, dp + c_p p \,.\, dv$. If therefore the specific heat c_p and the constant R are known, the equation (c) enables us to determine the specific heat for any change of state in the vp-plane. The specific heat has an infinite number of values for a given state in the vp-plane, depending on the direction in which this change of state takes place.

The expressions (a), (b), (c) show a noteworthy peculiarity. If one of them, say (a), is divided by θ, we obtain by the use of the fundamental equation $pv = R\theta$, $J \,.\, dQ/\theta = Jc_v \,.\, d\theta/\theta + R \,.\, dv/v$. If, for example, the gas passes from the state A (Fig. 133) to the state B, and if the temperatures and volumes at these points are θ_1, v_1 and θ_2, v_2 respectively, we have by integration,

(d) $$J \,.\, \int dQ/\theta = J \,.\, c_v \,.\, \log(\theta_2/\theta_1) + R \,.\, \log(v_2/v_1).$$

Therefore, while the integral $\int dQ$ depends on the path on which the gas passes from one state to another, the integral $\int dQ/\theta$ does not depend on this path.

Clausius called the quantity $S = J \,.\, \int dQ/\theta$ the *entropy*. This concept is of great importance in the theory of heat. *If a body passes from one state to another the change of the entropy is determined by the coordinates of the initial and final points.* This theorem is here proved only for a gas, but holds also for all bodies.

If the change of state of a gas occurs along an isothermal curve, we have from (a) $J \,.\, dQ = p \,.\, dv$. Using the equation of state and integrating, we obtain

(e) $$JQ = \int_1^2 p \,.\, dv = R\theta \,.\, \log(v_2/v_1).$$

All the heat communicated is therefore used in keeping the temperature constant. If we set v_2 equal to μv_1, $\mu^2 v_1$, $\mu^3 v_1$, etc., in succession, where μ is any number, the corresponding values for Q are

$$Q = R\theta/J \,.\, \log\mu, \quad 2R\theta/J \,.\, \log\mu, \quad 3R\theta/J \,.\, \log\mu, \text{ etc.}$$

If the change of state occurs along an isothermal curve, and if the quantities of heat introduced are in arithmetical progression, the volumes, according to equation (e), are in geometrical progression; at the same time the pressure changes proportionally to the density.

If *the change of state occurs along an adiabatic curve*, we have from (c) $c_v \log p + c_p \log v = c_1$. Setting $c_p/c_v = k$ we obtain (f) $pv^k = c$, where c is constant. The equation (f) is the *equation of the adiabatic curves.* Combining this with the equation $pv = R\theta$, we have from (f) $R\theta v^{k-1} = c$. If we introduce in this formula the density $\delta = M/v$ of the gas, where M denotes its mass, it follows that its temperature is proportional to the $(k-1)$ power of the density when the state of the gas changes along an adiabatic curve.

Further we obtain the relation [CX. (b)] $\int_1^2 p.dv = U_1 - U_2$. The work is therefore done at the expense of the internal energy, if the change of state is adiabatic.

Section CXII. Cyclic Processes.

A simple reversible cycle is one in which all changes occur in such a way that if reversed they may be effected under the same circumstances. The body which performs the cycle is called the working body. In the performance of a simple reversible cycle the working body must be associated with two others, one which communicates heat to it, and another which receives heat from it. In a gas engine the working body is the gas in the cylinder; in a steam engine it is the water or steam. The gases of the fire and the walls of the boiler give up heat, the water in the condenser receives heat. The gas or steam passes through a series of states and, at least in some machines, returns to its original state; it is then in condition to repeat the same process. Since the value of the internal energy U at the beginning and end of the process is the same, we have [CX. (a)]

(a) $$JQ = W,$$

where Q is the difference between the heat received and the heat given up.

The quantity of heat received by the working body and not given up to the colder body is the equivalent of the work done. If the working body is a gas, we have for the cycle $JQ = \int pdv$. The entropy of a gas depends only on the coordinates and therefore has the same value at the beginning and end of the process. If S_1 denotes the entropy at the starting point, the entropy at any instant during the process is equal to $S_1 + \int dQ/\theta$. If the integration is extended over the whole cycle, the entropy returns again to its value S_1, and we have therefore $\int dQ/\theta = 0$.

We will discuss more particularly a special case, the so-called *Carnot's cycle*, which is of great importance in the theory of heat.

Suppose a gram of gas to be in the state represented by the point B (Fig. 134) in the vp-plane. The curve representing the cycle is in this case composed of two isothermal curves BC and ED and of two adiabatic curves CD and BE. The gas first expands at the constant temperature θ_1. This is accomplished by keeping it in contact with the infinitely great body M_1 at the temperature θ_1, and by so regulating the external pressure on the gas that it passes to the state C along the path BC. During the change of state BC the quantity of heat Q_1 is absorbed and the work represented by the surface $BCC'B'$ is done. The gas then expands adiabatically, in the manner represented by the adiabatic curve CD, and its temperature falls to θ_2. Then the gas is brought in contact with an infinitely great body M_2 at the temperature θ_2 and compressed; during this process it gives up to M_2 the quantity of heat Q. Its state is represented by E. Finally the gas is further compressed without communication of heat until it returns to the original state B. The integral $\int p\,.\,dv$, extended over the whole cycle, equals the area $BCDE$, and represents the work done by the gas. We have [CX. (a)] (c) $J(Q_1 - Q_2) = W$.

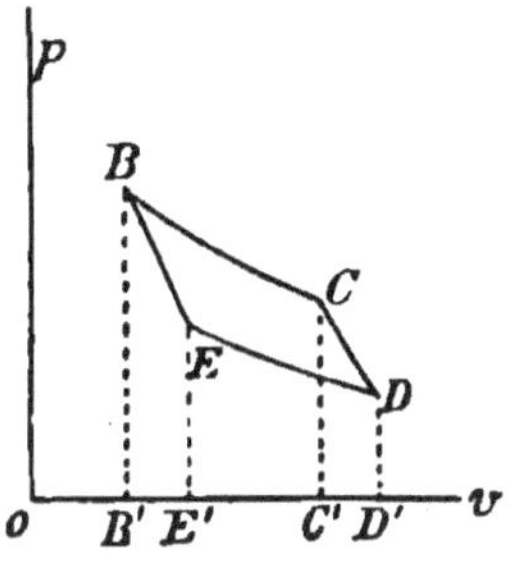

FIG. 134.

When the gas expands from B to C, its entropy is increased by Q_1/θ_1; it remains constant along the path from C to D, is diminished by Q_2/θ_2 along the path from D to E, and again remains constant along the path from E to B. Since the gas on its return to B has the same entropy as at the outset we have

(d) $$Q_1/\theta_1 - Q_2/\theta_2 = 0 \quad \text{or} \quad Q_1/\theta_1 = Q_2/\theta_2.$$

From (c) and (d) it follows that

(e) $$W = J(Q_1 - Q_2) = JQ_1(\theta_1 - \theta_2)/\theta_1.$$

Therefore the work done by this cyclic process is proportional to the quantity of heat Q_1 absorbed and to the difference of temperature $\theta_1 - \theta_2$, and is inversely proportional to the absolute temperature θ_1, at which the heat is absorbed. The heat received from the source M_1 is not wholly transformed into work, but is divided into two parts, one of which is transformed into work, and the other transferred to M_2.

The *efficiency* ζ of the Carnot's cycle is the ratio of the heat transformed into work to that communicated to the gas. We have

$$\zeta = W/JQ_1 = (\theta_1 - \theta_2)/\theta_1.$$

Hence the efficiency depends only on the temperatures θ_1 and θ_2 of the sources of heat. If we consider the reversed cycle, the state of the gas first changes along BE; along the path ED it receives a certain quantity of heat from M_2, and has a certain quantity of external work done upon it along the path DC. The heat received and the work done transformed into heat are given up by the gas to the body M_1 along the path CB. In the case of the cyclic process first considered heat is transformed into work; in the reverse process work is transformed into heat.

SECTION CXIII. CARNOT'S AND CLAUSIUS' THEOREM.

It was shown in the preceding section that $\int dQ/\theta = 0$ for any reversible cycle, when the body describing the cycle is a gas. We will now see if this theorem holds when any other body is used as the working body instead of a gas. Let us take the simple case in which the process is carried out along two isothermal lines BC and ED (Fig. 134), and two adiabatic lines CD and BE. Suppose the change of state to take place in the sense given by the letters $BCDE$. If the body at the temperature θ_1 expands from B to C, it receives the quantity of heat Q_1; when it is compressed from D to E it gives up the quantity of heat Q_2. Along the paths CD and EB heat will neither be received nor rejected. In this process the total quantity of heat received by the body is $Q_1 - Q_2$. Since it returns to its original state, the quantity of heat $Q_1 - Q_2$ is equivalent to the work done, which is therefore $J(Q_1 - Q_2)$.

S. Carnot published, in 1824, a work on the motive power of heat, in which he proposed an important theorem on the connection between heat and work. He was of the opinion that heat was a fundamental substance whose quantity remained invariable in nature. Applying this view to explain the action of the steam-engine, he supposed that the steam gave up a quantity of heat Q_1 at the higher temperature θ_1, that this heat was transferred to the condenser at the lower temperature θ_2, and that the motive power of the heat was due to its passage from the higher to the lower temperature. The work thus done by this passage of heat from a higher to a lower temperature was considered analogous to that done by a falling fluid or by any falling body. This latter is proportional to the weight of the falling body and to the distance which it falls. Hence for the work done by the heat Carnot proposed the expression

$KQ_1(\theta_1 - \theta_2)$ where K is a function of the absolute temperatures θ_1 and θ_2. This conclusion of Carnot was confirmed by experiment, but did not agree with the mechanical theory of heat in so far as it regarded heat as an invariable quantity. If for the present we disregard this error, we have for the cycle just described

(a) $$J(Q_1 - Q_2) = KQ_1(\theta_1 - \theta_2).$$

Since K must be independent of the nature of the body doing the work we have, if the body is a gas [CXII. (e)], (b) $K = J/\theta_1$. It therefore follows that $(Q_1 - Q_2)/(\theta_1 - \theta_2) = Q_1/\theta_1$ and hence $Q_1/\theta_1 = Q_2/\theta_2$, that is, *if a body traverses a Carnot's cycle any number of times, by being placed alternately in contact with two infinite sources of heat, the quantities of heat which it receives from one source and gives up to the other are in the same ratio as the temperatures of the sources.*

There can be no doubt that this theorem holds for a cycle of the kind considered. The application of the theorem in many departments of physics and chemistry has led to no results which are as yet contradicted by experiment. Several attempts were made to give a direct proof of the theorem, the first and most important of which is due to Clausius, whose method may be presented in the following way:

Suppose a gas to traverse the cycle $BCDE$ (Fig. 135) composed of the isothermal curves BC and DE, which correspond to the absolute temperatures θ_1 and θ_2, and of the adiabatic curves CD and BE. During its expansion from B to C, the gas takes the quantity of heat Q_1 from an infinitely great source M_1 whose temperature θ_1 is constant. It then expands from C to D without communication of heat. It is then brought in communication with the infinitely great source of heat M_2 whose temperature θ_2 is constant, and by compression is made to give up to it the quantity of heat Q_2. Finally it is brought back to its original state B. During the cycle the gas has received from the source M_1 the quantity of heat Q_1, which is divided into two parts. One of these parts is transferred as a quantity of heat Q_2 to the source M_2, the other is transformed into work and is represented in amount by the area $BCDE$.

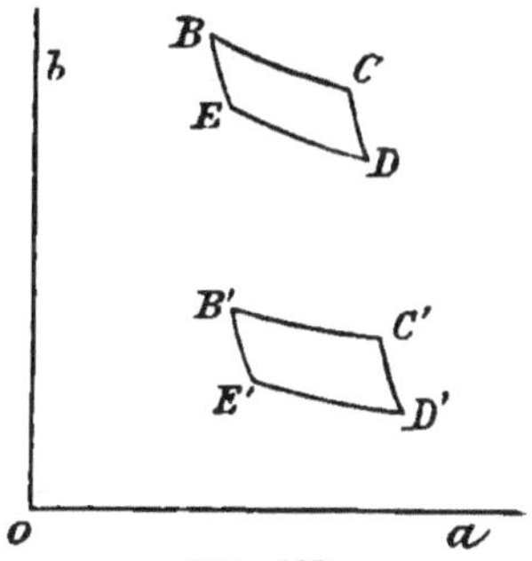

FIG. 135.

Suppose $B'C'$ and $E'D'$ (Fig. 135) to be the two isothermal curves corresponding to the temperatures θ_1 and θ_2 for another body, say for water vapour. $C'D'$ and $B'E'$ are two adiabatic curves so chosen that the surface $B'C'D'E'$ equals the surface $BCDE$. If the water-vapour is subjected to a process similar to that just described for the gas, the heat which it will receive while in contact with the source M_1 is Q_1+q, and during its passage from D' to E', while in contact with the source M_2, it gives up to that source the quantity of heat Q_2'. The work done by the vapour is equal to that done by the gas, because the surface $BCDE$ is equal to the surface $B'C'D'E'$, and we therefore have $Q_1+q-Q_2'=Q_1-Q_2$, and therefore $Q_2'=Q_2+q$. The vapour in expanding along $B'C'$ receives the quantity of heat Q_1+q, and gives up the quantity Q_2+q along the path $D'E'$.

The cycle described can also be performed in the opposite sense. For example, the water-vapour can expand along the isentropic curve $B'E'$; it may then be brought in contact with the source of heat M_2, and expand from E' to D', during which expansion the quantity of heat Q_2+q is taken from M_2. It may then be compressed along the isentropic curve $D'C'$, and lastly along the path $C'B'$, while in contact with the source of heat M_1. During this compression it gives up to M_1 the quantity of heat Q_1+q. To carry out this process, a quantity of work must be done which is equivalent to the heat

$$Q_1+q-(Q_2+q)=Q_1-Q_2,$$

this work is represented in Fig. 135 by the surface $B'C'D'E'$.

We consider finally two engines, one of which is a gas engine, in which the gas performs the cycle $BCDE$, and the other a steam-engine, in which the steam performs the reversed cycle $B'C'D'E'$. The work done by the one is equal to that supplied to the other, if we neglect friction and other resistances. The gas engine in each revolution takes from the source M_1 the quantity of heat Q_1 and gives up to the source M_2 the quantity Q_2; at the same time the steam-engine takes from M_2 the quantity of heat Q_2+q and gives up to M_1 the quantity Q_1+q. Hence, in these circumstances, the source of heat at higher temperature receives during each revolution the quantity of heat q, while the source at lower temperature M_2 gives up the same quantity of heat; this transfer of heat q from the lower to the higher temperature being effected without the doing of work.

By this process, therefore, heat can be transferred from a colder to a hotter body. This Clausius declares to contradict experience. While heat invariably tends to flow from hotter to colder bodies, in the process described above the opposite occurs. The objection has been raised to this conception of Clausius that a thermoelectric circuit, in which one junction is at the temperature 100° and the other at 0°, can produce a current which will heat a platinum wire red hot, so that heat passes from a colder to a hotter body, that is, to the red hot platinum. Clausius answered this objection by asserting that this transfer of heat to a higher temperature is compensated for by the heat generated at the points of contact.

Clausius therefore proposed this theorem: *Heat can never pass from a colder to a hotter body without the expenditure of work or the occurrence of some change of state.* Hence, by this principle of Clausius, $q=0$, and therefore for any cycle of the sort described, whatever body is used in it, we have (a) $Q_1/\theta_1 = Q_2/\theta_2$.

We can now show that a similar theorem holds for a cycle of any sort. Suppose that the change of state of a body proceeds from B along the curve BC (Fig. 136). If the isothermal curve BD passes through B and the adiabatic curve CD through C, we may replace the path BC by the path BDC, that is, the body may first expand at constant temperature along BD and then at constant entropy, that is, without communication of heat, along DC. If BC is infinitesimal, BD and DC are so also, and the change of state BC may be replaced by the two changes BD and DC. On both paths the body receives

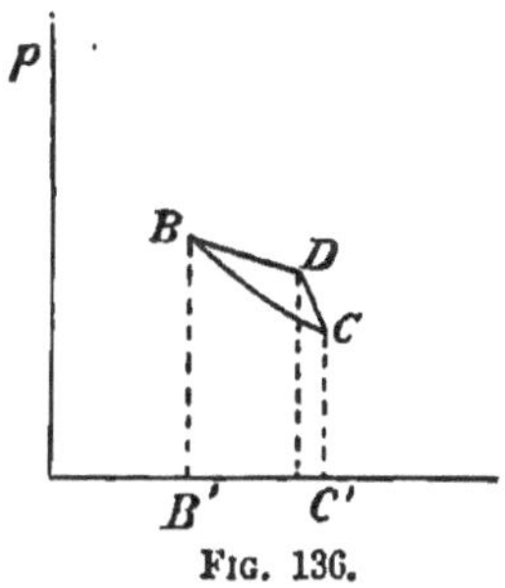

FIG. 136.

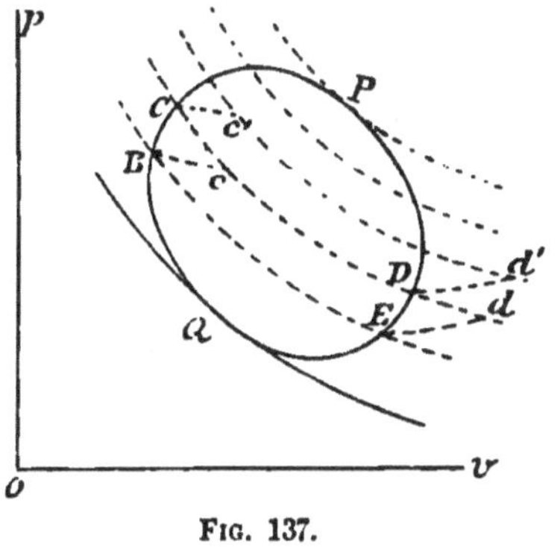

FIG. 137.

the same increment dU of internal energy. The external work is in the one case $BCC'B$, in the other $BDCC'B$. But since $B'C'=dv$ is infinitely small, while $B'B=p$ remains constant, the surface BDC vanishes in comparison with the surface $BCC'B$. If dQ represents the quantity of heat supplied, we have $J.dQ=dU+p.dv$ along the path BC, as well as along the path BDC.

Let $BCPDEQ$ (Fig. 137) be any cycle, Bc, Cc', Ed, Dd' isothermal curves, and BE, CD, etc., adiabatic curves. Let the body receive the quantity of heat dQ along BC, and give up the quantity dQ_2 along DE. As we have already seen, the body would receive by a change of state along Bc the same quantity of heat dQ_1, and give up along dE the same quantity dQ_2. Therefore we have for the cycle $BcdE$, $dQ_1/\theta_1 = dQ_2/\theta_2$, if θ_1 and θ_2 are the absolute temperatures corresponding to the isothermals Bc and Ed. In the same way we have for Cc' and Dd', etc., $dQ_1'/\theta_1' = dQ_2'/\theta_2'$, $dQ_1''/\theta_1'' = dQ_2''/\theta_2''$, etc. If Q and P denote the points at which the cycle touches two isothermal curves, we have by addition (b) $\int dQ_1/\theta_1 = \int dQ_2/\theta_2$, where dQ_1 is the quantity of heat received along an element of $QBCP$ and θ_1 the corresponding temperature, dQ_2 the quantity given up along an element of $PDEQ$ and θ_2 the corresponding temperature. *If the heat received is considered positive and that given up negative, the sum of all infinitesimal quantities of heat received during the performance of a reversible cycle, each divided by the absolute temperature at which it is received, equals zero*, that is, (c) $\int dQ/\theta = 0$. This is the *second law of thermodynamics*.

The theorem (c) which Clausius first expressed in this form may be given in another way. Let $ABCD$ (Fig. 138) represent a cycle, so that $\int_{ABCDA} dQ/\theta = 0$. We divide the integral into two parts,

$$_B\int_A^C dQ/\theta + {}_D\int_C^A dQ/\theta = 0,$$

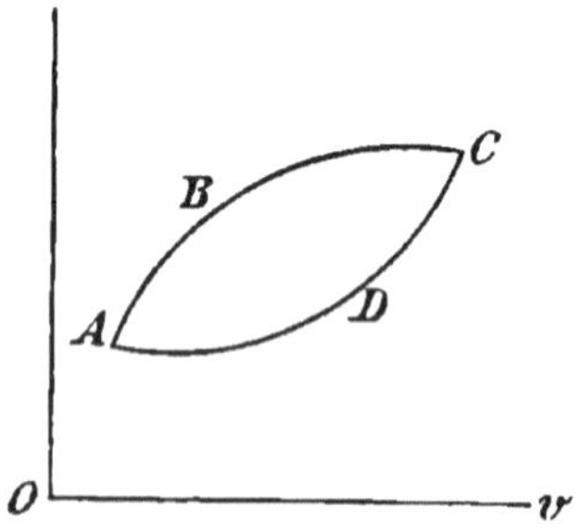

FIG. 138.

of which the first is extended from A over B to C, the second from C over D to A, and have

$$_B\int_A^C dQ/\theta = -{}_D\int_C^A dQ/\theta = {}_D\int_A^C dQ/\theta.$$

If, therefore, the body passes from the state A to the state C, the value of the integral $\int dQ/\theta$ is independent of the path. If θ_1 and v_1 are the coordinates of the point A, and θ_2 and v_2 those of the point C, we have

$$\int_1^2 dQ/\theta = f(\theta_2, v_2) - f(\theta_1, v_1).$$

Clausius introduced a special symbol for the entropy by setting $dS = J.dQ/\theta$, from which $J\int_1^2 dQ/\theta = S_2 - S_1$. The function S represents the entropy of the body; it depends only on the state of the body at any instant, and is independent of all previous states.

SECTION CXIV. APPLICATION OF THE SECOND LAW.

We have already obtained [CX. (a)] the equation

(a) $$J\,.\,dQ = dU + p\,.\,dv.$$

If the state of the body is determined only by the independent variables θ and v, equation (a) may take the form

(b) $$J\,.\,dQ = (\partial U/\partial\theta)_v\,.\,d\theta + \big((\partial U/\partial v)_\theta + p\big)dv,$$

where the indices attached indicate that the quantities which they represent remain constant during differentiation. If S denotes the entropy, we have

$$dS = J\,.\,dQ/\theta = 1/\theta\,.\,(\partial U/\partial\theta)_v\,.\,d\theta + \big(1/\theta\,.\,(\partial U/\partial v)_\theta + p/\theta\big)dv.$$

Since S is here a function of v and θ, we may set

$$(\partial S/\partial\theta)_v = 1/\theta\,.\,(\partial U/\partial\theta)_v; \quad (\partial S/\partial v)_\theta = 1/\theta\,.\,(\partial U/\partial v)_\theta + p/\theta.$$

But for the same reason we have also

$$\partial(\partial S/\partial\theta)_v/\partial v = \partial(\partial S/\partial v)_\theta/\partial\theta,$$

and further $$\partial(\partial S/\partial\theta)_v/\partial v = 1/\theta\,.\,\partial(\partial U/\partial\theta)_v/\partial v,$$

$$\partial(\partial S/\partial v)_\theta/\partial\theta = 1/\theta\,.\,\partial(\partial U/\partial v)_\theta/\partial\theta - 1/\theta^2\,.\,(\partial U/\partial v)_\theta + \partial(p/\theta)_v/\partial\theta.$$

Whence it follows that (c) $(\partial U/\partial v)_\theta = \theta^2\,.\,\partial(p/\theta)_v/\partial\theta$, since U is also a function of θ and v only, and

$$\partial(\partial U/\partial v)_\theta/\partial\theta = \partial(\partial U/\partial\theta)_v/\partial v.$$

The internal energy must satisfy the differential equation (c). The second law furnishes the means of determining the internal energy. It follows from equations (c) and (b) that

(d) $$J\,.\,dQ = (\partial U/\partial\theta)_v\,.\,d\theta + \big(\theta^2\partial(p/\theta)_v/\partial\theta + p\big)dv.$$

Hence if the equation of state and the specific heat $c_v = 1/J\,.\,(\partial U/\partial\theta)_v$ are known, the quantity of heat required for a given change in the state of the body may be determined by equation (d).

The quantity of heat which a body has received is not determined by the state of the body at any given instant, and therefore cannot be considered as a function of the coordinates. We can, however, set

(e) $$J(\partial Q/\partial\theta)_v = (\partial U/\partial\theta)_v, \quad J(\partial Q/\partial v)_\theta = (\partial U/\partial v)_\theta + p,$$

since $(\partial Q/\partial\theta)_v\,.\,d\theta$ is the quantity of heat which is used in raising the temperature by $d\theta$, while the volume remains constant, and $(\partial Q/\partial v)_\theta$ is the quantity which is absorbed during an increase in volume by dv at constant temperature. But $\partial^2 Q/\partial\theta\,\partial v$ is not equal to $\partial^2 Q/\partial v\,\partial\theta$. From equation (c) we obtain

$$\partial(\partial Q/\partial\theta)_v/\partial v - \partial(\partial Q/\partial v)_\theta/\partial\theta = -1/J\,.\,(\partial p/\partial\theta)_v.$$

The differential equation (c) is applied to the relations of an ideal gas, for which $p/\theta = R/v$. From this relation $(\partial U/\partial v)_\theta = 0$, which agrees with the results in CXI.

If the energy of a body at constant temperature is independent of its volume, its equation of state, from (c), will have the form

$$p/\theta = f(v).$$

SECTION CXV. THE DIFFERENTIAL COEFFICIENTS.

As a rule the equation of state of a body is unknown. There are, however, many bodies for which, within narrow limits, we know approximately the relations of volume, pressure, and temperature. Within these limits, therefore, an equation of state may be constructed of the form (a) $f(v, p, \theta) = 0$. *If the pressure p is constant*, we have

$$f(p, v + dv, \theta + d\theta) = 0 \text{ and } \partial f/\partial v \,.\, dv + \partial f/\partial \theta \,.\, d\theta = 0.$$

The ratio between dv and $d\theta$ is written in the form $(\partial v/\partial \theta)_p$. The volume v of the body is generally given in the form

$$v = v_0(1 + \alpha t + \beta t^2 + \ldots),$$

where $t = \theta - 273$. Hence we have

$$(\partial v/\partial \theta)_p = v\big(\alpha + 2\beta(\theta - 273) + \ldots\big)/\big(1 + \alpha(\theta - 273) + \ldots\big).$$

If β is very small, we obtain (b) $(\partial v/\partial \theta)_p = v\alpha$. This formula can be used even when α is a function of θ.

If *the temperature θ is constant* in equation (a), we have

$$\partial f/\partial p \,.\, dp + \partial f/\partial v \,.\, dv = 0.$$

From this equation the change of volume due to change of pressure at constant temperature can be determined. Since the volume always diminishes when the pressure increases, $(\partial v/\partial p)_\theta$ is negative. From the theory of elasticity (cf. XXIX.) we have found that

$$\lambda \partial v/v = -\partial p \text{ and } \Theta = \partial v/v = -(1 - 2k)3\partial p/E,$$

in the case of fluids and solids respectively. Therefore

(c) $$(\partial v/\partial p)_\theta = -v/\lambda.$$

It follows from the equation of state of an ideal gas that

$$(\partial v/\partial p)_\theta = -v/p,$$

and hence $\lambda = p$.

If the volume of the body is constant, the pressure is increased by dp by the rise of temperature $d\theta$; we obtain in this way a third quantity $(\partial p/\partial \theta)_v$. Besides these differential coefficients we must also notice three others, $\partial\theta/\partial v$, $\partial p/\partial v$, and $\partial\theta/\partial p$, which are connected with those already mentioned by the following relations:

(d) $(\partial v/\partial\theta)_p \,.\, (\partial\theta/\partial v)_p = 1$, $(\partial v/\partial p)_\theta \,.\, (\partial p/\partial v)_\theta = 1$, $(\partial p/\partial\theta)_v \,.\, (\partial\theta/\partial p)_v = 1$.

Let LM and NP (Fig. 139) represent two isothermal curves corresponding to the temperatures θ and $\theta + d\theta$. We then have

$$(\partial p/\partial v)_\theta = \operatorname{tg} AEv,$$

if AE is the tangent at the point A of the isothermal curve whose parameter is θ. If BAD is perpendicular to Ov, and AC parallel to Ov, we have

$$\operatorname{tg} AEv = -AD/DE = -AB/AC,$$

and hence

$$(\partial p/\partial v)_\theta = -AB/AC.$$

Further, we have

$$(\partial v/\partial\theta)_p = AC/d\theta,$$
$$(\partial\theta/\partial p)_v = d\theta/AB,$$

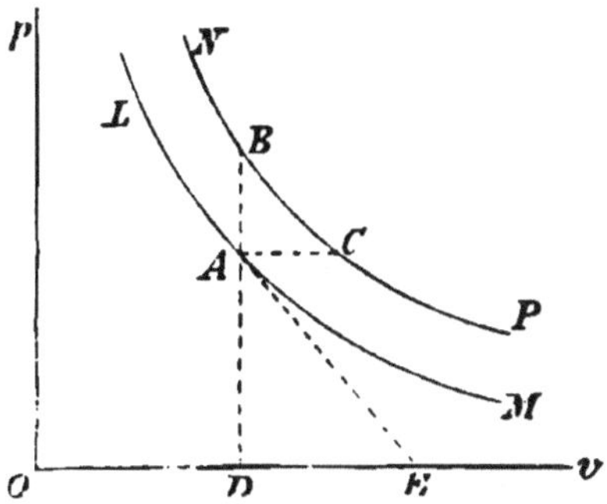

FIG. 139.

and hence we obtain (e) $(\partial p/\partial v)_\theta \,.\, (\partial v/\partial\theta)_p \,.\, (\partial\theta/\partial p)_v = -1$.

Equations (d) and (e) show that if we know two of these differential coefficients which are independent, the others are also known. Equation (e) may be derived in the following way: If p is considered a function of v and θ, we have $dp = (\partial p/\partial v)_\theta \,.\, dv + (\partial p/\partial\theta)_v \,.\, d\theta$. Assuming the pressure constant, so that $dp = 0$, we have

$$dv/d\theta = -(\partial p/\partial\theta)_v/(\partial p/\partial v)_\theta.$$

The quantity $dv/d\theta$ in this equation is that which has already been designated by $(\partial v/\partial\theta)_p$; we therefore again obtain equation (e).

For gases we have $(\partial p/\partial v)_\theta = -p/v$; $(\partial v/\partial\theta)_p = R/p$; $(\partial\theta/\partial p)_v = v/R$. These values satisfy equation (e).

For liquids and solids we have $(\partial v/\partial\theta)_p = av$, $(\partial v/\partial p)_\theta = -v/\lambda$, and hence by equation (e), (f) $(\partial p/\partial\theta)_v = a\lambda$.

SECTION CXVI. LIQUIDS AND SOLIDS.

If θ and v are the independent variables, we have [CXI.]

(a) $$J \,.\, dQ = (\partial U/\partial\theta)_v \,.\, d\theta + \big((\partial U/\partial v)_\theta + p\big) \,.\, dv.$$

From the second law of thermodynamics we obtain as in CXIV. (c), (b) $(\partial U/\partial v)_\theta = \theta^2 \,.\, \partial(p/\theta)_v/\partial\theta = \theta \,.\, (\partial p/\partial\theta)_v - p$. Hence equation (a) takes the form (c) $J.dQ = (\partial U/\partial\theta)_v \,.\, d\theta + \theta \,.\, (\partial p/\partial\theta)_v \,.\, dv$. If we designate the specific heat at constant volume by c_v we have (d) $J.c_v=(\partial U/\partial\theta)_v$. Equation (c) then becomes [CXV. (f)] (e) $dQ=c_v.d\theta+\theta a\lambda.dv/J$. It follows from equations (b) and (d) that

$$\text{(f)} \qquad J.\partial c_v/\partial v=\partial^2 U/\partial v\partial\theta=\theta.\partial(\partial p/\partial\theta)_v/\partial\theta,$$

and therefore from CXV. (f) that (g) $J.\partial c_v/\partial v=\theta.\partial(a\lambda)/\partial\theta$. This may be obtained also from equation (e) by the use of the second law. Equation (g) shows that c_v is independent of the volume if $a\lambda$ is independent of the temperature.

In order to express the dependence of the quantity of heat communicated to a body on its pressure and temperature, we set

$$dv=(\partial v/\partial\theta)_p.d\theta+(\partial v/\partial p)_\theta.dp,$$

and then obtain from (c)

$$J.dQ=\{(\partial U/\partial\theta)_v+\theta.(\partial p/\partial\theta)_v.(\partial v/\partial\theta)_p\}d\theta+\theta.(\partial p/\partial\theta)_v.(\partial v/\partial p)_\theta.dp,$$

or [CXV. (e)], (h) $J.dQ=\{Jc_v-\theta.(\partial v/\partial\theta)_p^2/(\partial v/\partial p)_\theta\}d\theta-\theta.(\partial v/\partial\theta)_p.dp$.

If c_p is the specific heat at constant pressure, we have

$$c_p=c_v-\theta/J.(\partial v/\partial\theta)_p^2/(\partial v/\partial p)_\theta.$$

Now since $\partial v/\partial p$ is always negative, c_p is greater than c_v, so long as $\partial v/\partial\theta$ is not equal to zero; the case in which $\partial v/\partial\theta=0$ is exhibited by water at 4° C.

Introducing the values for the differential coefficients found in CXV., we obtain

$$\text{(i)} \quad J.dQ=\{Jc_v+a^2\lambda v\theta\}d\theta-av\theta.dp \quad \text{and} \quad \text{(k)} \quad c_p=c_v+a^2\lambda v\theta/J.$$

The temperature of a fluid or of a solid is changed by compression. If we set $dQ=0$ in equation (i) the rise of temperature $d\theta$ due to the increase of pressure dp is $d\theta=+av\theta/c_pJ.dp$, that is, *the temperature rises with increasing pressure if a is positive, that is, if the body expands when heated; if a is negative the temperature falls with increasing pressure.*

Section CXVII. The Development of Heat by Change of Length.

If the pressure p is exerted on each unit of surface of the ends of a solid cylinder, each unit of length of the cylinder is shortened by p/ϵ, where ϵ is a constant. If l denotes the original length of the cylinder at 0° C., its length L at the temperature θ and under the pressure p, if the limits of elasticity are not exceeded, is

(a) $$L = l.(1 - p/\epsilon).(1 + \beta(\theta - 273)),$$

where β is the *coefficient of expansion*.

If the length of the body increases by dL and its temperature rises by $d\theta$, the quantity of heat dQ must be supplied to it; the work done by the elongation is $A.p.dL$, if A is the cross-section of the cylinder. If M represents the mass of the body when the temperature is θ and the length L, we have

(b) $$J.dQ = M.\partial U/\partial\theta.d\theta + (M.\partial U/\partial L + Ap).dL.$$

Applying the second law to this expression we obtain

$$\partial(M/\theta.\partial U/\partial\theta)/\partial L = \partial(M/\theta.\partial U/\partial L + Ap/\theta)/\partial\theta$$

or $$M.(\partial U/\partial L)_\theta = \theta^2.A.\partial(p/\theta)_L/\partial\theta.$$

Hence $$J.dQ = M.(\partial U/\partial\theta)_L.d\theta + \theta.A.(\partial p/\partial\theta)_L.dL.$$

If θ and p are taken as the independent variables, we have

$$dL = (\partial L/\partial\theta)_p.d\theta + (\partial L/\partial p)_\theta.dp \text{ and}$$

$$J.dQ = [M.(\partial U/\partial\theta)_L + \theta.A.(\partial p/\partial\theta)_L.(\partial L/\partial\theta)_p]d\theta$$
$$+ \theta.A.(\partial p/\partial\theta)_L.(\partial L/\partial p)_\theta.dp.$$

Since the deformation is very small, we have, representing by c_p the specific heat at constant pressure,

(c) $$J.dQ = JMc_p.d\theta - \theta A.(\partial L/\partial\theta)_p.dp,$$

since by analogy with CXV. (e) $(\partial p/\partial\theta)_L.(\partial\theta/\partial L)_p.(\partial L/\partial p)_\theta = -1$. If the pressure on the ends is increased by dp, so that the total pressure is $A.dp = P$, and if there is no communication of heat, the temperature of the body increases by $d\theta = \theta PL\beta/JMc_p$, or, if m is the mass of unit length, by $d\theta = \theta\beta P/Jmc_p$. If the cylinder is stretched by the force P a corresponding cooling will occur.

SECTION CXVIII. VAN DER WAAL'S EQUATION OF STATE.

The equation of state of an ideal gas is $pv = R\theta$, and its isothermal curve is therefore a rectangular hyperbola. Real gases, however, at low temperatures and under high pressures, do not conform to this equation. Suppose that a certain quantity of gas at a given temperature has the volume OC' (Fig. 140) and is under the pressure CC'. If the pressure is increased while the temperature remains constant, the volume will be diminished. At last the space in which the gas is contained becomes *saturated* with it; let the corresponding pressure be DD'. DD' is then the *pressure of the saturated vapour* or the *vapour pressure* at the given temperature. If the volume is still further diminished, the pressure remains constant, while a part of the vapour passes over into the liquid state. At last all the vapour is transformed into liquid; let the corresponding volume be OF'. *So long as the vapour and liquid are in the same space, the isothermal curve is a straight line parallel to the axis Ov.* If the volume of the liquid is now diminished, the pressure increases very rapidly; the corresponding isothermal curve is represented by FG (Fig. 140). Andrews found, by experimenting with carbon-dioxide, that, as the temperature rises, the line DF becomes shorter, and that, at a certain temperature, which he called the *critical temperature*, it disappears altogether. If the temperature of the carbon-dioxide remains constant, its state changes along curves which are represented in Fig. 141. The abscissas represent volumes, the ordinates pressures. For example, let us examine the isothermal curve $ABCD$, which corresponds to the temperature 15·1° C.; at the point A the carbon-dioxide is still in the gaseous state; at B it may be considered as saturated vapour. If the compression is continued, condensation begins, and the pressure remains constant until the substance has become liquid, that is, until its state is represented by C. From C on, the pressure increases very rapidly as the volume is diminished. At the temperature 21·5° C. the condensation begins at B, and the horizontal part of the curve is shorter. At 31·1° C. the horizontal part of the isothermal curve vanishes; the critical temperature has now been reached. Isothermals corresponding to higher temperatures are continuous curves; it is therefore impossible to reduce carbon-dioxide to the liquid state at a temperature higher than 31·1° C. At higher temperatures than this there is no apparent difference between its liquid and gaseous states. The liquid, at the critical temperature, has the same density as the saturated vapour. *A gas can be reduced*

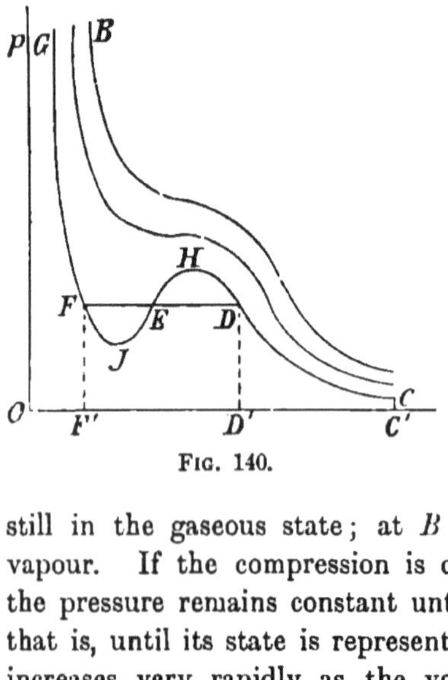

FIG. 140.

to the fluid state by compression only when its temperature is lower than the critical temperature.

James Thomson substituted for the isothermal curve here described a continuous curve *CDHEJFG* (Fig. 140); the part *DHEJF* corresponds to an unstable state. It appears from various investigations on the relations between vapours and their liquids at the boiling point, that it is possible to obtain a vapour in the states represented

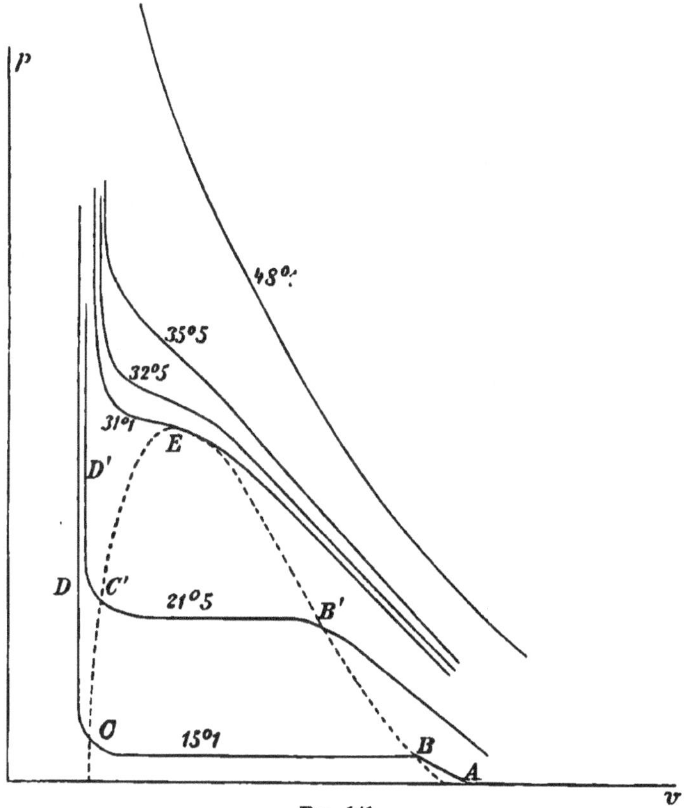

FIG. 141.

by *DH* and *FJ*, while the states represented by *HEJ* are always unstable, since in these states the pressure and volume change in the same sense.

J. Clerk Maxwell called attention to an important peculiarity of these isothermals which may be deduced by applying the laws of thermo-dynamics. If a gas traverses the cycle *FEDHEJF*, in passing along the straight line from *F* to *D* it receives the quantity of heat

L, and in passing along the curve $DHEJF$ gives up the quantity L'; the temperature is the same along both paths. Since the gas has traversed a complete cycle, we have $\int dQ/\theta = L/\theta - L'/\theta = 0$, and therefore $L = L'$. Therefore, since no heat is used in this cycle, no work can be done, and hence the surface FJE is equal to the surface DHE. Thus, if the isothermal curve $CDHEJFG$ is given, the maximum pressure of the vapour can be determined, by determining the line FD so that the surfaces FJE and DHE are equal.

Van der Waals has proposed an equation of state for gases, which represents, more exactly than the simple one, the behaviour of the gas and which permits the calculation of the critical temperature. The volume of a gas is determined not only by the external pressure but also by the attraction of its molecules; we may think of this attraction as replaced by a pressure p' added to the external pressure p. Since the attracting and attracted molecules approach one another as the density increases, p' must be directly proportional to the square of the density, and therefore inversely proportional to the square of the volume. Hence we set $p' = a/v^2$, so that the total pressure of the gas is $p + a/v^2$. Further, the molecules are not free to move everywhere in the region v, for they themselves occupy part of the region. Van der Waals assumed that the volume of a fluid cannot fall below a certain limit without the particles losing their freedom of motion. In place of the apparent volume v, he used, as the true or effective volume, $v - b$, where b is a very small quantity, though much greater (about 4 – 8 times) than the volume of all the molecules of the gas. We thus obtain the equation of state

(a) $$(p + a/v^2)(v - b) = R\theta.$$

If the volume v is very great, this equation becomes the equation of state of ideal gases.

The positions of the points H and J (Fig. 140), at which the tangents to the isothermal curves are parallel to the v-axis, are obtained from the equation $dp/dv = 0$ or (b) $p + a/v^2 - 2a(v - b)/v^3 = 0$. This is the equation of the curve which passes through all points at which the tangents to the isothermal curves are parallel to the axis Ov. All these isothermal curves correspond to temperatures at which the body can be either liquid or gaseous. When the two points coincide we reach the critical state. But since the two coincident points must have a line joining them parallel to the axis Ov, we introduce the condition for the critical state by setting dp/dv, obtained by differentiating the foregoing equation (b), equal to zero, or

(c) $$6a(v - b)/v^4 - 4a/v^3 = 0.$$

If v_1 denotes the critical volume, that is, the volume of unit mass of the gas or liquid at the critical temperature, we obtain from the last equation (d) $v_1 = 3b$. The critical temperature θ_1, and the critical pressure p_1, which must exist in order that the fluid shall not boil at a temperature which is lower by an infinitesimal than the critical temperature, are (e) $\theta_1 = 1/R \, . \, 8a/27b$, $p_1 = a/27b^2$. These values are obtained by introducing the value of v_1 in (b) and (a).

Choosing v_1, θ_1, and p_1 as units of volume, temperature, and pressure, we may set

$$v = Vv_1 = V \, . \, 3b, \quad p = Pp_1 = P \, . \, a/27b^2, \quad \theta = T\theta_1 = T \, . \, 8a/27bR,$$

and thus give to the equation of state the form

$$\text{(f)} \qquad (P + 3/V^2)(3V - 1) = 8T.$$

From this equation we obtain the following law: *If the critical pressure is taken as the unit of pressure, the critical volume as the unit of volume, and the absolute critical temperature as the unit of temperature, the isothermals of all bodies are the same.* We will now consider some applications of formula (f).

(*a*) Equation (f) may be written in the form

$$PV = \frac{8VT}{3V - 1} - 3/V.$$

The following table shows how the product PV depends on P when the temperature is constant:

$\frac{1}{V}$	$T=1\cdot0$		$T=1\cdot5$		$T=2\cdot0$		$T=2\cdot5$		$T=3\cdot0$	
	PV	P	PV	P	PV	P	PV	P	PV	P
0·0	2·67	0·00	4·00	0·00	5·33	0·00	6·67	0·00	8·00	0·00
0·2	2·26	0·45	3·69	0·74	5·11	1·02	6·54	1·31	7·97	1·59
0·4	1·88	0·75	3·42	1·37	4·96	1·98	6·50	2·60	8·03	3·21
0·6	1·53	0·92	3·20	1·92	4·87	2·92	6·54	3·92	8·20	4·92
0·8	1·24	0·99	3·06	2·44	4·87	3·90	6·69	5·35	8·51	6·81
1·0	1·00	1·00	3·00	3·00	5·00	5·00	7·00	7·00	9·00	9·00
1·2	0·85	1·01	3·07	3·68	5·29	6·35	7·51	9·01	9·73	11·68
1·4	0·80	1·12	3·30	4·62	5·80	8·12	8·30	11·62	10·80	15·12
1·6	0·91	1·46	3·77	6·03	6·63	10·60	9·49	15·38	12·34	19·75
1·8	1·27	2·28	4·60	8·28	7·93	14·28	11·27	20·28	14·60	26·28
2·0	2·00	4·00	6·00	12·00	10·00	20·00	14·00	28·00	18·00	36·00

The results of this table may be represented by a figure, in which P is the abscissa and PV the ordinate; in the neighbourhood of the critical temperature $T=1$, PV is very variable, and the departures from the ordinary laws are in consequence very considerable. As the temperatures rise the relations change rather rapidly, PV becoming nearly constant. The product PV has a minimum value, which in some cases is reached only when the pressure is negative. This minimum is determined by $dPV/dV = 3/V^2 - 8T/(3V-1)^2 = 0$. PV is therefore a minimum if $3V - 1/V = \sqrt{8T/3}$.

The corresponding values of P and V, which we may designate by P_m and V_m, are, by the use of the equation of state, equal to

$$P_m = 3(2\sqrt{8T/3} - 3)(3 - \sqrt{8T/3}), \quad V_m = 1/(3 - \sqrt{8T/3}),$$

and hence $P_m V_m = 6\sqrt{8T/3} - 9$.

Hence we obtain for

$T=$	1·0	1·5	2·0	2·5	3·0.
$P_m=$	1·09	3·00	3·35	2·72	1·37.
$P_m V_m=$	0·80	3·00	4·86	6·49	7·97.

As an example of the application of the relations here developed, the following table has been calculated for 18° C. or $\theta=291$:

	Crit. Temp. θ_1	Crit. Press. P_1	T	P_m	$p=P_m p_1$
Hydrogen,	33	—	7·3	neg.	neg.
Nitrogen,	127	33	2·29	3·08	102
Carbon-monoxide,	132	36	2·20	3·19	115
Oxygen,	155	50	1·88	3·37	169
Nitrogen-monoxide,	179	71	1·63	3·21	228
Methane,	191	55	1·52	3·05	168
Ethylene,	263	51	1·11	1·68	86

The product of pressure and volume therefore diminishes as the pressure increases, if the pressure is less than 102 atmospheres. If the pressure is greater than this the product increases with the pressure.

(*b*) The coefficient a_v which is called the *coefficient of pressure* (formerly called the coefficient of expansion at constant volume), and which denotes the change of pressure for a rise of temperature of 1° C. at constant volume of the gas, is determined in the following way: We have

$$a_v = 1/p \,.\, \partial p/\partial\theta = 1/\theta_1 P \,.\, \partial P/\partial T = 1/\theta_1 \,.\, 8/(3PV - P)$$
$$= 1/\theta_1 \,.\, 1/(T - \tfrac{9}{8} \,.\, 1/V + \tfrac{3}{8} \,.\, 1/V^2).$$

For very great values of V, $a_v = 1/\theta$, corresponding to an ideal gas. Further $a_v \cdot \theta_1 = 1/(T - \frac{9}{8} \cdot 1/V + \frac{3}{8} \cdot 1/V^2)$. Van der Waals defines *corresponding states* as those in which the volumes, temperatures and pressures of both gases are in the same ratio to the same quantities in the critical state, that is, in which

$$v/v_1 = v'/v_1' = V, \quad p/p_1 = p'/p_1' = P, \quad \theta/\theta_1 = \theta'/\theta_1' = T,$$

where the quantities v', v_1', p', etc., refer to the second gas; thus we have $a_v\theta_1 = a_v'\theta_1'$ or $a_v/a_v' = \theta_1'/\theta_1$. *Hence the coefficients a_v and a_v' in corresponding states are inversely as the critical temperatures.*

(*c*) The coefficient of expansion a_p, which represents the increase of volume for a rise of temperature of 1° C. at constant pressure, is defined in the following way:

$$a_p = 1/v \cdot \partial v/\partial\theta = 1/\theta_1 V \cdot \partial V/\partial T = 1/\theta_1 \cdot 1/(T + \tfrac{1}{8}P - \tfrac{9}{4} \cdot 1/V + \tfrac{9}{8} \cdot 1/V^2).$$

With increasing values of V, a_p approaches the value a_v. If the bodies are in corresponding states, we have $a_p\theta_1 = a_p'\theta_1'$. If the basis for this method is sound it should find application in the expansion of liquids by heat. Since changes of pressure have only slight effect on the volume of liquids, it is sufficient to compare the coefficients of expansion at corresponding temperatures. The calculations made by van der Waals have shown that this law is in essentials correct for liquids whose critical temperatures are known.

(*d*) *The pressure of saturated vapours.*—To determine the pressure of saturated vapours we use the above-mentioned theorem of Maxwell, which states that the surface $F'FEDD'$ (Fig. 140) is equal to the surface $F'FJEHDD'$, that is, setting

$$OF' = V_1, \quad OD' = V_2, \quad \text{and} \quad F'F = D'D = P_1,$$

we have
$$P_1(V_2 - V_1) = \int_{V_1}^{V_2} P \, dV.$$

Effecting the integration and using the equation of state (f) it follows that

(g) $$\tfrac{8}{3} \cdot T \cdot \log\left((3V_2 - 1)/(3V_1 - 1)\right) = (P_1 + 3/V_1V_2)(V_2 - V_1).$$

For the points F and D the following equations hold:

$$(P_1 + 3/V_1^2)(3V_1 - 1) = 8T \quad \text{and} \quad (P_1 + 3/V_2^2)(3V_2 - 1) = 8T.$$

If V_1 and V_2 are eliminated from these equations, a relation is obtained between the pressure P_1 of the saturated vapour and the temperature T. We may therefore conclude with van der Waals that: *If for different bodies the absolute temperature is the same multiple of the critical temperature, the pressure of their saturated vapours is also the same multiple of their critical pressures.* Similar laws hold for the relation between the volume of the saturated vapour and its pressure and temperature.

On the basis of van der Waals' investigations, Clausius has presented a slightly altered form of the equation of state, viz.,

$$\left(p + a/\theta(v+\beta)^2\right)(v-b) = R\theta.$$

This equation represents the actual relations better than the other, but leads to essentially the same results. Starting from the equation (h) $J\,.\,dQ = (\partial U/\partial\theta)_v d\theta + \left((\partial U/\partial v)_\theta + p\right)dv$, and applying the second law we obtain the relation [CXIV. (c)], $(\partial U/\partial v)_\theta = \theta^2\,.\,\partial(p/\theta)_v/\partial\theta$. From Clausius' equation $p/\theta = R/(v-b) - a/\theta^2(v+\beta)^2$, and hence

(i) $$(\partial U/\partial v)_\theta = 2a/\theta(v+\beta)^2.$$

If the temperature increases by $d\theta$ and the volume by dv, the internal energy increases by

(k) $$dU = Jc_v\,.\,d\theta + 2a/\theta(v+\beta)^2\,.\,dv,$$

where $Jc_v = (\partial U/\partial\theta)_v$ is an unknown function of θ and v. If the changes of temperature are slight, c_v may be considered constant, as is shown by observation. If the temperature increases from θ to $\theta + \Delta\theta$, while v increases from v_1 to v_2, the increment ΔU of the internal energy is approximately equal to

(l) $$\Delta U = Jc_v\,.\,\Delta\theta + 2a/\theta\,.\,(1/v_1 - 1/v_2).$$

If the gas expands without resistance the internal energy remains constant. Hence, if we set $\Delta U = 0$ in equation (l), we have

(m) $$\Delta\theta = -2a/Jc_v\theta\,.\,(1/v_1 - 1/v_2).$$

In this case the temperature falls.

Section CXIX. Saturated Vapours.

If the volume of a gram of a certain substance, or its *specific volume*, is called v_1 when it is a vapour and v_2 when it is a liquid, the volume v occupied by a gram of liquid and vapour is (a) $v = v_1 x + v_2(1-x)$, if the volume contains x grams of vapour and therefore $(1-x)$ grams of liquid. If U_1 is the internal energy of the vapour, U_2 that of the liquid, the internal energy of the mixture of the two is

(b) $$U = U_1 x + U_2(1-x).$$

So long as the vapour is saturated, its internal energy and pressure depend only on the temperature; the pressure and volume of the liquid are also determined by it. Hence, for a mixture of liquid and saturated vapour, we have (c) $p = f(\theta)$, *where p is the pressure of the saturated vapour at the temperature θ, and consequently θ is the boiling point of the liquid under the pressure p.*

That quantity of heat which is needed to transform one gram of liquid into vapour at constant temperature θ and under the corresponding pressure p is called the *heat of vaporization L.* This heat is partly used in increasing the internal energy, partly in doing external work. If U_2 denotes the energy of the liquid, U_1 that of the vapour, at the temperature θ, the internal energy is increased by $U_1 - U_2$ by the transformation. Since the pressure p is constant, the external work is $\int_{v_2}^{v_1} p dv = p(v_1 - v_2)$. The work needed to evaporate a gram of liquid is (d) $J.L = U_1 - U_2 + p(v_1 - v_2)$.

If a mixture of liquid and vapour receives heat and also changes its volume, the increase of internal energy, since θ and x both vary, is

(e) $$dU = (U_1 - U_2)dx + (x.\partial U_1/\partial\theta + (1-x).dU_2/\partial\theta)d\theta.$$

Since, from equation (a),

(f) $$dv = (v_1 - v_2)dx + (x.\partial v_1/\partial\theta + (1-x).\partial v_2/\partial\theta)d\theta,$$

it follows from the equation $J.dQ = dU + p.dv$ that the heat imparted to the mixture is determined by

(g) $$\begin{cases} J.dQ = \{x(dU_1/\partial\theta + p.\partial v_1/\partial\theta) + (1-x)(\partial U_2/\partial\theta + p.\partial v_2/\partial\theta)\}d\theta \\ \qquad\qquad + (U_1 + pv_1 - U_2 - pv_2)dx. \end{cases}$$

If in this equation we set $d\theta = 0$, and integrate from $x = 0$ to $x = 1$, we obtain equation (d). If equation (g) is brought into the form $J.dQ = \Theta.d\theta + X.dx$, it follows, from the Carnot-Clausius theorem, that $\partial(\Theta/\theta)/\partial x = \partial(X/\theta)/\partial\theta$. But we have

$$\partial(\Theta/\theta)/\partial x = 1/\theta.(\partial U_1/\partial\theta + p.\partial v_1/\partial\theta) - 1/\theta.(\partial U_2/\partial\theta + p.\partial v_2/\partial\theta);$$

$$\partial(X/\theta)/\partial\theta = 1/\theta.(\partial U_1/\partial\theta + p.\partial v_1/\partial\theta + v_1.\partial p/\partial\theta)$$
$$- 1/\theta.(\partial U_2/\partial\theta + p.\partial v_2/\partial\theta + v_2.\partial p/\partial\theta) - 1/\theta^2.(U_1 + pv_1 - U_2 - pv_2).$$

Hence it follows that (h) $U_1 - U_2 = (v_1 - v_2).\theta.\partial p/\partial\theta - p(v_1 - v_2)$, and thus the difference between the internal energy of the vapour and that of the fluid is determined.

We obtain from equations (d) and (h), (i) $JL = (v_1 - v_2)\theta.\partial p/\partial\theta$. Now we know by observation that $v_1 > v_2$, so that $\partial p/\partial\theta$ is positive. *The boiling point is therefore higher, the higher the pressure.*

We may also apply equation (i) to the process of *melting.* In that case v_1 denotes the volume of the liquid, v_2 that of the solid. We must here distinguish between two kinds of substances, those like wax, whose volume increases during melting, and those like ice, whose volume diminishes during melting. For the former $v_1 > v_2$, and therefore $\partial p/\partial\theta$ is positive; for the latter $v_1 < v_2$, and therefore $\partial p/\partial\theta$ is negative. *For those substances whose volume increases during melting, the melting temperature rises as the pressure increases. For those substances whose volume diminishes during melting, the melting temperature falls as the pressure increases.*

If the volume is always filled with saturated vapour only, it follows from (g), since $x=1$, that (k) $J.dQ=(\partial U_1/\partial\theta+p.\partial v_1/\partial\theta)d\theta$. Hence dQ is the quantity of heat which must be imparted to the vapour that its temperature shall increase by $d\theta$ while it remains saturated.

From equation (d) we have $U_1-U_2=JL-p(v_1-v_2)$. If c denotes the specific heat of the liquid, and if k is a constant, we may set $U_2=Jc\theta+k$. If we consider v_2 as constant, it follows that

$$\partial U_1/\partial\theta=Jc+J.dL/d\theta-(v_1-v_2).\partial p/\partial\theta-p.\partial v_1/\partial\theta.$$

From equations (i) and (k) we then have $dQ=(dL/d\theta-L/\theta+c)d\theta$. Designating by h the quantity of heat which must be used in raising the temperature of the vapour by 1°C. while it remains saturated, we have (l) $$h=dL/d\theta-L/\theta+c.$$

SECTION CXX. THE ENTROPY.

The methods which have here been applied to the discussion of the equilibrium of a fluid and its vapour may be used to advantage in many other cases, especially in connection with chemical problems. All the methods are based on the equation $\int dQ/\theta=0$ for a cyclic process. M. Planck has given general formulas by which treatment of such questions is much facilitated. The bodies whose chemical equilibrium is to be investigated are contained in the volume V at the temperature θ, and are subjected to an external pressure. A change in the chemical composition, or in the proportions of the mixture, is accompanied by a change of volume dV and a change of temperature $d\theta$, and at the same time the quantity of heat dQ is received from surrounding bodies. If S denotes the entropy and U the internal energy of the system, we have (a) $dS=(dU+P.dV)/\theta$.

The state of the system of bodies is determined by the pressure P, the temperature θ, and certain other variables n, n_1, n_2, etc. If, for example, the space contains water and saturated water vapour and if the whole mass equals M, we may call the quantity of vapour Mn and the quantity of liquid Mn_1 where $n+n_1=1$. If we are dealing with a case of *dissociation*, we may use n for the number of molecules of the original gas, while n_1 and n_2 are the numbers of the dissociated molecules. Hence the state of a system depends generally on the quantities θ, P, n, n_1, $n_2\ldots$, and we have

$$\text{(b)} \quad \begin{cases} dU=\partial U/\partial\theta\,.\,d\theta+\partial U/\partial P\,.\,dP+\partial U/\partial n\,.\,dn+\partial U/\partial n_1\,.\,dn_1\ldots; \\ dV=\partial V/\partial\theta\,.\,d\theta+\partial V/\partial P\,.\,dP+\partial V/\partial n\,.\,dn+\partial V/\partial n_1\,.\,dn_1\ldots \\ dS=\partial S/\partial\theta\,.\,d\theta+\partial S/\partial P\,.\,dP+\partial S/\partial n\,.\,dn+\partial S/\partial n_1\,.\,dn_1+\ldots. \end{cases}$$

From the definition (a) of the entropy we have

$$\text{(b}') \quad \begin{cases} \partial S/\partial\theta=1/\theta\,.(\partial U/\partial\theta+P\,.\,\partial V/\partial\theta), \\ \partial S/\partial V=1/\theta\,.(\partial U/\partial V+P\,.\,\partial V/\partial P), \end{cases}$$

since θ and P are independent. It follows from (a) and (b), since θ and P do not depend on n, n_1, $n_2\ldots$, that

$$\partial S/\partial n\,.\,dn+\partial S/\partial n_1\,.\,dn_1+\ldots$$
$$=\partial\big((U+PV)/\theta\big)/\partial n+\partial\big((U+PV)/\theta\big)/\partial n_1+\ldots.$$

If we set (c) $\Phi=S-(U+PV)/\theta$, this equation takes the form

$$\text{(d)} \quad \partial\Phi/\partial n\,.\,dn+\partial\Phi/\partial n_1\,.\,dn_1+\partial\Phi/\partial n_2\,.\,dn_2+\ldots=0.$$

If the quantities n, n_1, $n_2\ldots$ are independent of each other, we have

$$\partial\Phi/\partial n=\partial S/\partial n-1/\theta\,.\,(\partial U/\partial n+P\,.\,\partial V/\partial n)$$

and analogous equations, which may also in this case be obtained directly from (a). In general, there will be some relation among the quantities n, n_1, $n_2\ldots$. As an example of this method, we will consider the problem of the change of state. If a quantity of vapour Mn and a quantity of liquid Mn_1 are enclosed in a given volume, we have as above $n+n_1=1$. Then the following equations hold

$$S=Mns+Mn_1s_1,\quad U=Mnu+Mn_1u_1,\quad V=Mnv+Mn_1v_1,$$

where s, u, v denote the entropy, the internal energy and the volume of the vapour respectively, while s_1, u_1, and v_1 denote the same quantities for the liquid. From equation (c) we then have

$$\Phi=Mn\big(s-(u+Pv)/\theta\big)+Mn_1\big(s_1-(u_1+Pv_1)/\theta\big)$$

and $$0=M\big(s-(u+Pv)/\theta\big)dn+M\big(s_1-(u_1+Pv_1)/\theta\big)dn_1.$$

In addition, we have $dn+dn_1=0$. Hence equilibrium exists between the vapour and the liquid if (e) $s\theta-u-Pv=s_1\theta-u_1-Pv_1$. Since the quantities in this equation depend only on P and θ, it may take the form $P=f(\theta)$. Hence the equation (e) states the way in which the pressure of the saturated vapour depends on the temperature.

If the unit mass of the substance is transformed from liquid to vapour at the temperature θ, the entropy increases by (f) $s - s_1 = JL/\theta$. We therefore obtain from equation (e) $JL = u - u_1 + P(v - v_1)$, as in CXIX. (d). If equation (e) is differentiated with respect to θ, and P considered as a function of θ, and if we use the equations

$$\partial s/\partial\theta = 1/\theta\,.\,(\partial u/\partial\theta + P\,.\,\partial v/\partial\theta);$$
$$\partial s_1/\partial\theta = 1/\theta\,.\,(\partial u_1/\partial\theta + P\,.\,\partial v_1/\partial\theta)$$

we obtain the relation $J\,.\,L = (v - v_1)\theta\,.\,\partial P/\partial\theta$.

We may consider the method here described as having its basis in a certain tendency in nature. Almost all natural processes are accompanied by the development of heat; hence energy seems to be especially inclined to assume the form of heat, and heat tends to pass from bodies of a higher to those of a lower temperature. In all such transformations the total energy remains unchanged, but it loses more and more the capacity of transforming itself into kinetic energy. A body which contains a quantity of heat Q_1 and whose temperature is θ_1, while the temperature of all surrounding bodies is θ_2, may yield [CXII. (e)] the kinetic energy $Q_1(\theta_1 - \theta_2)/\theta_1$. Hence the lower θ_1, the less the kinetic energy yielded, if θ_2 remains constant.

Clausius expressed this principle in the statement that *the entropy always increases and tends toward a maximum.* For example, if a quantity of heat Q passes by conduction or radiation from a body at the higher temperature θ_1 to one at the lower temperature θ_2, the increase of entropy is $\Delta S = Q/\theta_2 - Q/\theta_1$. The entropy remains unchanged only in the case of a cycle, in which the bodies receiving the heat have the same temperature as those giving it up, and in which the whole system is in neutral equilibrium, since after the performance of this cycle $\int dQ/\theta = 0$ or $\Delta S = 0$. This is also the case at any instant during the cycle, if we take into account not only the entropy of the working body, but also that of surrounding bodies; if the first receives the quantity of heat dQ, the second gives up the same quantity; since the temperature of both bodies is the same, the entropy remains unchanged. This holds, however, only for ideal cycles; in any actual movement of heat, the heat passes from a higher to a lower temperature, and the entropy must therefore increase.

Hence the condition for a change in the state of a system is an increase of the entropy; changes in which the entropy diminishes are impossible. The state of a body in equilibrium is such that if it undergoes a small change, the entropy will either increase or remain constant.

We thus return to the conditions of equilibrium (a). By the communication of the quantity of heat dQ the entropy of the body considered diminishes by dS; hence the increase of the total entropy is $dS - dQ/\theta$, and this must be equal to zero. Since $dQ = dU + P.dV$ we obtain $dS - 1/\theta.(dU + P.dV) = 0$.

Section CXXI. Dissociation.

If a compound gas is separated into two or more constituent gases, either by heating or by diminution of pressure, it is said to be dissociated; the extent of the dissociation depends on the pressure and temperature. In order to determine it, we must determine the function (a) $\Phi = S - (U + PV)/\theta$. If n denotes the number of molecules of the original gas, n_1, n_2... the numbers of molecules of the products of dissociation, we have, as an expression for the total internal energy U, $U = nu + n_1u_1 + n_2u_2 + \ldots$, where u, u_1, u_2 denote the energies of the separate molecules. If m is the mass of a molecule, and c_v its specific heat at constant volume, we may set the internal energy at $\theta°$ C. equal to $Jmc_v\theta + h$, when h is a constant. If we represent Jmc_v by c, we have (b) $U = n(c\theta + h) + n_1(c_1\theta + h_1) + \ldots$. If v denotes the volume of a molecule of a gas and p its pressure, we have $pv = R\theta$, where R does not depend on the nature of the gas. For the sake of simplicity Planck sets $R = 1$. Since the gases are uniformly distributed throughout the whole volume, we have $nv = n_1v_1 = \ldots = V$, and therefore $pV = n\theta$, $p_1V = n_1\theta$, etc., (c) $PV = (n + n_1 + n_2 + \ldots)\theta$. By this equation V is given as a function of P, θ, n, n_1....

The entropy of the system is equal to the sum of the entropies of all the gases, so that $S = ns + n_1s_1 + n_2s_2 + \ldots$, if s denotes the entropy of the molecule. Using c with the meaning given above, the entropy of a molecule equals [CXI. (d)] $c.\log\theta + \log v + k$. Here $nv = V$, and therefore by the use of equation (c),

$$s = c.\log\theta + \log\left(\theta/P.(n + n_1 + n_2 + \ldots)/n\right) + k.$$

If we set

$$C = n/(n + n_1 + n_2 + \ldots),\quad C_1 = n_1/(n + n_1 + n_2 + \ldots),\ \text{etc.},\quad C + C_1 + \ldots = 1,$$

we have (d)

$$S = n[(c+1)\log\theta - \log P - \log C + k] + n_1[(c_1 + 1)\log\theta - \log P - \log C_1 + k_1] + \ldots$$

By the use of equations (b), (c), and (d), (a) takes the form

$$\text{(e)}\quad \begin{cases} \Phi = n[(c+1)(\log\theta - 1) - \log P - \log C + k - h/\theta] \\ \quad + n_1[(c_1+1)(\log\theta - 1) - \log P - \log C_1 + k_1 - h_1/\theta] + \ldots. \end{cases}$$

But we have $\partial(n \,.\, \log C + n_1 \,.\, \log C_1 + \ldots)/\partial n = \log C$, and further

$$\text{(f)}\qquad \partial\Phi/\partial n = (c+1)(\log\theta - 1) - \log P - \log C + k - h/\theta.$$

Similar expressions hold for the other differential coefficients. Introducing these values in the condition of equilibrium

$$\partial\Phi/\partial n \,.\, dn + \partial\Phi/\partial n_1 \,.\, dn_1 + \ldots = 0,$$

it follows that the problem may be solved if we know the relations existing among the quantities n, n_1, $n_2 \ldots$.

If we investigate, for example, the dissociation of hydriodic acid into iodine and hydrogen, the proportions of the gases in the mixture can be represented in the following way: nJH, n_1H_2, n_2J_2. By dissociation two molecules of hydriodic acid form one molecule of hydrogen and one of iodine, hence the ratio between dn, dn_1, dn_2 is $-2:1:1$. If we set generally $dn : dn_1 : dn_2 : \ldots = \nu : \nu_1 : \nu_2 : \ldots$, the condition of equilibrium becomes

$$\text{(g)}\qquad \nu \,.\, \partial\Phi/\partial n + \nu_1 \,.\, \partial\Phi/\partial n_1 + \nu_2 \,.\, \partial\Phi/\partial n_2 + \ldots = 0.$$

From equations (f) and (g) we obtain

$$\text{(h)}\qquad \Sigma[\nu(c+1)(\log\theta - 1) - \nu\log P - \nu\log C + \nu k - \nu h/\theta] = 0.$$

To simplify the calculation Planck assumed that the atomic heat is constant even in the compound gas, and that the molecular heat is equal to the sum of the atomic heats; experiment shows that this is approximately true in all cases. If a, a_1, a_2 denote the number of atoms in the molecule of each gas, we may set

$$c = \gamma a, \quad c_1 = \gamma a_1, \quad c_2 = \gamma a_2 \ \ldots$$

Since the whole number of atoms is unchanged by dissociation, the sum $na + n_1a_1 + n_2a_2 + \ldots$ is constant, and hence

$$a \,.\, dn + a_1 \,.\, dn_1 + a_2 \,.\, dn_2 + \ldots = 0.$$

Consequently also $\nu a + \nu_1 a_1 + \nu_2 a_2 + \ldots = 0$ and $\nu c + \nu_1 c_1 + \nu_2 c_2 + \ldots = 0$. Further if we set

$$\nu + \nu_1 + \nu_2 + \nu_3 + \ldots = \nu_0; \quad \nu h + \nu_1 h_1 + \nu_2 h_2 + \ldots = -\log h_0;$$
$$\nu(k-1) + \nu_1(k_1 - 1) + \ldots = \log k_0,$$

it follows from (h) that

$$\nu\log C + \nu_1\log C_1 + \nu_2\log C_2 + \ldots = \log k_0 + (\log h_0)/\theta + \nu_0 \,.\, \log(\theta/P),$$

or (i)
$$C^{\nu} \,.\, C_1^{\nu_1} \,.\, C_2^{\nu_2} \ldots = k_0 \,.\, h_0^{1/\theta} \,.\, (\theta/P)^{\nu_0}.$$

Hence for hydriodic acid we have

$$\nu_0 = 0, \text{ and } C_1 C_2 / C'^2 = k_0 \,.\, h_0^{1/\theta}.$$

If no hydrogen or iodine is present except that set free by dissociation, we have $n_1 = n_2$, and therefore $C_1 = C_2$ and $C_1/C = \sqrt{k_0 \,.\, h_0^{1/\theta}}$. Now $C_1/C = n_1/n$. Hence the degree of dissociation is independent of the pressure, but increases with the temperature. In this case, however, the dissociation can never become complete, since, for $\theta = \infty$, we have $C_1/C = \sqrt{k_0}$.

From equation (i) the pressure has no influence on the degree of dissociation if the total volume remains unchanged by the dissociation. This is the case when $\nu_0 = 0$. If, on the other hand, the volume increases during the dissociation, any increase of the pressure will lessen the degree of dissociation. This occurs in the case of nitrogen-dioxide, N_2O_4, in which one molecule is broken up by the dissociation into two molecules NO_2. Hence $\nu = -1$, $\nu_1 = 2$, and therefore

$$C_1^2 / C = k_0 \,.\, h_0^{1/\theta} \,.\, \theta / P.$$

This equation, together with $C_1 + C = 1$, determines the degree of dissociation.

In order to occasion the dissociation determined by the quantities dn, dn_1, dn_2, at constant temperature and at constant pressure, the quantity of heat dQ is required, which is determined by

$$J \,.\, dQ = dU + P \,.\, dV,$$

or, from equations (b) and (c), by $J \,.\, dQ = \Sigma(c\theta + h)dn + \theta \,.\, \Sigma dn$. Since $\Sigma\nu c = 0$, the quantity of heat required for the dissociation determined by the quantities ν, ν_1, $\nu_2 \ldots$ is determined by

(k) $$J \,.\, Q = \nu h + \nu_1 h_1 + \nu_2 h_2 + \ldots + \nu_0 \theta = \nu_0 \theta - \log h_0.$$

We reach the same result from the equation $J \,.\, dQ = \theta \,.\, dS$ together with the relations (d) and (h).

CHAPTER XIV.

CONDUCTION OF HEAT.

SECTION CXXII. FOURIER'S EQUATION.

IF the temperatures of the different parts of a body are different, a gradual change goes on until the temperature of all parts of the body is the same, that is, until *equilibrium of temperature* has been reached. In this statement it is assumed that the body neither receives heat from surrounding bodies nor gives up heat to them. The rate at which the condition of equilibrium is reached depends upon the facility with which the body conducts heat. Without making any assumptions on the nature of heat, we may say that heat flows in a body until a state of equilibrium is reached. We define the *rate of flow of heat* in any direction, as that quantity of heat which passes in unit time through unit area perpendicular to that direction.

Hence, if Q represents the rate of flow of heat through an area dS within the body, the quantity of heat which will pass through that area in the time dt is $Q.dS.dt$. If U, V, W are the components of flow in the directions of the coordinate axes, the quantities of heat which pass through the elementary areas $dy.dz$, $dx.dz$, $dx.dy$, in the time dt, are $U.dy.dz.dt$, $V.dx.dz.dt$, and $W.dx.dy.dt$ respectively.

Using the general equations (XIV.) of fluid motion, we obtain for Q (a) $Q=lU+mV+nW$, where l, m, n are the direction cosines of the normal to the elementary area dS.

Let OO' (Fig. 142) be a rectangular parallelepiped, whose edges $OA=a$, $OB=b$, and $OC=c$ are parallel to the coordinate axes. If U, V, W represent the components of flow at the point O, those at A are $U+\partial U/\partial x.a$, $V+\partial V/\partial x.a$, $W+\partial W/\partial x.a$ respectively, supposing a, b, c so small that only the first terms in the expansion need be retained. The parallelepiped receives the quantity of heat $U.dt.bc$ in the time dt through the surface $OBA'C$, and loses the quantity $(U+\partial U/\partial x.a)bc.dt$, which flows out through the surface $ACO'B'$ in the same time. The parallelepiped gains, on the whole, the quantity $-\partial U/\partial x.a.bc.dt$. If we take account of the other surfaces, the quantity of heat which remains in the parallelepiped is $-(\partial U/\partial x+\partial V/\partial y+\partial W/\partial z).abc.dt$, or, if we set $a.b.c=dv$,

$$-(\partial U/\partial x+\partial V/\partial y+\partial W/\partial z)dvdt.$$

This quantity of heat raises the temperature of the parallelepiped by $d\theta$, which, if c denotes the specific heat of the body, and ρ its density, is determined by the following equation,

(b) $$c\rho.d\theta=-(\partial U/\partial x+\partial V/\partial y+\partial W/\partial z)dt.$$

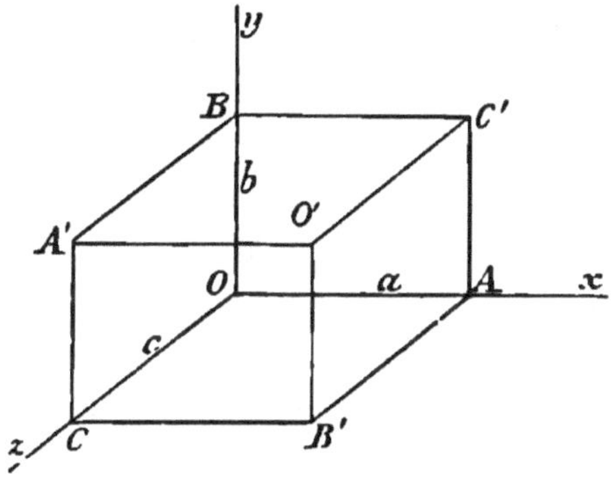

FIG. 142.

This equation holds only if the heat received is used solely in causing change of temperature and does not produce any change in the state of aggregation or any chemical change. Sometimes, too, heat exists in the interior of a body which has not penetrated into it in the form of heat, but is produced by friction or by an electrical current in the body; and to this the above equation does not apply.

The components of flow U, V, W depend on the distribution of heat in the body and on the nature of the body. If the body conducts heat equally well in all directions, that is if it is *isotropic*, we may determine the rate of flow in the following way. Let A and B be two points within the body infinitely near each other, in which the temperatures are respectively θ and θ'. If $d\nu$ denotes the distance between the points A and B, and k the conductivity of the body for heat, the rate of flow of heat in the direction AB is given by $Q=k(\theta-\theta')/d\nu$. Hence the *conductivity* is the quantity of heat which flows in unit time through unit area of a surface in the body, parallel with and between two surfaces whose temperatures differ by 1°, and which are distant from each other by one centimetre.

Now since $\theta'=\theta+d\theta/d\nu\,.\,d\nu$, we have also (c) $Q=-k\,.\,d\theta/d\nu$. We obtain in like manner for the components of flow U, V, W, the expressions (d) $U=-k\,.\,\partial\theta/\partial x$, $V=-k\,.\,\partial\theta/\partial y$, $W=-k\,.\,\partial\theta/\partial z$. In actual cases the conductivity k is a function of θ; but for the sake of simplicity we will assume that k is constant. We obtain from (b) and (d)

(e) $$c\rho\,.\,\partial\theta/\partial t=k(\partial^2\theta/\partial x^2+\partial^2\theta/\partial y^2+\partial^2\theta/\partial z^2).$$

This equation was first given by Fourier, and is therefore called *Fourier's equation*. The specific heat c, the density ρ, and the conductivity k are functions of θ; we will, however, consider them constant. Fourier's equation may also take the form

(f) $$\partial\theta/\partial t=\kappa^2(\partial^2\theta/\partial x^2+\partial^2\theta/\partial y^2+\partial^2\theta/\partial z^2),$$

where (g) $\kappa^2=k/c\rho$.

In the following table the values of k and κ for several metals at the temperatures 0° and 100° C. are given from the experiments of L. Lorenz:

	k_0	k_{100}	κ_0	κ_{100}
Copper, . .	0,7198	0,7226	0,909	0,873
Tin, . . .	0,1598	0,1423	0,392	0,344
Iron, . . .	0,1665	0,1627	0,202	0,179
Lead, . . .	0,0836	0,0764	0,242	0,222.

SECTION CXXIII. STEADY STATE.

The state of the body with respect to heat is called *steady*, if the temperatures of the different parts of the body are different, but do not change with the time. In this case each particle gives up on the one side as much heat as it receives on the other, and the temperature is independent of the time t and dependent only on the coordinates x, y, z. For the steady state, equation CXXII. (f) becomes (a) $\partial^2\theta/\partial x^2+\partial^2\theta/\partial y^2+\partial^2\theta/\partial z^2=\nabla^2\theta=0$. The components of flow are expressed by equations CXXII. (d).

Flow of heat in a plate.—We will consider a thin plate whose faces L and M are parallel to the yz-plane. The temperatures of the faces are respectively θ_1 and θ_2. This being so, the flow of heat is parallel to the x-axis, and the temperature θ in the vicinity of the x-axis depends only on x, so that from (a) we have $d^2\theta/dx^2=0$. Hence $\theta=px+q$. If the distances of the faces L and M from the yz-plane are a and b respectively, we have $\theta_1=pa+q$, $\theta_2=pb+q$, and further

$$\theta=(b\theta_1-a\theta_2)/(b-a)-(\theta_1-\theta_2)x/(b-a).$$

If we represent the distance $b-a$ between the faces by e, the rate of flow of heat U between them is (c) $U=k(\theta_1-\theta_2)/e$.

Every integral of equation (a) corresponds to a steady state of heat. If $\theta=f(x, y, z)$ is an integral of (a), $\theta_1=f(x, y, z)$ and $\theta_2=f(x, y, z)$ are the equations of two surfaces of constant temperatures, or of two *isothermal surfaces*, where θ_1 and θ_2 are constant. If the body is bounded by the surfaces which are determined by θ_1 and θ_2, and if θ is a temperature which lies between θ_1 and θ_2, $\theta=f(x, y, z)$ is the equation of any isothermal surface.

The flow of heat in a sphere.—If m and c are constant, and if

$$r^2=x^2+y^2+z^2, \quad \theta=m/r+c$$

is a solution of (a). Therefore setting $\theta_1=m/r_1+c$, $\theta_2=m/r_2+c$, we have (d)

$$\theta=\frac{(\theta_1-\theta_2)r_1r_2}{r(r_2-r_1)}-\frac{r_1\theta_1-r_2\theta_2}{r_2-r_1}$$

as the equation of the system of isothermal surfaces, which in this case are spheres. For the rate of flow of heat U in the direction r, we have (e) $U=-k\,.\,d\theta/dr=k(\theta_1-\theta_2)r_1r_2/r^2(r_2-r_1)$. The temperature and flow of heat in a hollow sphere, whose internal and external surfaces are at the temperatures θ_1 and θ_2 respectively, are also given by equations (d) and (e). The total quantity of heat which flows out through the hollow sphere is $4\pi r^2U=4\pi k(\theta_1-\theta_2)r_1r_2/(r_2-r_1)$.

The flow of heat in a tube.—If c and c' are constants and if $r^2=x^2+y^2$, we have, from XV., $\theta=c\log r+c'$ as an integral of (a). Therefore, if we set $\theta_1=c\,.\log r_1+c'$, $\theta_2=c\,.\log r_2+c'$, we obtain

(f) $\theta=(\theta_1-\theta_2)\log r/(\log r_1-\log r_2)+(\theta_1\log r_2-\theta_2\log r_1)/(\log r_2-\log r_1)$.

The rate of flow U in the direction r is $U=k(\theta_1-\theta_2)/r(\log r_2-\log r_1)$. The quantity of heat which flows out through a unit length of the tube is (g) $2\pi rU=2\pi k(\theta_1-\theta_2)/(\log r_2-\log r_1)$.

SECTION CXXIV. THE PERIODIC FLOW OF HEAT IN A GIVEN DIRECTION.

If the temperature of the body depends only on one coordinate, say on x, Fourier's equation becomes (a) $\partial\theta/\partial t=\kappa^2\,.\,\partial^2\theta/\partial x^2$. We will hereafter investigate in what way this equation can be integrated.

For the present we will consider special integrals which correspond to simple yet important cases.

The temperature of the earth changes during the year; it rises and falls with the temperature of the air. The time at which the maximum or minimum temperature at any point is reached is later as the point lies further below the surface. In the following discussion we will not take into account the internal heat of the earth.

If the temperature at the earth's surface is given by (b) $\theta = \sin at$, we may express the temperature at a point in the interior of the earth by (c) $\theta = P . \sin at + Q . \cos at$, where P and Q are functions of the distance x of that point from the earth's surface. If we substitute for θ in (a) the expression (c) we have

$$Pa . \cos at - Qa . \sin at = \kappa^2(\sin at . d^2P/\partial x^2 + \cos at . d^2Q/dx^2).$$

Hence we must have $\kappa^2 . d^2P/dx^2 = -Qa$ and $\kappa^2 . d^2Q/dx^2 = Pa$. Now if we set $\epsilon^2 = a/\kappa^2$, we have (d), (e), $d^4P/dx^4 = -\epsilon^4 P$ and $Q = -1/\epsilon^2 . d^2P/dx^2$. In order to integrate equation (d) we set $P = Ae^{px}$, and obtain $p = \epsilon\sqrt[4]{-1}$. The integral of equation (d) then takes the form

$$P = Ae^{(1+\sqrt{-1})\epsilon x/\sqrt{2}} + Be^{(1-\sqrt{-1})\epsilon x/\sqrt{2}} + Ce^{(-1+\sqrt{-1})\epsilon x/\sqrt{2}} + De^{(-1-\sqrt{-1})\epsilon x/\sqrt{2}}.$$

Since θ must equal 0 when $x = \infty$, we have $A = B = 0$, and hence

$$P = Ce^{(-1+\sqrt{-1})\epsilon x/\sqrt{2}} + De^{(-1-\sqrt{-1})\epsilon x/\sqrt{2}}.$$

We obtain from equation (e)

$$Q = (Ce^{(-1+\sqrt{-1})\epsilon x/\sqrt{2}} - De^{(-1-\sqrt{-1})\epsilon x/\sqrt{2}}) . \sqrt{-1}.$$

But from equations (b) and (c) we have $P = 1$ and $Q = 0$ when $x = 0$, and therefore $C = D = \frac{1}{2}$. Hence we obtain

$$P = e^{-\epsilon x/\sqrt{2}} . \cos(\epsilon x/\sqrt{2}); \quad Q = -e^{-\epsilon x/\sqrt{2}} . \sin(\epsilon x/\sqrt{2});$$

and $$\theta = e^{-\epsilon x/\sqrt{2}} . \sin(at - \epsilon x/\sqrt{2}).$$

Substituting for ϵ its value, we have

$$\theta = e^{-x\sqrt{\frac{1}{2}a/\kappa}} . \sin(at - x\sqrt{\tfrac{1}{2}a/\kappa}).$$

The difference between the highest and lowest temperatures at the depth x below the surface is therefore $2e^{-x\sqrt{\frac{1}{2}a/\kappa}}$. This difference depends on the value of a. The faster the temperature changes at the surface the smaller the influence of this change on the temperature in the interior. For example, if we set the temperature at the surface equal to $\theta = \sin(2\pi t/T)$, the difference between the highest and lowest temperatures is equal to $2e^{-x\sqrt{\pi/T/\kappa}}$, and this is very much greater when T is a year than when it is a day.

The temperature relations within the earth are actually different from those here described, because the temperature at the surface cannot be expressed in any simple way. The main features of the phenomena, however, are similar to those deduced in this discussion.

SECTION CXXV. A HEATED SURFACE.

Let the temperature in an infinite body at the time $t=0$ be everywhere zero, except in a plane in which each unit of area contains the quantity of heat σ. Fourier showed that at the time t the temperature θ at a point at the distance x from the heated plane is given by

(a) $$\theta=\frac{1}{\sqrt{\pi}}\frac{\sigma\kappa}{2k\sqrt{t}}e^{-\frac{x^2}{4\kappa^2 t}},$$

where k is the conductivity and κ the quantity defined in CXXII. (g). We will now examine whether this expression for θ satisfies all the conditions of the problem. We will first consider the differential equation $\partial\theta/\partial t=\kappa^2.\partial^2\theta/\partial x^2$. From (a) we obtain

(b), (c), $$\partial\theta/\partial t=(-1/2t+x^2/4\kappa^2t^2)\theta,\quad \partial\theta/\partial x=-x\theta/2\kappa^2 t,$$

(d) $$\partial^2\theta/\partial x^2=(-1/2\kappa^2 t+x^2/4\kappa^4t^2)\theta.$$

It follows from (b) and (d) that the differential equation is satisfied. Since the function ze^{-z^2} approaches zero as its limit if z becomes infinitely great, it follows that for $t=0$ we have $\theta=0$, for all values of x, with the exception of the value $x=0$. If θ is determined by the equation (a), we can further show that each unit of area of the heated surface S contains the quantity of heat σ at the time $t=0$. The total quantity of heat which is present in the body is given by the expression

$$\int_{-\infty}^{+\infty}S\rho c\theta\,.\,dx=\frac{S\sigma}{\sqrt{\pi}}\int_{-\infty}^{+\infty}\frac{1}{2\kappa\sqrt{t}}e^{-x^2/4\kappa^2t}\,.\,dx.$$

But because (e) $$\int_{-\infty}^{+\infty}e^{-q^2}dq=\sqrt{\pi},$$

the quantity of heat present at any time must be $S\sigma$, and since this quantity is present on the infinite surface S at the time $t=0$, the unit of surface at that time must contain the quantity σ.

It follows from (a) that $\theta=0$ for $t=0$ as well as for $t=\infty$; therefore there must be a certain time at which θ is a maximum. This time is found from (f) $\dot{\theta}=0$, which gives $t=x^2/2\kappa^2$. The corresponding value of θ is (g) $\theta=1/\sqrt{2\pi e}\,.\,\sigma/c\rho x$. It appears from equation

(a) that the heat is propagated with an infinite velocity, since θ is everywhere different from zero as soon as t has a finite value.

We will now determine the temperature at any time in a region in which the original distribution of heat depends only on one of the coordinates. Let $\theta = f(a)$ when $t=0$, where a is the distance from the yz-plane. The part S of the region which is bounded by two parallel planes for which $x=a$ and $x=a+da$, contains the quantity of heat $S\sigma = S \cdot da \cdot \rho c \cdot f(a)$. Therefore the quantity σ which is present in the unit of area of this sheet is $\sigma = da \cdot \rho c \cdot f(a)$.

If the temperature of the rest of the region is zero, the heat flows out from this sheet on both sides, and at a point whose distance from the yz-plane equals x, and which therefore is at the distance $x-a$ from the shell, the temperature by (a) is

$$d\theta = \frac{1}{\sqrt{\pi}} \frac{\rho c \kappa}{2k\sqrt{t}} e^{-(x-a)^2/4\kappa^2 t} \cdot f(a) da.$$

All other similar sheets emit heat according to the same law, and we therefore have

$$\text{(h)} \qquad \theta = \frac{1}{\sqrt{\pi}} \frac{1}{2\kappa\sqrt{t}} \cdot \int_{-\infty}^{+\infty} e^{-(x-a)^2/4\kappa^2 t} f(a) da.$$

If we set (i) $q = (a-x)/2\kappa\sqrt{t}$, we have

$$\theta = 1/\sqrt{\pi} \cdot \int_{-\infty}^{+\infty} e^{-q^2} f(x + 2\kappa q \sqrt{t}) dq.$$

The expressions (h) and (i) contain the complete solution of the problem before us. If we set $t=0$ in equation (i) and make use of (e), it follows at once that $\theta = f(x)$.

For example, if the initial temperature is constant and equal to θ_0 within the portion determined by $-l < x < +l$, but equal to zero outside these limits, the integration in (h) is effected between these limits, so that

$$\text{(k)} \qquad \theta = \frac{1}{\sqrt{\pi}} \cdot \frac{\theta_0}{2\kappa\sqrt{t}} \int_{-l}^{+l} e^{-(x-a)^2/4\kappa^2 t} da.$$

Section CXXVI. The Flow of Heat from a Point.

Let us suppose that, at the time $t=0$, the temperature in an infinitely great body is everywhere equal to zero, except at one point, in which the quantity of heat m is concentrated. We will investigate the distribution of heat in the body at any subsequent time t. This problem was first handled by Fourier, who found that the temperature θ at a point whose distance from m is r is given by

(a) $$\theta = m\kappa^2/k \, . \, (1/2\kappa\sqrt{\pi t})^3 \, . \, e^{-r^2/4\kappa^2 t}.$$

We can show that this expression satisfies all the conditions of the problem. If the point which contains the quantity of heat m is at the origin of coordinates, Fourier's equation

$$\partial\theta/\partial t = \kappa^2(\partial^2\theta/\partial x^2 + \partial^2\theta/\partial y^2 + \partial^2\theta/\partial z^2)$$

takes the form (XV.) (b) $\partial\theta/\partial t = \kappa^2(\partial^2\theta/\partial r^2 + 2/r \, . \, \partial\theta/\partial r)$, because θ is a function of r only. This equation may be given the form

(c) $$\partial(r\theta)/\partial t = \kappa^2\partial^2(r\theta)/\partial r^2.$$

We obtain from (a)

(d) $$\begin{cases} \partial(r\theta)/\partial t = (-3/2t + r^2/4\kappa^2 t^2) \, . \, r\theta, \\ \partial(r\theta)/\partial r = (1/r - r/2\kappa^2 t) \, . \, r\theta, \\ \partial^2(r\theta)/\partial r^2 = (-3/2\kappa^2 t + r^2/4\kappa^4 t^2) \, . \, r\theta, \end{cases}$$

and by the use of these values prove first that Fourier's equation is satisfied. Further, $\theta = 0$ when $t = 0$. The quantity of heat originally present is m, since the total quantity of heat at any time is given by

$$\int_0^\infty 4\pi r^2 \, . \, dr \, . \, \rho c\theta = \int_0^\infty 4\pi r^2 \, . \, dr \, . \, m(1/2\kappa\sqrt{\pi t})^3 \, . \, e^{-r^2/4\kappa^2 t}.$$

If we set $q = r/2\kappa\sqrt{t}$, the integral takes the form

$$4m/\sqrt{\pi} \, . \int_0^\infty e^{-q^2} q^2 dq.$$

We find by integration by parts, and by the use of CXXV. (e), that the value of the integral is m.

The time t, at which θ reaches its maximum value, is obtained from the equation $\dot{\theta} = 0$, and is from (d), $t = r^2/6\kappa^2$. The corresponding maximum value of θ is $\theta = (1/\sqrt{\tfrac{2}{3}\pi e})^3 \, . \, m/c\rho r^3$.

SECTION CXXVII. THE FLOW OF HEAT IN AN INFINITELY EXTENDED BODY.

We will now investigate, with the aid of the results already obtained, the flow of heat in an infinitely extended body, when the distribution of heat at a particular time is given. Let $\theta = f(a, b, c)$ at the time $t = 0$, where a, b, c are the coordinates of a point referred to a system of rectangular axes. The quantity of heat contained by a volume-element $da \, . \, db \, . \, dc$ is $dm = f(a, b, c) \, . \, k/\kappa^2 \, . \, da db dc$.

If this quantity is propagated through the body, it produces the rise of temperature [CXXVI. (a)],

$$d\theta = (1/2\kappa\sqrt{\pi t})^3 e^{-[(x-a)^2+(y-b)^2+(z-c)^2]/4\kappa^2 t} . f(a, b, c) da\, db\, dc.$$

If we take the sum of all increments of temperature which arise from the distribution of heat considered, we obtain for the temperature θ at the point x, y, z,

$$\text{(a)} \quad \theta = (1/2\kappa\sqrt{\pi t})^3 \int_{-\infty}^{+\infty}\int_{-\infty}^{+\infty}\int_{-\infty}^{+\infty} e^{-[(x-a)^2+(y-b)^2+(z-c)^2]/4\kappa^2 t} . f(a, b, c) da\, db\, dc.$$

This expression for θ is an integral of the differential equation

$$\text{(b)} \quad \partial\theta/\partial t = \kappa^2(\partial^2\theta/\partial x^2 + \partial^2\theta/\partial y^2 + \partial^2\theta/\partial z^2).$$

We notice that the integration of this equation depends on that of the simpler equation (c) $\partial X/\partial t = \kappa^2 \partial^2 X/\partial x^2$. For if X is a function of x and t, which satisfies equation (c), and if Y and Z are functions of y, t and z, t respectively, which satisfy equations analogous to (c) for y and z, the equation $\theta = XYZ$ satisfies (b). We have

$$YZ\dot{X} + XZ\dot{Y} + XY\dot{Z} = \kappa^2(YZ . \partial^2 X/\partial x^2 + XZ . \partial^2 Y/\partial y^2 + XY . \partial^2 Z/\partial z^2).$$

It follows from (c) and the analogous equations for y and z that this equation is satisfied, from CXXV. (a), by $X = 1/\sqrt{t} . e^{-(x-a)^2/4\kappa^2 t}$, hence the expression

$$1/\sqrt{t} . e^{-(x-a)^2/4\kappa^2 t} . 1/\sqrt{t} . e^{-(y-b)^2/4\kappa^2 t} . 1/\sqrt{t} . e^{-(z-c)^2/4\kappa^2 t}$$

is an integral of equation (b). Therefore, also,

$$\theta = C\int_{-\infty}^{+\infty}\int_{-\infty}^{+\infty}\int_{-\infty}^{+\infty}(1/\sqrt{t})^3 . e^{-[(x-a)^2+(y-b)^2+(z-c)^2]/4\kappa^2 t} . f(a, b, c,) da\, db\, dc$$

is an integral of equation (b). C is a constant, and $f(a, b, c)$ an arbitrary function of a, b, c. If we set

$$\alpha = (a-x)/2\kappa\sqrt{t}, \quad \beta = (b-y)/2\kappa\sqrt{t}, \quad \gamma = (c-z)/2\kappa\sqrt{t},$$

it follows that

$$\theta = (2\kappa)^3 C\int_{-\infty}^{+\infty}\int_{-\infty}^{+\infty}\int_{-\infty}^{+\infty} e^{-\alpha^2-\beta^2-\gamma^2} f(x + 2\kappa\alpha\sqrt{t},\ y + 2\kappa\beta\sqrt{t},$$
$$z + 2\kappa\gamma\sqrt{t}) d\alpha\, d\beta\, d\gamma.$$

If we now assume $t = 0$, we obtain by the help of CXXV. (e),

$$\theta = (2\kappa)^3 . C(\sqrt{\pi})^3 f(x, y, z).$$

If $f(x, y, z)$ is an expression for the temperature when $t = 0$, we set $C = 1/(2\kappa\sqrt{\pi})^3$, and obtain

$$\text{(d)} \quad \begin{cases} \theta = (1/\sqrt{\pi})^3 \int_{-\infty}^{+\infty}\int_{-\infty}^{+\infty}\int_{-\infty}^{+\infty} e^{-\alpha^2-\beta^2-\gamma^2} f(x + 2\kappa\alpha\sqrt{t}, \\ \qquad y + 2\kappa\beta\sqrt{t},\ z + 2\kappa\gamma\sqrt{t}) d\alpha\, d\beta\, d\gamma. \end{cases}$$

The expressions (a) and (d) are identical, as may be shown by the substitution already employed.

SECTION CXXVIII. THE FORMATION OF ICE.

Suppose that the temperature of a mass of water is everywhere $\theta=0$, and that the surface of the mass is in contact with another surface whose temperature is $-\theta_0$. θ_0 may be either constant or variable, but must be always below zero. A sheet of ice will be formed under this surface, whose thickness ϵ is a function of the time t. The temperature θ of the mass of ice is itself a function of t and of the distance x from the surface. For $x=\epsilon$, θ is always equal to zero. The equation (a) $\partial\theta/\partial t=\kappa^2\partial^2\theta/\partial x^2$ holds everywhere within the mass of ice. New ice will form continually on the bounding surface of the ice and water. The quantity of heat which flows outward through unit area of the lowest sheet of ice is given by $k\partial\theta/\partial x\,.\,dt$. During the same time a sheet of ice, whose thickness is $d\epsilon$, is formed, and the quantity of heat set free thereby is $L\rho d\epsilon$, where L represents the heat of fusion of ice and ρ its density. When $x=\epsilon$, we have

(b) $$k\partial\theta/\partial x=L\rho d\epsilon/dt, \text{ or } \partial\theta/\partial x=L/c\kappa^2\,.\,d\epsilon/dt.$$

We may write for θ the expression

(c)* $$-c\theta/L=\frac{1}{1\,.\,2}\cdot\frac{d(\epsilon-x)^2}{\kappa^2\,.\,dt}+\frac{1}{1\,.\,2\,.\,3\,.\,4}\cdot\frac{d^2(\epsilon-x)^4}{\kappa^4\,.\,dt^2}+\ldots.$$

As may easily be seen, this expression satisfies equation (a). It also satisfies the condition that $\theta=0$ when $x=\epsilon$. In order to find whether it satisfies the condition contained in (b), we differentiate (c) with respect to x, and obtain

$$-c/L\,.\,\partial\theta/\partial x=-\frac{d(\epsilon-x)}{\kappa^2\,.\,dt}-\frac{1}{1\,.\,2\,.\,3}\cdot\frac{d^2(\epsilon-x)^3}{\kappa^4\,.\,dt^2}-\ldots.$$

When $x=\epsilon$ this becomes equation (b).

Since, at the surface, $\theta=-\theta_0$, it follows from (c) that

(d) $$c\theta_0/L=\frac{1}{1\,.\,2}\cdot\frac{d\epsilon^2}{\kappa^2dt}+\frac{1}{1\,.\,2\,.\,3\,.\,4}\cdot\frac{d^2\epsilon^4}{\kappa^4dt^2}+\ldots.$$

If the thickness of the sheet of ice is given as a function of the time t, θ_0 may be easily determined; on the other hand, if θ_0 is given, it is in general difficult to determine ϵ.

* This solution was communicated to the author by L. Lorenz. See also Stefan, *Wied. Ann.*, Bd. XLII., S. 269.

If θ_0 is constant, the right side of equation (d) must also be constant. This condition is fulfilled if $\epsilon^2/\kappa^2 = 2p^2 t$, where p is constant. From (d) we then obtain the equation

$$\text{(e)} \qquad c\theta_0/L = \frac{p^2}{1} + \frac{p^4}{1\,.\,3} + \frac{p^6}{1\,.\,3\,.\,5} + \ldots,$$

which serves to determine p. In order to put the series in (e) into a finite form, we form from (e)

$$d(c\theta_0/Lp)/dp = 1 + \frac{p^2}{1} + \frac{p^4}{1\,.\,3} + \frac{p^6}{1\,.\,3\,.\,5} + \ldots,$$

and thus obtain $d(c\theta_0/Lp)/dp = 1 + c\theta_0/L$. The integral of this equation is (f) $c\theta_0/L = p\int_0^p e^{-(\alpha^2-p^2)/2} d\alpha$. If the thickness of the sheet of ice increases in direct ratio with the time t, that is, if $\epsilon = q\kappa t$, where q is a new constant, it follows from (d) that

$$c\theta_0/L = q^2 t + \frac{q^4 t^2}{1\,.\,2} + \frac{q^6 t^3}{1\,.\,2\,.\,3} + \ldots, \text{ or (g) } c\theta_0/L = \epsilon^{q^2 t} - 1.$$

If ϵ is very small, we obtain from (d) $c\theta_0/L = 1/2\kappa^2 \,.\, d\epsilon^2/dt$, and hence (h) $\epsilon^2 = 2k/L\rho \,.\, \int_0^t \theta_0 dt$. This result also follows if we set the flow of heat upward equal to $k\theta_0/\epsilon$, in which, however, we assume that the temperature in the ice increases uniformly from its upper surface downward. On this assumption, the quantity of heat $k\theta_0 \,.\, dt/\epsilon$ flows upward through the ice in the time dt. In the same time a sheet of ice, whose thickness is $d\epsilon$, is formed, and the quantity of heat $L\rho \,.\, d\epsilon$ is set free. Hence we have $k\theta_0 \,.\, dt/\epsilon = L\rho \,.\, d\epsilon$. This equation leads to the result we have already obtained. If θ_0 is constant, it follows that (i) $\epsilon = \sqrt{2\theta_0 kt/L\rho}$.

SECTION CXXIX. THE FLOW OF HEAT IN A PLATE WHOSE SURFACE IS KEPT AT A CONSTANT TEMPERATURE.

It is in general very difficult to determine the variations of temperature in a limited body. We will discuss a few cases in which it is possible to solve this problem. Suppose that the temperature in the interior of a plate bounded by parallel plane faces is $\theta = f(x)$, where x denotes the distance of the point considered from one of the faces of the plate. From the time $t = 0$ on, the surfaces are supposed to be in contact with a mixture of ice and water, or to be so conditioned that their temperature is kept at zero. The law according to which the temperature changes in the interior of the plate is to be determined. Designating the thickness of the plate by a, we have

$$\text{(a)}\qquad \begin{cases} \text{for } t=0,\ \theta=f(x); & \text{for } t=\infty,\ \theta=0; \\ \text{for } x=0,\ \theta=0; & \text{for } x=a,\ \theta=0. \end{cases}$$

The rate at which the temperature changes at the surface is infinitely great; just outside the surface it is equal to zero, while just within it, at one face, it is equal to $f(0)$. At the other face the temperature outside the plate is also zero, and within it $f(a)$. The function θ must satisfy not only these conditions, but also the differential equation (b) $\partial\theta/\partial t = \kappa^2\partial^2\theta/\partial x^2$. An integral of this equation is

$$\text{(c)}\qquad \theta = e^{-m^2\kappa^2 t}(A\sin mx + B\cos mx).$$

From (a) $B=0$, so that (d) $\theta = Ae^{-m^2\kappa^2 t}\sin mx$. This value of θ satisfies not only equation (b), but also vanishes for $x=0$. Since θ is also zero when $x=a$, we must have $\sin ma=0$, and therefore $ma=\pm p\pi$, where p is a whole number. Hence we have

$$\text{(e)}\qquad \theta = Ae^{-p^2\pi^2\kappa^2 t/a^2}\,.\,\sin(px\pi/a).$$

If we notice further that $\theta=f(x)$ when $t=0$, we have

$$\text{(f)}\qquad f(x) = A\sin(p\pi x/a).$$

In general, the function $f(x)$ can not be represented by this expression. To solve the problem we use the following method.

Since the expression (e) is an integral of Fourier's equation, the complete integral is obtained by taking the sum of the similar expressions, which are obtained by giving p all values between 1 and ∞. The terms which correspond to a negative value of p differ from those terms for which p is positive only in sign, and can therefore be considered as contained in the latter. Hence we set

$$\text{(g)}\qquad \theta = A_1\sin(\pi x/a)\,.\,e^{-\pi^2\kappa^2 t/a^2} + A_2\sin(2\pi x/a)\,.\,e^{-2^2\pi^2\kappa^2 t/a^2} + \dots.$$

When $t=0$, we have $\theta=f(x)$, so that for $0<x<a$

$$\text{(h)}\qquad f(x) = A_1\sin(\pi x/a) + A_2\sin(2\pi x/a) + \dots.$$

We will now investigate whether a function $f(x)$, which is arbitrary within the given limits, can be represented by a trigonometrical series of this form. For this purpose we choose instead of the infinite series (h) another series with $(n-1)$ coefficients $A_1, A_2, \dots A_{n-1}$, which coincides with $f(x)$ at $(n-1)$ points, namely, at the points

$$x=a/n,\quad x=2a/n,\quad \dots\quad x=(n-1)a/n.$$

We then obtain the following $n-1$ equations:

$$f(a/n)=A_1\sin(\pi/n)+A_2\sin(2\pi/n)+\dots+A_{n-1}\sin\big((n-1)\pi/n\big),$$
$$f(2a/n)=A_1\sin(2\pi/n)+A_2\sin(2.2\pi/n)+\dots+A_{n-1}\sin\big(2(n-1)\pi/n\big),$$
$$f\big((n-1)a/n\big)=A_1\sin\big((n-1)\pi/n\big)+A_2\sin\big((n-1)2\pi/n\big)+\dots$$
$$+A_{n-1}\sin\big((n-1)(n-1)\pi/n\big).$$

If we multiply the first of these equations by $\sin(\pi/n)$, the second by $\sin(2\pi/n)$, etc., and add the right and left sides of the equations thus obtained, we have

$$f(a/n)\sin(\pi/n)+f(2a/n).\sin(2\pi/n)+\dots+f\big((n-1)a/n\big)\sin\big((n-1)\pi/n\big)$$
$$=A_1\big[\sin^2(\pi/n)+\sin^2(2\pi/n)+\dots+\sin^2\big((n-1)\pi/n\big)\big]$$
$$+A_2\big[\sin(\pi/n).\sin(2\pi/n)+\sin(2\pi/n).\sin(2.2\pi/n)+\dots$$
$$+\sin\big((n-1)\pi/n\big)\sin\big((n-1)2\pi/n\big)\big]+\dots.$$

Now

$$\sin^2\alpha+\sin^2 2\alpha+\dots+\sin^2\big((n-1)\alpha\big)$$
$$=\tfrac{1}{2}\Big[n-1-\Big(\cos 2\alpha+\cos 4\alpha+\dots+\cos\big((2n-2)\alpha\big)\Big)\Big].$$

But because

$$\cos\beta+\cos 2\beta+\dots+\cos\big((n-1)\beta\big)=\cos(\tfrac{1}{2}n\beta).\sin\big(\tfrac{1}{2}(n-1)\beta\big)/\sin\tfrac{1}{2}\beta,$$

we have

$$\sin^2\alpha+\sin^2 2\alpha+\dots+\sin^2\big((n-1)\alpha\big)=\tfrac{1}{2}[n-1-\cos n\alpha.\sin(n-1)\alpha/\sin\alpha].$$

Substituting for α its value π/n, we have

$$\sin^2(\pi/n)+\sin^2(2\pi/n)+\dots\sin^2\big((n-1)\pi/n\big)=n/2.$$

Now we can give the factor of A_2 the form

$$\tfrac{1}{2}\big[\cos(\pi/n)-\cos(3\pi/n)+\cos(2\pi/n)-\cos(6\pi/n)\dots$$
$$+\cos\big((n-1)\pi/n\big)-\cos\big((n-1)3\pi/n\big)\big].$$

Applying the above given summation formula, we find that this factor is equal to zero. In the same way the factors of A_3, A_4, etc., vanish, and we obtain finally

$$A_1=2/n.\big[f(a/n)\sin(\pi/n)+f(2a/n).\sin(2\pi/n)+\dots$$
$$+f\big((n-1)a/n\big)\sin\big((n-1)\pi/n\big)\big].$$

In general we have, for $0<m<n$,

$$\text{(i)}\qquad \begin{cases} A_m=2/n.\big[f(a/n).\sin(m\pi/n)+f(2a/n).\sin(2m\pi/n) \\ \qquad +\dots+f\big((n-1)a/n\big).\sin\big((n-1)m\pi/n\big)\big]. \end{cases}$$

Hence it is possible to so determine the coefficients A_1, A_2, ... that $f(x)$ and the trigonometrical series coincide for $(n-1)$ values of x between 0 and a. The greater the value of n, the more values will the two functions have in common, and when $n=\infty$, one function may be replaced by the other between the limits considered. The two functions are not, however, necessarily identical, for their differential coefficients may be entirely different. One of them is related to the other in the same way as a straight line to a zig-zag line, whose irregularities are infinitely small.

We will now assume that $n=\infty$ and write the equation (i),

$$A_m = 2/\pi \,.\, \pi/n \,.\, [f(a/n) \,.\, \sin(m\pi/n) + f(2a/n)\sin(2m\pi/n) + \ldots \\ + f(ra/n)\sin(rm\pi/n) + \ldots].$$

Setting $r\pi/n=\gamma$, $\pi/n=d\gamma$, $ra/n=a\gamma/\pi$, it follows that

$$\text{(k)} \qquad A_m = 2/\pi \,.\, \int_0^\pi f(a\gamma/\pi) \,.\, \sin(m\gamma)d\gamma.$$

Further, if we set $x=a\gamma/\pi$, we have

$$\text{(l)} \qquad A_m = 2/a \,.\, \int_0^a f(x)\sin(m\pi x/a)dx.$$

The same result is obtained in another way in XXXVII. (c).

Therefore, within given limits, we can replace the function $f(x)$ by a trigonometrical series, and set, for $0<x<a$,

$$\text{(m)} \qquad \begin{cases} f(x) = 2/a \,.\, [\sin(\pi x/a) \,.\, \int_0^a f(x)\sin(\pi x/a)dx \\ \qquad + \sin(2\pi x/a) \,.\, \int_0^a f(x)\sin(2\pi x/a)dx + \ldots]. \end{cases}$$

Introducing the values for A_1, A_2... contained in (g), the problem is solved, and we obtain

$$\text{(n)} \qquad \begin{cases} \tfrac{1}{2}a\theta = \sin(\pi x/a)e^{-\pi^2\kappa^2 t/a^2} \,.\, \int_0^a f(x)\sin(\pi x/a)dx \\ \qquad + \sin(2\pi x/a)e^{-2^2\pi^2\kappa^2 t/a^2} \,.\, \int_0^a f(x)\sin(2\pi x/a)dx + \ldots. \end{cases}$$

For example, if the initial temperature of the plate is constant and equal to θ_0, we have

$$\int_0^a \theta_0 \sin(m\pi x/a)dx = (1-\cos m\pi)a\theta_0/m\pi,$$

and therefore

$$\text{(o)} \qquad \tfrac{1}{4}\pi\theta = \theta_0 \,.\, \sin(\pi x/a)e^{-(\pi\kappa/a)^2 . t} + \tfrac{1}{3}\theta_0 \,.\, \sin(3\pi x/a) \,.\, e^{-(3\pi\kappa/a)^2 . t} + \ldots.$$

When $t=0$, we obtain, for $0<x<a$,

$$\text{(p)} \qquad \tfrac{1}{4}\pi = \sin(\pi x/a) + \tfrac{1}{3} \,.\, \sin(3\pi x/a) + \tfrac{1}{5} \,.\, \sin(5\pi x/a) + \ldots$$

SECTION CXXX. THE DEVELOPMENT OF FUNCTIONS IN SERIES OF SINES AND COSINES.

As shown in the foregoing paragraph, we may always set

(a) $$f(x) = A_1 \sin(\pi x/a) + A_2 \sin(2\pi x/a) + \dots,$$

in which [CXXIX. (1)]

(b) $$A_m = 2/a \,.\int_0^a f(\alpha)\sin(m\pi\alpha/a)d\alpha,$$

where x is replaced by α. This development holds only for $0 < x < a$; it does not hold for the limits 0 and a, except when $f(x)$ itself is equal to zero for these limits. The right side of (a) is an odd function, which changes its sign with x. The series (a) holds then within the limits $-a < x < 0$, when $f(x)$ is also odd. Setting $f(x) = x$, we have

$$A_m = 2/a \,.\int_0^a \alpha \sin(m\pi\alpha/a)d\alpha = -2a\cos(m\pi)/m\pi,$$

and further

(c) $$\tfrac{1}{2}\,.\,\pi x/a = \sin(\pi x/a) - \tfrac{1}{2}\,.\sin(2\pi x/a) + \tfrac{1}{3}\,.\sin(3\pi x/a) - \dots.$$

Since x is an odd function, the series holds for negative values of x if it holds for positive values. Further, since the series holds for $x = 0$, it holds within the limits $-a < x < +a$. Setting $\pi x/a = y$, we have for $-\pi < y < +\pi$,

(d) $$\tfrac{1}{2}y = \sin y - \tfrac{1}{2}\,.\sin 2y + \tfrac{1}{3}\,.\sin 3y - \dots.$$

Further, if we set (e) $f(x) = B_0 + B_1\cos(\pi x/a) + B_2\cos(2\pi x/a) + \dots$, multiply both sides of this equation by $\cos(m\pi x/a)$, and then integrate from 0 to a, it follows, if m and n are whole numbers, that

$$\int_0^a \cos(m\pi x/a)\cos(n\pi x/a)dx = 0 \quad \text{and} \quad \int_0^a \cos^2(m\pi x/a)dx = \tfrac{1}{2}a.$$

Hence for $m > 0$ we obtain

(f) $$B_m = 2/a \,.\int_0^a f(x)\,.\cos(m\pi x/a)dx \quad \text{and} \quad B_0 = 1/a \,.\int_0^a f(x)dx.$$

We obtain B_0 by multiplying both sides of (e) by dx and integrating from 0 to a. If $f(x)$ is an even function, the series holds within the limits $-a < x < a$, since the cosine series does not change its sign with x. But if $f(x)$ is an odd function, the series (e) holds only within the limits 0 and a.

We therefore obtain the result (g)

$$\tfrac{1}{2}a.f(x) = \sin(\pi x/a)\int_0^a f(\alpha)\sin(\pi\alpha/a)d\alpha + \sin(2\pi x/a)\int_0^a f(\alpha)\sin(2\pi\alpha/a)d\alpha + \dots,$$

and (h) $\begin{cases} \frac{1}{2}a\,.\,f(x)=\frac{1}{2}\,.\int_0^a f(\alpha)d\alpha+\cos(\pi x/a)\,.\int_0^a f(\alpha)\cos(\pi\alpha/a)d\alpha \\ \qquad +\cos(2\pi x/a)\int_0^a f(\alpha)\cos(2\pi\alpha/a)d\alpha+\ldots. \end{cases}$

An arbitrary function $f(x)$ can also be developed in a series of sines and cosines, so that the development holds within the limits $-a<x<a$. To effect this, we set

$$f(x)=\tfrac{1}{2}\,.\,[f(x)+f(-x)]+\tfrac{1}{2}\,.\,[f(x)-f(-x)],$$

in which $\frac{1}{2}[f(x)+f(-x)]$ is an even function, because it remains unchanged when x is replaced by $-x$. This function $\frac{1}{2}[f(x)+f(-x)]$ can therefore be represented by a cosine series. The coefficient of $\cos(m\pi x/a)$ is

$$\tfrac{1}{2}\,.\int_0^a[f(\alpha)+f(-\alpha)]\cos(m\pi\alpha/a)d\alpha$$
$$=\tfrac{1}{2}\,.\int_0^a f(\alpha)\cos(m\pi\alpha/a)d\alpha+\tfrac{1}{2}\,.\int_0^a f(-\alpha)\cos(m\pi\alpha/a)d\alpha.$$

If in the last integral α is replaced by $-\alpha$, the integral is transformed into $-\frac{1}{2}\,.\int_0^{-a} f(\alpha)\cos(m\pi\alpha/a)d\alpha$, and the coefficient sought becomes $\frac{1}{2}\int_{-a}^{+a} f(\alpha)\cos(m\pi\alpha/a)d\alpha$. Hence we obtain

(i) $\begin{cases} \frac{1}{2}a\,.\,[f(x)+f(-x)]=\frac{1}{2}\,.\int_{-a}^{+a} f(\alpha)d\alpha+\cos(\pi x/a)\int_{-a}^{+a} f(\alpha)\cos(\pi\alpha/a)d\alpha \\ \qquad +\cos(2\pi x/a)\int_{-a}^{+a} f(\alpha)\cos(2\pi\alpha/a)d\alpha+\ldots. \end{cases}$

On the other hand, the function $\frac{1}{2}\,.\,[f(x)-f(-x)]$ is an odd function, because it changes its sign with x; therefore, by using (g), we can represent this function by a sine series. The coefficient of $\sin(m\pi x/a)$ is

$$\tfrac{1}{2}\,.\int_0^a[f(\alpha)-f(-\alpha)]\sin(m\pi\alpha/a)d\alpha$$
$$=\tfrac{1}{2}\,.\int_0^a f(\alpha)\sin(m\pi\alpha/a)d\alpha-\tfrac{1}{2}\,.\int_0^a f(-\alpha)\sin(m\pi\alpha/a)d\alpha.$$

If we replace $-\alpha$ by α in the last integral, it is transformed into $-\frac{1}{2}\,.\int_0^{-a} f(\alpha)\sin(m\pi\alpha/a)d\alpha$, and the coefficient becomes

$$\tfrac{1}{2}\,.\int_{-a}^{+a} f(\alpha)\sin(m\pi\alpha/a)d\alpha.$$

We therefore obtain

(k) $\begin{cases} \frac{1}{2}a\,.\,[f(x)-f(-x)]=\sin(\pi x/a)\,.\int_{-a}^{+a} f(\alpha)\sin(\pi\alpha/a)d\alpha \\ \qquad +\sin(2\pi x/a)\int_{-a}^{+a} f(\alpha)\sin(2\pi\alpha/a)d\alpha+\ldots. \end{cases}$

By the addition of equations (i) and (k) we obtain finally

(l)
$$\begin{cases} a\,.f(x) = \tfrac{1}{2}\,.\int_{-a}^{+a} f(\alpha)d\alpha + \int_{-a}^{+a} f(\alpha)\cos\big(\pi(x-\alpha)/a\big)d\alpha \\ \qquad + \int_{-a}^{+a} f(\alpha)\cos\big(2\pi(x-\alpha)/a\big)d\alpha + \ldots, \end{cases}$$

or, for $-a<x<a$,

(m) $$f(x) = 1/a\,.\int_{-a}^{+a}\left[\tfrac{1}{2} + \cos\big(\pi(x-\alpha)/a\big) + \cos\big(2\pi(x-\alpha)/a\big) + \ldots\right]f(\alpha)d\alpha.$$

This series is due to Fourier. It may also be expressed in the form

(n) $$f(x) = 1/2a\,.\int_{-a}^{+a} f(\alpha)d\alpha + 1/a\,.\sum_{m=1}^{m=\infty}\int_{-a}^{+a} f(\alpha)\cos\big(m\pi(x-\alpha)/a\big)d\alpha.$$

We may now ascribe any value to a. If a is infinitely great, and if $\int_{-a}^{a} f(\alpha)d\alpha$ is finite, the first term on the right side of the equation (n) vanishes, and we obtain

$$f(x) = 1/\pi\,.\sum_{m=1}^{m=\infty}\pi/a\,.\int_{-a}^{a} f(\alpha)\cos\big(m\pi(x-\alpha)/a\big)d\alpha.$$

Now setting $m\pi/a = \lambda$, and therefore $\pi/a = d\lambda$, it follows that

(o) $$f(x) = 1/\pi\,.\int_{0}^{\infty} d\lambda\int_{-\infty}^{+\infty} f(\alpha)\cos\big(\lambda(x-\alpha)\big)d\alpha,$$

where $-\infty < x < \infty$. Instead of this equation we may often use one of the two which are obtained from (g) and (h). The general term in (g) is

$$\sin(m\pi x/a)\,.\int_{0}^{a} f(\alpha)\sin(m\pi\alpha/a)d\alpha,$$

and hence

$$f(x) = 2/\pi\,.\,\pi/a\,.\sum_{m=1}^{m=\infty}\sin(m\pi x/a)\int_{0}^{a} f(\alpha)\sin(m\pi\alpha/a)d\alpha.$$

Now if we set $m\pi/a = \lambda$, and therefore $\pi/a = d\lambda$, we have for $0 < x < \infty$,

(p) $$f(x) = 2/\pi\,.\int_{0}^{\infty} d\lambda\,.\,\sin(\lambda x)\int_{0}^{\infty} f(\alpha)\sin(\lambda\alpha)d\alpha.$$

From (h) we obtain in the same way for $0 < x < \infty$,

(q) $$f(x) = 2/\pi\,.\int_{0}^{\infty} d\lambda\,\cos(\lambda x)\,.\int_{0}^{\infty} f(\alpha)\cos(\lambda\alpha)d\alpha.$$

SECTION CXXXI. THE APPLICATION OF FOURIER'S THEOREM TO THE CONDUCTION OF HEAT.

If the temperature in a certain region depends only on the x-coordinate, the temperature θ must satisfy Fourier's equation

(a) $$\partial\theta/\partial t = \kappa^2\partial^2\theta/\partial x^2.$$

From CXXIX. (c) $\theta = e^{-\lambda^2\kappa^2 t}(A \sin\lambda x + B\cos\lambda x)$ is an integral of equation (a) where λ, A, B are constants. We may also give the expression for θ the form $\theta = e^{-\lambda^2\kappa^2 t}\cos\big(\lambda(x-a)\big).f(a)$, where $f(a)$ is an arbitrary function of a, and λ and a are constants, which may take all possible values. Any sum of such terms satisfies the equation, and as the integral

(b) $$\theta = 1/\pi.\int_0^\infty d\lambda e^{-\lambda^2\kappa^2 t}\int_{-\infty}^{+\infty} f(a)\cos\big(\lambda(x-a)\big)da$$

is such a sum, it will also satisfy the equation. But when $t=0$, we have

$$\theta = 1/\pi.\int_0^\infty d\lambda\int_{-\infty}^{+\infty} f(a)\cos\big(\lambda(x-a)\big)da,$$

from which, by comparison with CXXX. (o), we obtain $\theta = f(x)$.

The formula (b) contains the solution of the problem, to determine the temperature in a body at any time t, when the temperature is given, at the time $t=0$, by $\theta = f(x)$. This problem has already been solved in another way in CXXV. (h) and (i). We proceed to show that the solution here given is identical with the former one.

Since (c) $$\theta = 1/\pi.\int_{-\infty}^{+\infty} f(a)da\int_0^\infty e^{-\lambda^2\kappa^2 t}\cos\big(\lambda(x-a)\big)d\lambda,$$

we first determine the value of the integral

$$\int_0^\infty e^{-\lambda^2\kappa^2 t}\cos\big(\lambda(x-a)\big)d\lambda.$$

If we develop $\cos\big(\lambda(x-a)\big)$ in a series, this integral is represented by

$$\sum_{n=0}^{n=\infty}(-1)^n\frac{(x-a)^{2n}}{[2n]}.\int_0^\infty e^{-\lambda^2\kappa^2 t}\lambda^{2n}.\ d\lambda.$$

It follows by integration by parts that

$$\int_0^\infty e^{-\lambda^2\kappa^2 t}\lambda^{2n}d\lambda = (2n-1)/2\kappa^2 t.\int_0^\infty e^{-\lambda^2\kappa^2 t}\lambda^{2n-2}d\lambda,$$

and by continued reduction

$$\int_0^\infty e^{-\lambda^2\kappa^2 t}\lambda^{2n}d\lambda = \frac{(2n-1)(2n-3)\dots 3.1}{(2\kappa^2 t)^n}.\int_0^\infty e^{-\lambda^2\kappa^2 t}d\lambda.$$

But because
$$\int_0^\infty e^{-q^2}dq = \tfrac{1}{2}\sqrt{\pi},$$
we obtain
$$\int_0^\infty e^{-\lambda^2\kappa^2 t}\lambda^{2n}d\lambda = \frac{(2n-1)(2n-3)\ldots 3.1}{(2\kappa^2 t)^n}\,\frac{\sqrt{\pi}}{2\kappa\sqrt{t}}.$$
The value of the integral sought is therefore
$$\frac{\sqrt{\pi}}{2\kappa\sqrt{t}}\sum_{n=0}^{n=\infty}(-1)^n\frac{(x-a)^{2n}}{[2n]}\cdot\frac{(2n-1)(2n-3)\ldots 3.1}{(2\kappa^2 t)^n} = \frac{\sqrt{\pi}}{2\kappa\sqrt{t}}\sum_0^\infty\frac{(-1)^n}{[n]}\frac{(x-a)^{2n}}{(4\kappa^2 t)^n},$$
or
$$\frac{\sqrt{\pi}}{2\kappa\sqrt{t}}\cdot e^{-(x-a)^2/4\kappa^2 t}.$$
If we replace the integral in (c) by this value, we obtain

(d)
$$\theta = \frac{1}{\sqrt{\pi}}\cdot\frac{1}{2\kappa\sqrt{t}}\cdot\int_{-\infty}^{+\infty}f(a)e^{-(x-a)^2/4\kappa^2 t}da.$$

This expression for θ is identical with that given in CXXV. (h).

We will now apply Fourier's theorem to find the law of penetration of heat into a body. For this purpose we will consider the simple case in which the body is in contact over a plane bounding surface F with another body whose temperature θ_0 is constant and given. Let the original temperature of the cold body be zero.

If we proceed as before and use CXXX. (p), we obtain
$$\theta = \theta_0 + 2/\pi\,.\int_0^\infty d\lambda\sin(\lambda x)e^{-\lambda^2\kappa^2 t}\,.\int_0^\infty f(a)\sin(\lambda a)da.$$
This expression for θ satisfies all conditions if only we have, when $t = 0$,
$$0 = \theta_0 + 2/\pi\,.\int_0^\infty d\lambda\sin(\lambda x)\,.\int_0^\infty f(a)\sin(\lambda a)da.$$
This condition is fulfilled [CXXX. (p)] when $f(a) = -\theta_0$. Hence the solution of the proposed problem is contained in

(e)
$$\theta = \theta_0 - 2\theta_0/\pi\,.\int_0^\infty d\lambda\sin(\lambda x)e^{-\lambda^2\kappa^2 t}\,.\int_0^\infty\sin(\lambda a)da.$$

Using the same method of reduction as that by which (c) is transformed into (d), we obtain
$$\theta = \theta_0 - \frac{\theta_0}{2\kappa\sqrt{\pi t}}\cdot\left\{\int_0^\infty e^{-(x-a)^2/4\kappa^2 t}da - \int_0^\infty e^{-(x+a)^2/4\kappa^2 t}da\right\},$$
from which
$$\theta = \theta_0 - \frac{\theta_0}{\sqrt{\pi}}\cdot\left\{\int_{-x/2\kappa\sqrt{t}}^\infty e^{-q^2}dq - \int_{x/2\kappa\sqrt{t}}^\infty e^{-q^2}dq\right\},$$
and therefore
$$\theta = \theta_0\left\{1 - 1/\sqrt{\pi}\,.\int_{-x/2\kappa\sqrt{t}}^{+x/2\kappa\sqrt{t}}e^{-q^2}dq\right\}.$$

Since e^{-q^2} is an even function, we obtain

$$\theta = \theta_0 \left\{ 1 - 2/\sqrt{\pi} \, . \int_0^{x/2\kappa\sqrt{t}} e^{-q^2} dq \right\},$$

or by the help of CXXV. (e),

(f) $$\theta = 2\theta_0/\sqrt{\pi} \, . \int_{x/2\kappa\sqrt{t}}^{\infty} e^{-q^2} dq.$$

Let A and B be two points within the body, whose distances from the surface F are x_1 and x_2 respectively. The temperature θ' which A attains after the lapse of the time t_1 is $\theta' = 2\theta_0/\sqrt{\pi} \, . \int_{x_1/2\kappa\sqrt{t_1}}^{\infty} e^{-q^2} dq$. B attains the same temperature after the lapse of the time t_2, given by the equation $\theta' = 2\theta_0/\sqrt{\pi} \, . \int_{x_2/2\kappa\sqrt{t_2}}^{\infty} e^{-q^2} dq$. Comparing the two integrals, it appears that $x_1/\sqrt{t_1} = x_2/\sqrt{t_2}$ or $t_2/t_1 = x_2^2/x_1^2$, that is, *the times required for two points to attain the same temperature are proportional to the squares of the distances of the points from the heated surface F.*

We will now determine the quantity of heat which flows into the cooler body through unit area in unit time. For this purpose we give the equation (f) the form $\theta = 2\theta_0/\sqrt{\pi} \, . [f(\infty) - f(x/2\kappa\sqrt{t})]$, from which follows, by (f), $-k\partial\theta/\partial x = k\theta_0/\kappa\sqrt{\pi t} \, . \, e^{-x^2/4\kappa^2 t}$. Setting $x = 0$, we find the quantity of heat U desired, (g) $U = k\theta_0/\kappa\sqrt{\pi t}$.

By the help of equation (g) we may solve an important problem. Two bodies L and L' are in contact over a plane surface, the temperature of one of these bodies being T, that of the other T'. If the two bodies are brought in contact, one of them is heated and the other is cooled. We can also determine the temperature T_0 of the surface of contact. Assuming that T_0 is constant, the quantity of heat which L receives in unit time is, from (g), given by

$$U = k(T_0 - T)/\kappa\sqrt{\pi t}.$$

In the same time, L' receives the quantity of heat

$$U' = k'(T_0 - T')/\kappa'\sqrt{\pi t},$$

where k' and κ' have the same meaning for L' as k and κ for L. But since the infinitely thin bounding surface can contain no heat, $U + U'$ must equal zero, or $k/\kappa \, . (T_0 - T) = k'/\kappa' \, . (T' - T_0)$, from which follows (h)* $T_0 = (T\sqrt{kc\rho} + T'\sqrt{k'c'\rho'})/(\sqrt{kc\rho} + \sqrt{k'c'\rho'})$. It is thus shown that the assumption is correct, that the temperature in the bounding surface between two bodies which meet in a plane surface is constant. Strictly speaking, the bodies in contact must both be infinitely large,

* L. Lorenz, *Lehre von der Wärme*. S. 178. Kopenhagen, 1877.

but the formula (h) may also be applied to small bodies if we only consider them shortly after they are brought in contact. We may show from (h) that the temperature of a heated solid is very little diminished by contact with the air; this holds for the metals and for good conductors in general. It follows from equation (h) that

$$(T - T_0)/(T_0 - T') = \sqrt{k'c'\rho'/kc\rho}.$$

If T represents the temperature of the solid, ρ is always very much greater than ρ'. Hence $T_0 - T'$ is very much greater than $T - T_0$, especially since k is also greater than k', while c and c' are not very different from each other.

SECTION CXXXII. THE COOLING OF A SPHERE.

Let us suppose that the temperature at a point in the interior of a sphere depends only on the distance of that point from the centre of the sphere. In this case [CXXVI. (c)] Fourier's equation takes the form (a) $\partial(r\theta)/\partial t = \kappa^2\partial^2(r\theta)/\partial r^2$. If m, A, B are arbitrary constants, an integral of equation (a) is

$$r\theta = e^{-m^2\kappa^2 t}.\left(A\sin(mr) + B\cos(mr)\right).$$

But since this equation leads to the conclusion that $\theta = \infty$ when $r = 0$, B must equal zero, and we obtain as the integral of equation (a)

(b) $$r\theta = A\,.\,e^{-m^2\kappa^2 t}.\sin(mr).$$

We will first consider the case of the sphere immersed in a mixture of ice and water, or so situated that its surface is kept at the temperature 0° by any means. If we represent the radius of the sphere by R, we have $\theta = 0$ when $r = R$, and therefore $\sin(mR) = 0$. Hence, if p is an arbitrary whole number, we must have $mR = \pm p\pi$. We can now set

(c) $$r\theta = A_1 e^{-(\pi\kappa/R)^2 t}.\sin(\pi r/R) + A_2 e^{-(2\pi\kappa/R)^2 t}.\sin(2\pi r/R) + \dots.$$

The constants A_1, A_2 ... are determined by the help of the temperatures of the different parts of the sphere at the time $t = 0$. Let these temperatures be given by $f(r)$. We then have

$$r.f(r) = A_1\sin(\pi r/R) + A_2\sin(2\pi r/R) + \dots.$$

From CXXIX. (l) we obtain for A_m

(d) $$A_m = 2/R.\int_0^R rf(r)\sin(m\pi r/R)dr.$$

If the temperature is constant and equal to θ_0 at the time $t=0$, we have

$$A_m = 2\theta_0/R . \int_0^R r \sin(m\pi r/R) dr, \text{ and hence } A_m = -2\theta_0 R/m\pi . \cos(m\pi).$$

Using these values we obtain finally

$$\text{(e)} \quad \begin{cases} \theta = 2R\theta_0/\pi r . [\sin(\pi r/R) . e^{-(\pi\kappa/R)^2 t} \\ \qquad - \tfrac{1}{2} . \sin(2\pi r/R) e^{-(2\pi\kappa/R)^2 t} + \tfrac{1}{3} . \sin(3\pi r/R) e^{-(3\pi\kappa/R)^2 t} - \ldots]. \end{cases}$$

The mean temperature θ' is

$$\text{(f)} \qquad \theta' = 6\theta_0/\pi^2 . \{e^{-(\pi\kappa/R)^2 t} + \tfrac{1}{4} . e^{-(2\pi\kappa/R)^2 t} + \tfrac{1}{9} e^{-(3\pi\kappa/R)^2 t} + \ldots\}.$$

This equation may be applied to a thermometer which is immersed in a fluid cooler than itself. The temperature of the thermometer is then given, to a close approximation, by the first term of the above equation. The rate of cooling is (g) $-d\theta'/dt = \pi^2 k\theta'/c\rho R^2$.

We will now consider another important case, that of a sphere in vacuo losing heat by radiation. We suppose the temperature of the region, or rather of its boundary, to be 0°. We suppose the radiation to take place according to Newton's law, and therefore to be proportional to the temperature on the surface of the sphere. From (b) the integral takes the form (h) $r\theta = \Sigma A_m e^{-m^2\kappa^2 t} . \sin(mr)$. If E represents the coefficient of radiation, the quantity of heat which radiates in unit time from an element dS of the surface is $dS . E\theta$. We will assume that E is constant. The same surface-element dS receives from the interior of the sphere in the same time the quantity of heat $-k . dS . d\theta/dr$. Since the quantity of heat which dS receives must be equal to that which it emits, we have (i) $-k . d\theta/dr = E\theta$ or $-d\theta/dr = h\theta$, where, for brevity, we set $h = E/k$. Hence, for $r = R$,

$$\Sigma A_m e^{-m^2\kappa^2 t} . (m \cos(mR)/R - \sin(mR)/R^2) = -h\Sigma A_m e^{-m^2\kappa^2 t} \sin(mR)/R,$$

or

$$\Sigma A_m e^{-m^2\kappa^2 t} . [mR \cos(mR) - (1 - hR) \sin(mR)] = 0.$$

If this equation is to hold for every value of t, we must have

$$\text{(k)} \qquad mR . \cos(mR) = (1 - hR) . \sin(mR).$$

This equation must be solved for m. We set $mR = x$ and obtain (l) $\operatorname{tg} x = x/(1 - hR)$. If we further set $y_1 = \operatorname{tg} x$, y_1 and x may be considered as the rectangular coordinates of a curve (Fig. 143). This curve has an infinite number of branches, oa, πb, $2\pi c \ldots$, to which the straight lines $x = \frac{1}{2}\pi$, $x = \frac{3}{2}\pi \ldots$ are asymptotes. Further, if we set $y_2 = x/(1 - hR)$ this equation represents a straight line, such as opq, which passes through the origin of coordinates.

The constant h is positive, and its value must lie between 0 and ∞. First consider the case where $h=0$; then $y_2=x$, and this equation represents the straight line opq which touches the curve oa at the point o, and cuts πb at p, $2\pi c$ at q, etc. The abscissas 0, x_1, x_2... are roots of equation (1). We have further

$$\pi<x_1<3\pi/2\,;\quad 2\pi<x_2<5\pi/2\,;\quad \ldots n\pi<x_n<(n+\tfrac{1}{2})\pi.$$

As n increases x_n approaches the superior limit $(n+\frac{1}{2})\pi$. Besides its positive roots equation (1) has also negative roots, which are equal in absolute value to the positive.

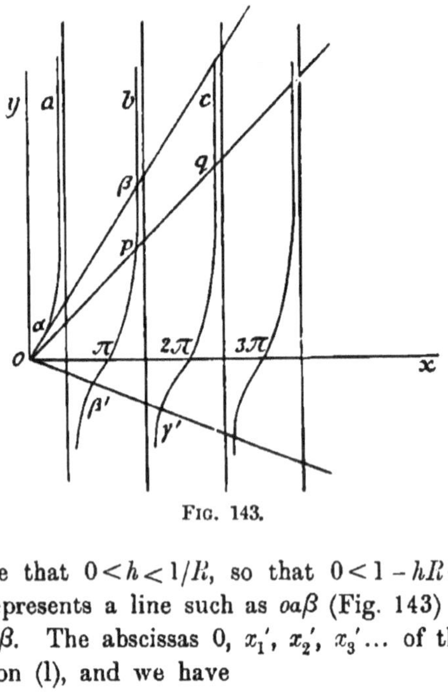

FIG. 143.

Next suppose that $0<h<1/R$, so that $0<1-hR<1$, and hence $y_2>x$. This represents a line such as $o\alpha\beta$ (Fig. 143) which cuts the curve at o, α, β. The abscissas 0, x_1', x_2', x_3'... of these points are roots of equation (1), and we have

$$0<x_1'<\tfrac{1}{2}\pi\,;\ \pi<x_2'<3\pi/2\,;\ 2\pi<x_3'<5\pi/2\ldots(n-1)\pi<x_n'<(n-\tfrac{1}{2})\pi.$$

As h increases, the angle $\alpha o\pi$ approaches the angle $\frac{1}{2}\pi$, and the roots approach their superior limits. If $h=1/R$, we have $x=0$, and the roots are 0, $\frac{1}{2}\pi$, $3\pi/2$, $5\pi/2,\ldots$. Now if $1/R<h<\infty$, we have $y_2=-x/(hR-1)$. The straight line then has the position $o\beta'\gamma'$. If in this case we represent the roots of equation (1) by 0, x_1'', $x_2''\ldots$, we have $\frac{1}{2}\pi<x_1''<\pi$; $3\pi/2<x_2''<2\pi\ldots$. As h increases, the roots approach their superior limits. And if $h=\infty$, we obtain $x_1''=\pi$; $x_2''=2\pi\ldots$.

We may now consider the roots of equation (l) as known and determine m from them. The values of m corresponding to the several roots are $0, m_1, m_2, \ldots$. We may neglect the negative roots, because the terms in (h) corresponding to them may be considered as included in those arising from the positive roots. We therefore set (m) $r\theta = A_1 e^{-m_1^2\kappa^2 t}\sin(m_1 r) + A_2 e^{-m_2^2\kappa^2 t}\sin(m_2 r) + \ldots$. If the temperature at the time $t=0$ is given by $\theta = f(r)$, we have

$$\text{(n)} \qquad r\,.\,f(r) = A_1 \sin(m_1 r) + A_2 \sin(m_2 r) + \ldots.$$

Now let m_a and m_b be two roots of equation (k), and multiply both sides of equation (n) by $\sin(m_a r)$. It follows by integration from 0 to R that

$$\text{(o)} \qquad \int_0^R r\,.\,f(r)\sin(m_a r)dr = \Sigma A_b \int_0^R \sin(m_b r)\sin(m_a r)dr.$$

Now

$$\text{(p)} \quad \begin{cases} \displaystyle\int_0^R \sin(m_b r)\sin(m_a r)dr = \tfrac{1}{2}\,.\int_0^R \{\cos[(m_b - m_a)r] - \cos[(m_b + m_a)r]\}dr \\ \quad = \tfrac{1}{2}\sin[(m_b - m_a)R]/(m_b - m_a) - \tfrac{1}{2}\,.\,\sin[(m_b + m_a)R]/(m_b + m_a) \\ \quad = [m_a \sin(m_b R)\cos(m_a R) - m_b \cos(m_b R)\sin(m_a R)]/(m_b^2 - m_a^2). \end{cases}$$

But because we have from (k)

$$m_a R = (1 - hR)\,\mathrm{tg}\,(m_a R), \quad m_b R = (1 - hR)\,\mathrm{tg}\,(m_b R),$$

and therefore

$$m_a \,\mathrm{tg}\,(m_b R) = m_b \,\mathrm{tg}\,(m_a R),$$

or

$$m_a\,.\,\sin(m_b R)\,.\,\cos(m_a R) = m_b\,.\,\sin(m_a R)\,.\,\cos(m_b R),$$

we have

$$\int_0^R \sin(m_b r)\,.\,\sin(m_a r)dr = 0,$$

whenever m_a and m_b are different from each other. But if they are equal, (p) is indeterminate. We find the value of the expression (p) in this case by setting $m_b = m_a + \epsilon$, where ϵ is a small quantity. We reach the same result more simply if we investigate the value of the integral

$$\int_0^R \sin^2(m_a r)dr = \tfrac{1}{2}\,.\int_0^R [1 - \cos(2m_a r)]dr = \tfrac{1}{2}[R - \sin(2m_a R)/2m_a].$$

We thus obtain

$$A_a = 2/R\,.\int_0^R r\,.\,f(r)\sin(m_a r)dr/[1 - \sin(2m_a R)/2m_a R].$$

Hence the complete solution of the problem is contained in

$$\text{(q)} \quad \begin{cases} \dfrac{Rr\theta}{2} = \dfrac{\sin(m_1 r)e^{-m_1^2\kappa^2 t}}{1 - \sin(2m_1 R)/2m_1 R}\cdot\displaystyle\int_0^R r\,.\,f(r)\sin(m_1 r)dr \\ \qquad + \dfrac{\sin(m_2 r)e^{-m_2^2\kappa^2 t}}{1 - \sin(2m_2 R)/2m_2 R}\cdot\displaystyle\int_0^R r\,.\,f(r)\sin(m_2 r)dr + \ldots \end{cases}$$

In the simple case in which the initial temperature of the sphere is everywhere equal, we have $f(r)=\theta_0$, and then find

$$\theta_0 \,.\int_0^R r\sin(mr)dr=\theta_0/m^2\,.\,[\sin(mR)-mR\cos(mR)],$$

or, by the use of (k),

$$\theta_0\int_0^R r\sin(mr)dr=hR\theta_0\sin(mR)/m^2.$$

We therefore get the result

$$\text{(r)}\quad \tfrac{1}{4}r\theta=hR\theta_0\,.\left(\frac{\sin(m_1R)\sin(m_1r)e^{-m_1^2\kappa^2t}}{m_1[2m_1R-\sin(2m_1R)]}+\frac{\sin(m_2R)\sin(m_2r)e^{-m_2^2\kappa^2t}}{m_2[2m_2R-\sin(2m_2R)]}+\dots\right).$$

If the coefficient of radiation E and therefore also h are very small, or if the radius of the sphere is small, the product hR is a small quantity. In this case, if we neglect the higher powers when the sine and cosine are developed in series, we obtain from (k),

$$1-\tfrac{1}{2}\,.\,m_1^2R^2=(1-hR)(1-\tfrac{1}{6}\,.\,m_1^2R^2),$$

from which follows $m_1^2=3h/R$.

The other values of m are so very much greater that the corresponding terms in (r) vanish in comparison with the first term. We therefore obtain $\theta=\theta_0\,.\,e^{-3h\kappa^2t/R}$, or, substituting the value of h,

$$\text{(s)}\qquad \theta=\theta_0e^{-3Et/\rho cR}.$$

We can derive this formula more simply. The quantity of heat which the sphere radiates in the time dt is $4\pi R^2E\theta dt$. Thus the temperature of the sphere increases by $-d\theta$; and the quantity of heat given up is $-4\pi/3\,.\,R^3c\rho d\theta$. Hence we have

$$4\pi R^2E\theta dt=-4\pi/3\,.\,R^3c\rho d\theta,$$

from which follows $\theta=\theta_0e^{-3Et/c\rho R}$, since the temperature of the sphere is θ_0 at the time $t=0$.

SECTION CXXXIII. THE MOTION OF HEAT IN AN INFINITELY LONG CYLINDER.

Let the cross-section S of the cylinder be so small that its temperature θ is constant, and let A and B be two cross-sections separated by the distance dx. The quantity of heat $-Sk\,.\,\partial\theta/\partial x\,.\,dt$ flows through A in the time dt, and the quantity $-Sk(\partial\theta/\partial x+\partial^2\theta/\partial x^2\,.\,dx)dt$ flows through B in the same time. Hence the part of the cylinder between A and B receives the quantity of heat $Sk\,.\,\partial^2\theta/\partial x^2\,.\,dxdt$. A part of this heat is given up to surrounding bodies by conduction or radiation. If P is the perimeter of the cylinder, E a constant, and if the temperature of the medium around the cylinder is 0, the heat given up by conduction or radiation is $PE\theta . dxdt$. Another portion of the heat received serves to heat the cylinder; this portion is $S . dx . c\rho . d\theta$. Hence we obtain the equation

$$Sk . \partial^2\theta/\partial x^2 = S\rho c . \partial\theta/\partial t + PE\theta, \text{ or (a) } \partial\theta/\partial t = \kappa^2 . \partial^2\theta/\partial x^2 - h\theta,$$

if we set $\kappa^2 = k/c\rho$ and $h = PE/S\rho c$.

If the state of the cylinder or rod has become steady, we have $\partial\theta/\partial t = 0$, and equation (a) takes the form $\kappa^2 . \partial^2\theta/\partial x^2 = h\theta$. From this it follows that (b) $\theta = Ae^{x\sqrt{h}/\kappa} + Be^{-x\sqrt{h}/\kappa}$. If the temperature of the rod is given at two points, we obtain from (b) the temperature of the intermediate points. We will assume that the temperature of a certain point in the rod is θ_0, and that at a very great distance from this point the temperature of the rod is 0. Then, for $x = \infty$, we will have $\theta = 0$ and therefore (c) $\theta = \theta_0 . e^{-x\sqrt{h}/\kappa}$. But if the rod is not in a steady state the equation (a) must be used. If we substitute in this equation $\theta = u . e^{-ht}$, we obtain $\partial u/\partial t = \kappa^2 . \partial^2 u/\partial x^2$. The integral of this differential equation has already been given. By the help of equations (c) and (d) we can determine the cooling of any rod which is heated in any manner.

We will here consider only the case in which one cross-section S of the rod has the temperature θ_0, while the temperature of all other parts of the rod is zero. The heat flows from S toward both parts of the rod, and after an infinitely long time the temperature at the distance x from S is given by $\theta = \theta_0 . e^{-x\sqrt{h}/\kappa}$. On the other hand, the temperature at the same place at the time t is given by

$$\text{(e)} \qquad \theta = \theta_0 . e^{-x\sqrt{h}/\kappa} + u . e^{-ht}.$$

In this case u must satisfy the following conditions:

(1) It must satisfy the equation $\partial u/\partial t = \kappa^2 . \partial^2 u/\partial x^2$;

(2) When $t = 0$ the equation holds $0 = \theta_0 . e^{-x\sqrt{h}/\kappa} + u$;

(3) When $x = 0$ we must have $u = 0$.

Conditions (1) and (3) are satisfied by

$$u = 2/\pi . \int_0^\infty d\lambda \sin(\lambda x) \int_0^\infty f(a) \sin(\lambda a) e^{-\lambda^2\kappa^2 t} da.$$

And as u also satisfies (2) we have

$$0 = \theta_0 . e^{-\sqrt{h}/\kappa} + 2/\pi . \int_0^\infty d\lambda \sin(\lambda x) \int_0^\infty f(a) \sin(\lambda a) da.$$

It is requisite for this that $f(a) = -\theta_0 e^{-a\sqrt{r/\kappa}}$. Hence the solution of the proposed problem is

$$\theta = \theta_0 . e^{-x\sqrt{h/\kappa}} - 2\theta_0/\pi . e^{-ht}\int_0^\infty d\lambda \sin(\lambda x) . \int_0^\infty \sin(\lambda a) e^{-\lambda^2\kappa^2 t - a\sqrt{h/\kappa}} da.$$

As in CXXXI., we can give the expression

$$2/\pi . \int_0^\infty \sin(\lambda x) . \sin(\lambda a) e^{-\lambda^2\kappa^2 t} d\lambda$$

the form

$$1/\pi . \int_0^\infty \cos[\lambda(x-a)] e^{-\lambda^2\kappa^2 t} d\lambda - 1/\pi . \int_0^\infty \cos[\lambda(x+a)] e^{-\lambda^2\kappa^2 t} d\lambda,$$

and this latter expression equals $1/2\kappa\sqrt{\pi t}.(e^{-(x-a)^2/4\kappa^2 t} - e^{-(x+a)^2/4\kappa^2 t})$. Substituting this expression in the equation for θ, it follows that

$$\theta = \theta_0\Big[e^{-x\sqrt{h/\kappa}} - \frac{e^{-ht}}{2\kappa\sqrt{\pi t}} . \int_0^\infty (e^{-(x-a)^2/4\kappa^2 t} - e^{-(x+a)^2/4\kappa^2 t}) . e^{-a\sqrt{h/\kappa}} da\Big].$$

To simplify this expression, we set $p = (a-x)/2\kappa\sqrt{t}$, $a = x + 2\kappa p\sqrt{t}$. It then follows that

$$\frac{e^{-ht}}{2\kappa\sqrt{\pi t}} . \int_0^\infty e^{-(x-a)^2/4\kappa^2 t} . e^{-a\sqrt{h/\kappa}} da = 1/\sqrt{\pi} . \int_{-x/2\kappa\sqrt{t}}^\infty e^{-p^2} . e^{-x\sqrt{h/\kappa} - 2p\sqrt{ht}} . e^{-ht} . dp$$

$$= \frac{e^{-x\sqrt{h/\kappa}}}{\sqrt{\pi}} . \int_{-x/2\kappa\sqrt{t}}^\infty e^{-(p+\sqrt{ht})^2} dp = \frac{e^{-x\sqrt{h/\kappa}}}{\sqrt{\pi}} . \int_{\sqrt{ht} - x/2\kappa\sqrt{t}}^\infty e^{-q^2} dq.$$

We obtain in the same way

$$\frac{e^{-ht}}{2\kappa\sqrt{\pi t}} . \int_0^\infty e^{-(x+a)^2/4\kappa^2 t} . e^{-a\sqrt{h/\kappa}} da = \frac{e^{x\sqrt{h/\kappa}}}{\sqrt{\pi}} . \int_{\sqrt{ht} + x/2\kappa\sqrt{t}}^\infty e^{-q^2} dq,$$

and therefore

$$\text{(f)} \qquad \theta = \theta_0\Big[e^{-x\sqrt{h/\kappa}} - \frac{e^{-x\sqrt{h/\kappa}}}{\sqrt{\pi}} . \int_{\sqrt{ht} - x/2\kappa\sqrt{t}}^\infty e^{-q^2} dq + \frac{e^{x\sqrt{h/\kappa}}}{\sqrt{\pi}} . \int_{\sqrt{ht} + x/2\kappa\sqrt{t}}^\infty e^{-q^2} dq\Big].$$

A careful examination of this expression shows that it represents the flow of heat through an infinitely long rod. For $t=0$ the lower limit of the first integral equals $-\infty$, and the value of the integral itself is then equal to $\sqrt{\pi}$; the lower limit of the second integral is in the same case ∞, and therefore the value of the second integral equals zero. Hence for $t=0$ we also obtain $\theta=0$, which should be the case for all cross-sections of the rod, except for the heated section. Both integrals have the same value for $x=0$, and therefore $\theta=\theta_0$. Both integrals vanish when $t=\infty$, and for the steady state of heat we have the evidently correct result $\theta = \theta_0 . e^{-x\sqrt{h/\kappa}}$. Because $h = PE/Sc\rho$, h is infinitely small, if the cross-section of the rod is infinitely great, or if the coefficient of radiation E is infinitely small. Setting $h=0$ in (f) we come back to a case already treated, since

$$\text{(g)} \qquad \theta=\theta_0\left(1-1/\sqrt{\pi}\,.\int_{-x/\kappa 2\sqrt{t}}^{\infty} e^{-q^2}dq+1/\sqrt{\pi}\,.\int_{x/2\kappa\sqrt{t}}^{\infty} e^{-q^2}dq\right).$$

This result is also found in CXXXI. The expression (g) gives the temperature in an infinitely extended body, having, at the time $t=0$, the temperature $\theta=0$ at all points, with the exception of the points on the surface $x=0$, for which $\theta=\theta_0$.

The solution (f) holds only for positive values of x; and that it shall hold for those parts of the rod which correspond to negative values of x, x must be replaced by $-x$ in (f).

SECTION CXXXIV. ON THE CONDUCTION OF HEAT IN FLUIDS.

Up to this point we have treated only the motion of heat in solids. The results which have thus been obtained cannot in general be applied to fluids, because any difference of temperature which causes a different expansion in different parts of the fluid, occasions so-called convection currents. In general, differences of temperature are more quickly equalized by these currents than by conduction alone. The relations are therefore very complicated. We will confine ourselves to developing the general equations of motion which will be applied in some simple cases.

We use the notation of hydrodynamics. The equation of continuity, which expresses that the quantity of matter is constant, becomes [cf. XLI. (d)]

$$\text{(a)} \qquad \partial\rho/\partial t+\partial(\rho u)/\partial x+\partial(\rho v)/\partial y+\partial(\rho w)/\partial z=0.$$

The momentum received by the unit of volume in the unit of time is equal to the force acting on that unit of volume. We have therefore from XLI.

$$\text{(b)} \qquad \left\{\begin{aligned} A&=\rho(\partial u/\partial t+u\partial u/\partial x+v\partial u/\partial y+w\partial u/\partial z)\\ &=\partial X_x/\partial x+\partial X_y/\partial y+\partial X_z/\partial z+\rho X,\\ B&=\rho(\partial v/\partial t+u\partial v/\partial x+v\partial v/\partial y+w\partial v/\partial z)\\ &=\partial Y_x/\partial x+\partial Y_y/\partial y+\partial Y_z/\partial z+\rho Y,\\ C&=\rho(\partial w/\partial t+u\partial w/\partial x+v\partial w/\partial y+w\partial w/\partial z)\\ &=\partial Z_x/\partial x+\partial Z_y/\partial y+\partial Z_z/\partial z+\rho Z.\end{aligned}\right.$$

The symbols A, B, C are introduced on account of the use to be subsequently made of them.

Suppose the fluid which is here considered to be a liquid and incompressible. In this case it contains energy only in the form of kinetic energy or heat. If the body, on the other hand, is gaseous, we suppose it to be an ideal gas, which conforms to the law of Boyle and Gay-Lussac. Such a gas can indeed be compressed, but the work done by the compression is transformed into heat, so that the energy contained by the gas is independent of the volume, and is determined only by its kinetic energy and temperature.

A volume-element of the fluid $d\omega = dxdydz$ contains, at the time t, a quantity of energy which is the sum of the kinetic energy and the quantity of heat contained in it. We multiply the latter by the mechanical equivalent J of the unit of heat. If E is the unit of volume, and if the unit of mass receives the quantity of heat Θ, we have, designating the velocity by h, $E = \frac{1}{2}\rho h^2 + J\rho\Theta$. During the time dt the volume element $d\omega$ receives the quantity of energy

(c) $$dE/dt \,.\, dtd\omega, \text{ where } dE/dt = \tfrac{1}{2}d(\rho h^2)/dt + J \,.\, d(\rho\Theta)/dt.$$

The increment of energy which the element $d\omega$ receives in the time dt proceeds from the following causes:

1. From the work done by the accelerating forces X, Y, Z.

2. From the kinetic energy which, in consequence of the flow of the fluid, passes into the volume-element $d\omega$ through its surface.

3. From the work done by the surface forces X_x, Y_y, ... on that part of the fluid which is situated on the surface of the element $d\omega$.

4. From the heat contained by that part of the fluid which flows through the surface-element $d\omega$.

5. From the heat which passes into the element $d\omega$ by conduction.

We will designate these quantities of energy in order by $e_1d\omega dt$, $e_2d\omega dt$, $e_3d\omega dt$, $e_4d\omega dt$, and $e_5d\omega dt$; e_1 is therefore the quantity of energy received by the unit of volume in the unit of time only through the influence of the accelerating forces. We will now investigate the values of e_1, e_2, ...

We determine the work done by the accelerating forces in the time dt in the following way. The volume-element contains the mass $\rho d\omega$ and moves in the time dt through the distance udt in the direction of the x-axis. Thus the force X does the work $\rho d\omega \,.\, Xudt$. The work done by the forces Y and Z is determined in the same way. The work considered is therefore $\rho(uX + vY + wZ)d\omega dt$. We have represented this quantity of work by $e_1d\omega dt$ and hence obtain

(d) $$e_1 = \rho(uX + vY + wZ).$$

The kinetic energy which $d\omega$ receives from that part of the fluid which flows in the time dt through the element $d\omega$, is determined thus. The mass which flows through the surface-element $dydz$ in the time dt is $\rho.u.dt.dydz$; the kinetic energy of this mass is therefore $\frac{1}{2}.\rho.udt.dydz.h^2$. But if we set $U=\frac{1}{2}\rho uh^2$, U is the component of flow of the kinetic energy in the direction of the x-axis. Let the corresponding components of flow with respect to the y- and z-axes be V and W respectively; we then have

$$V=\tfrac{1}{2}.\rho vh^2,\quad W=\tfrac{1}{2}.\rho wh^2.$$

By a method similar to that used in XIV., it may be shown that in the time dt the volume-element $d\omega$ receives the quantity of energy $-(\partial U/\partial x+\partial V/\partial y+\partial W/\partial z)d\omega dt$. We represent this quantity by $e_2 d\omega dt$ and hence obtain

$$e_2=-\tfrac{1}{2}[\partial(\rho uh^2)/\partial x+\partial(\rho vh^2)/\partial y+\partial(\rho wh^2)/\partial z],$$

or

$$\begin{aligned}e_2=&-\tfrac{1}{2}h^2.[\partial(\rho u)/\partial x+\partial(\rho v)/\partial y+\partial(\rho w)/\partial z]\\&-\rho u(u\partial u/\partial x+v\partial v/\partial x+w\partial w/\partial x)\\&-\rho v(u\partial u/\partial y+v\partial v/\partial y+w\partial w/\partial y)\\&-\rho w(u\partial u/\partial z+v\partial v/\partial z+w\partial w/\partial z).\end{aligned}$$

It follows from this by the help of equations (a) and (b) that

$$e_2=\tfrac{1}{2}h^2.d\rho/dt+\rho(u\partial u/\partial t+v\partial v/\partial t+w\partial w/\partial t)-(Au+Bv+Cw),$$

or more simply

$$e_2=\tfrac{1}{2}h^2 d\rho/dt+\tfrac{1}{2}\rho dh^2/dt-(Au+Bv+Cw),$$

(e) $$e_2=\tfrac{1}{2}.d(\rho h^2)/dt-(Au+Bv+Cw).$$

The quantity of energy which the surface forces X_x, Y_y ... impart to the element $d\omega$, may be determined in the following way. The force $-X_x dydz$ acts on the surface-element $dydz$ which bounds $d\omega$ on the side lying in the direction of the negative x-axis, in the direction of the x-axis. The fluid particles which flow in the time dt through the element $dydz$ traverse the path udt in the direction of the x-axis. Thus the force $-X_x$ does the work $-X_x dydz.udt$. But the fluid particles in the surface-element have also tangential motions. They traverse the path vdt in the direction of the y-axis under the influence of the force $-Y_x dydz$, by which the work $-Y_x dydz.vdt$ is done. The same particles also move in the direction of the z axis, so that the work $-Z_x dydz.wdt$ is done. The total work done by the forces in the time dt on the element $dydz$ is therefore

$$-(X_x u+Y_x v+Z_x w)dydzdt.$$

We will designate this flow of energy in the direction of the x-axis by $U'dydzdt$; let the corresponding flow in the direction of the y- and z-axes be $V'dxdzdt$ and $W'dxdydt$ respectively. We then have

$$U' = -(X_x u + Y_x v + Z_x w),\quad V' = -(X_y u + Y_y v + Z_y w),$$
$$W' = -(X_z u + Y_z v + Z_z w).$$

The quantity of energy which $d\omega$ thus receives is determined as in the foregoing case, and is equal to

$$e_3 d\omega dt = -(\partial U'/\partial x + \partial V'/\partial y + \partial W'/\partial z)d\omega dt.$$

Hence we obtain

$$e_3 = \partial(X_x u + Y_x v + Z_x w)/\partial x + \partial(X_y u + Y_y v + Z_y w)/\partial y$$
$$+ \partial(X_z u + Y_z v + Z_z w)/\partial z.$$

But using equations (b) it follows that

$$\text{(f)}\quad \begin{cases} e_3 = X_x \partial u/\partial x + Y_y \partial v/\partial y + Z_z \partial w/\partial z \\ \qquad + Z_y(\partial w/\partial y + \partial v/\partial z) + X_z(\partial u/\partial z + \partial w/\partial x) \\ \qquad + Y_x(\partial v/\partial x + \partial u/\partial y) + (A - \rho X)u + (B - \rho Y)v + (C - \rho Z)w. \end{cases}$$

The quantity of heat which the separate parts of the fluid contain is transferred with them by the flow. During the time dt the mass $\rho udt\,.\,dydz$ enters the element $d\omega$ through the surface $dydz$, and brings with it the quantity of heat $\rho udydzdt\Theta$ or the energy $J\rho udydzdt\Theta$. We determine in the same way the quantities of heat which enter the element $d\omega$ through the other bounding surfaces. If we use the method given above and set

$$U'' = J\rho u\Theta,\quad V'' = J\rho v\Theta,\quad W'' = J\rho w\Theta,$$

the quantity of heat $e_4 d\omega dt$ received by the parallelepiped $d\omega$ in the time dt, is given by

$$e_4 = -(\partial U''/\partial x + \partial V''/\partial y + \partial W''/\partial z)$$

or

$$e_4 = -J[\partial(\rho u\Theta)/\partial x + \partial(\rho v\Theta)/\partial y + \partial(\rho w\Theta)/\partial z].$$

Hence, by use of equation (a), we have

$$\text{(g)}\qquad e_4 = J\,.\,\partial(\rho\Theta)/\partial t - J\rho(\partial\Theta/\partial t + u\partial\Theta/\partial x + v\partial\Theta/\partial y + w\partial\Theta/\partial z).$$

Finally the element $d\omega$ receives heat by conduction. The components of flow of heat are [CXXII.]

$$-Jk\,.\,\partial\theta/\partial x,\quad -Jk\,.\,\partial\theta/\partial y,\quad -Jk\,.\,\partial\theta/\partial z.$$

If we set the quantity of energy thus received by $d\omega$ in the time dt, equal to $e_5 d\omega dt$, and assume the conductivity constant, we have

$$\text{(h)}\qquad e_5 = J\,.\,k(\partial^2\theta/\partial x^2 + \partial^2\theta/\partial y^2 + \partial^2\theta/\partial z^2).$$

The increase in energy which $d\omega$ receives in the time dt is given by $[\frac{1}{2} . d(\rho h^2)/dt + Jd(\rho\Theta)/dt]d\omega dt$. At the same time the quantity of energy $(e_1 + e_2 + e_3 + e_4 + e_5)\ d\omega dt$ enters the element $d\omega$, and we therefore have

(i) $$\frac{1}{2} . d(\rho h^2)/dt + Jd(\rho\Theta)/dt = e_1 + e_2 + e_3 + e_4 + e_5.$$

Introducing in this equation the values found for e_1, e_2, e_3, ... it follows that

(k) $$\left\{\begin{aligned} &J\rho(\partial\Theta/\partial t + u\partial\Theta/\partial x + v\partial\Theta/\partial y + w\partial\Theta/\partial z) - Jk . \nabla^2\theta \\ &\quad = X_x\partial u/\partial x + Y_y\partial v/\partial y + Z_z\partial w/\partial z + Z_y(\partial w/\partial y + \partial v/\partial z) \\ &\quad + X_z(\partial u/\partial z + \partial w/\partial x) + Y_x(\partial v/\partial x + \partial u/\partial y). \end{aligned}\right.$$

If internal friction exists in the fluid, we have from XLVII. (b)

$$X_x = -p + 2\mu . \partial u/\partial x - \tfrac{2}{3}\mu(\partial u/\partial x + \partial v/dy + \partial w/\partial z)$$

$$Z_y = \mu(\partial w/\partial y + \partial v/\partial z), \text{ etc.}$$

By the help of these relations we may give equation (k) the form

(l) $$\left\{\begin{aligned} &J\rho(\partial\Theta/\partial t + u\partial\Theta/\partial x + v\partial\Theta/\partial y + w\partial\Theta/\partial z) - Jk\nabla^2\theta \\ &\quad = -p(\partial u/\partial x + dv/\partial y + \partial w/\partial z) + 2\mu[(\partial u/\partial x)^2 \\ &\quad + (\partial v/\partial y)^2 + (\partial w/\partial z)^2 - \tfrac{1}{3}(\partial u/\partial x + \partial v/\partial y + \partial w/\partial z)^2] \\ &\quad + \mu[(\partial w/\partial y + \partial v/\partial z)^2 + (\partial u/\partial z + \partial w/\partial x)^2 + (\partial v/\partial x + \partial u/\partial y)^2]. \end{aligned}\right.$$

For the determination of the motion and temperature of the fluid we have the five equations given under (a), (b), and (l). These five equations are not sufficient to determine the seven unknown quantities u, v, w, ρ, p, Θ, and θ. We obtain two other equations in the following way. The total quantity of heat Θ contained by the unit of mass must depend on θ, and we assume that (m) $\Theta = c\theta$, where c is the specific heat, a constant. If the fluid considered is gaseous, c denotes the specific heat of constant volume.

The second equation must express the relation between density, pressure, and temperature. In the case of liquids, we may set approximately (n) $\rho = \rho_0/(1 + \alpha\theta)$, where ρ_0 is the density when $\theta = 0$ and α is a constant. But for gases, if V is the volume of the unit of mass at pressure p and temperature θ, V_0 the volume of the same mass at pressure p_0 and temperature 0°, we have $pV = p_0V_0(1 + \alpha\theta)$. Since $V\rho = 1$ and $V_0\rho_0 = 1$, we have (o) $p/\rho = p_0/\rho_0 . (1 + \alpha\theta)$. The equation (o) in connection with (a), (b), (l), and (m) serves to determine the unknown quantities. The complicated equations which determine temperature and motion in a fluid are very hard to integrate, so that up to this time no case has been completely solved.

SECTION CXXXV. THE INFLUENCE OF THE CONDUCTION OF HEAT ON THE INTENSITY AND VELOCITY OF SOUND IN GASES.

We have the following equations [CXXXIV.] for the determination of motion in a gas in which the temperature is variable:

1. The equation of continuity [CXXXIV. (a)], which may take the following form:

$$\partial\rho/\partial t + \rho(\partial u/\partial x + \partial v/\partial y + \partial w/\partial z) + u\partial\rho/\partial x + v\partial\rho/\partial y + w\partial\rho/\partial z = 0.$$

2. The equations of motion [CXXXIV. (b)]. We replace in these equations the forces X_x, X_y, ... by the values found in XLVII. (b) and (h), and obtain [cf. XLVIII. (a)]

$$\rho(\partial u/\partial t + u\partial u/\partial x + v\partial u/\partial y + w\partial u/\partial z)$$
$$= \rho X - \partial p/\partial x + \mu\nabla^2 u + \tfrac{1}{3}\mu\partial(\partial u/\partial x + \partial v/\partial y + \partial w/\partial z)/\partial x,$$

and analogous equations for y and z.

3. The condition for the conservation of energy [cf. CXXXIV. (l)].

4. The connection between the heat contained in the body and the temperature [cf. CXXXIV. (m)].

5. The Boyle-Gay-Lussac law [cf. CXXXIV. (o)].

Let the velocity and change of temperature be very small quantities; the same is then true of such differential coefficients as $\partial\rho/\partial x$, $\partial\Theta/\partial x$, etc., and we will therefore neglect the product of these quantities, that is, terms of the form $u\partial\rho/\partial x$, $u\partial u/\partial x$, $u\partial\Theta/dx$, etc. The equations 1—5 are then very much simplified. We obtain

(a) $$\partial(\log\rho)/\partial t + \partial u/\partial x + \partial v/\partial y + \partial w/\partial z = 0.$$

If we further set $\mu/\rho = \mu'$, equation (2), by use of (a), takes the form

(b) $$\partial u/\partial t + 1/\rho \,.\, \partial p/\partial x = \mu'\nabla^2 u - \tfrac{1}{3}\mu' \,.\, \partial^2(\log\rho)/\partial x\partial t.$$

Similar equations hold for u and v, if x is replaced by y and z respectively.

Eliminating Θ in equation CXXXIV. (l) by means of the relation $\Theta = c\theta$ and introducing the heat equivalent A of the unit of work for $1/J$, it follows that (c) $c\rho \,.\, \partial\theta/\partial t - k\nabla^2\theta = Ap\partial(\log\rho)/\partial t$. We have further the equation [CXXXIV. (o)], (d) $p/\rho = p_0/\rho_0 \,.\, (1 + a\theta)$. We consider μ', k, and c as constants. We substitute ρ_0 for ρ, if ρ or $1/\rho$ occurs as a coefficient; we also substitute p_0 for p in (c). In these substitutions we neglect only infinitely small quantities of the second order. Setting $\rho = \rho_0(1 + \sigma)$, we obtain (e) $\log\rho = \log\rho_0 + \sigma$, because σ is a small quantity. Hence equation (d) takes the form $p = p_0(1 + \sigma)(1 + a\theta)$, or, because θ is also a small quantity,

(f) $$p = p_0(1 + \sigma + a\theta).$$

Equation (c) now becomes $\partial\theta/\partial t - k/c\rho_0 \,.\, \nabla^2\theta = Ap_0/c\rho_0 \,.\, \partial\sigma/\partial t$, and if we set $\kappa^2 = k/c\rho_0$ and $\Theta = c\rho_0\theta/Ap_0$, we obtain from the last equation (g) $\partial\Theta/\partial t - \kappa^2\nabla^2\Theta = \partial\sigma/\partial t$.

By the use of (e) and (f) we may transform equation (b) into

$$\partial u/\partial t + p_0/\rho_0 \,.\, \partial\sigma/\partial x + p_0 a/\rho_0 \,.\, \partial\theta/\partial x = \mu' \,.\, \nabla^2 u - \tfrac{1}{3}\mu' \,.\, \partial^2\sigma/\partial t\,\partial x.$$

Introducing in place of θ the quantity Θ already defined, we have

(h) $\partial u/\partial t + p_0/\rho_0 . \partial\sigma/\partial x + p_0/\rho_0 \,.\, Ap_0 a/\rho_0 c . \partial\Theta/\partial x = \mu' . \nabla^2 u - \tfrac{1}{3}\mu' . \partial^2\sigma/\partial t\,\partial x$.

The heat required to raise the temperature of a gram of air under constant pressure from θ to $\theta + d\theta$ is equal to $C.d\theta$, if C is the specific heat at constant pressure. A part of this heat, namely $c.d\theta$, is used in raising the temperature, the other part is used in overcoming resistance during the expansion, by which the work $p.dV$ is done. We have therefore $C.d\theta = c.d\theta + Ap.dV$. It follows from the equation $pV = p_0V_0(1 + a\theta)$, because p is here constant, that $p.dV = p_0V_0 a.d\theta$, and therefore (i) $C = c + Ap_0 a/\rho_0$, since $V_0\rho_0 = 1$.

Finally, if we set $a^2 = p_0C/\rho_0 c$ and $b^2 = p_0/\rho_0$, the equations (a), (b), and (c) take the following forms:

$$\text{(k)}\quad \begin{cases} \partial\sigma/\partial t + \partial u/\partial x + \partial v/\partial y + \partial w/\partial z = 0, \\ \partial u/\partial t + b^2 \,.\, \partial\sigma/\partial x + (a^2 - b^2)\partial\Theta/\partial x = \mu' \,.\, \nabla^2 u - \tfrac{1}{3}\mu' \,.\, \partial^2\sigma/\partial t\,\partial x, \\ \partial v/\partial t + b^2 \,.\, \partial\sigma/\partial y + (a^2 - b^2)\partial\Theta/\partial y = \mu' \,.\, \nabla^2 v - \tfrac{1}{3}\mu' \,.\, \partial^2\sigma/\partial t\,\partial y, \\ \partial w/\partial t + b^2 \,.\, \partial\sigma/\partial z + (a^2 - b^2)\partial\Theta/\partial z = \mu' \,.\, \nabla^2 w - \tfrac{1}{3}\mu' \,.\, \partial^2\sigma/\partial t\,\partial z, \\ \partial\Theta/\partial t - \kappa^2\nabla^2\Theta = \partial\sigma/\partial t. \end{cases}$$

These equations are due to Kirchhoff.* Reference may be made to Kirchhoff's work for the application of these equations to the more difficult cases of the transmission of sound. We will here investigate only the influence of conduction and friction on the motion of plane sound waves. First, however, the physical significance of the constants (a) and (b) must be determined.

If there is neither conduction nor friction in the air, we have $\kappa = 0$ and $\mu' = 0$; further, if the vibrations occur in the direction of the x-axis, we also have $v = w = 0$. Under these circumstances the equations (k) become

$$\partial\sigma/\partial t + \partial u/\partial x = 0,\quad \partial u/\partial t + b^2 \,.\, \partial\sigma/\partial x + (a^2 - b^2) \,.\, \partial\Theta/\partial x = 0,\quad \partial\Theta/\partial t = \partial\sigma/\partial t.$$

If the second of these three equations is differentiated with respect to t, we obtain

$$\partial^2 u/\partial t^2 + b^2 \,.\, \partial^2\sigma/\partial x\,\partial t + (a^2 - b^2) \,.\, \partial^2\Theta/\partial x\,\partial t = 0.$$

* Kirchhoff, *Pogg. Ann.*, Vol. 134. 1868.

It thus follows, by the use of the first and last of these equations, that $\partial^2 u/\partial t^2 = a^2 . \partial^2 u/\partial x^2$. An integral of this equation is

$$u = \cos[2\pi/T . (t - x/a)],$$

and this expression represents a wave motion which proceeds with the velocity (l) $a = \sqrt{p_0 C/\rho_0 c} = b\sqrt{C/c}$. This value for the velocity of sound was found by Laplace. It differs from the value calculated in XXXV., which was originally found by Newton, and which in our present notation is $b = \sqrt{p_0/\rho_0}$. The difference between the two formulas is due to the fact that in the first we have taken into account the heating of the air by compression and its cooling by expansion. Since the ratio C/c has been determined by direct experiment, the true velocity of sound in the air may be calculated. For atmospheric air at 0° C., $C/c = 1,405$; hence $a = 33815$ cm. This value agrees very well with experiment.

Suppose that a plane-wave is propagated in the direction of the x-axis, and that κ and μ' are not zero. The vibrations are parallel to the x-axis, so that $v = 0$ and $w = 0$. Since u, Θ, and σ are then functions of x and t alone, equations (k) become

$$\text{(m)} \quad \begin{cases} \partial\sigma/\partial t + \partial u/\partial x = 0, \\ \partial u/\partial t + b^2 . \partial\sigma/\partial x + (a^2 - b^2)\partial\Theta/\partial x = \mu' . \partial^2 u/\partial x^2 - \frac{1}{3}\mu' . \partial^2\sigma/\partial t\,\partial x, \\ \partial\Theta/\partial t - \kappa^2 . \partial^2\Theta/\partial x^2 = \partial\sigma/\partial t. \end{cases}$$

The unknown quantities u, Θ, and σ are periodic functions of t. We will represent by h a real magnitude, and by u', Θ', and σ' three magnitudes which are functions of x alone. It is then admissible to make the assumptions (n) $u = u' . e^{hit}$, $\Theta = \Theta' . e^{hit}$, $\sigma = \sigma' . e^{hit}$, where $i = \sqrt{-1}$. By the help of these equations we obtain from (m)

$$hi\sigma' + du'/dx = 0,$$
$$hiu' + b^2 . d\sigma'/dx + (a^2 - b^2)d\Theta'/dx = \mu' . d^2u'/dx^2 - \tfrac{1}{3}\mu' hi . d\sigma'/dx,$$
$$hi\Theta' - \kappa^2 . d^2\Theta'/dx^2 = hi\sigma'.$$

We eliminate σ' from these equations, and then have

$$\text{(o)} \quad \begin{cases} -h^2u' + hi(a^2 - b^2) . d\Theta'/dx = (b^2 + \tfrac{4}{3}\mu' hi) . d^2u'/dx^2, \\ du'/dx = \kappa^2 d^2\Theta'/dx^2 - hi\Theta'. \end{cases}$$

If the first of these equations (o) is differentiated with respect to x, u' may be eliminated, and we obtain the following differential equation:

$$\text{(p)} \quad \kappa^2(b^2 + \tfrac{4}{3}\mu' hi) . d^4\Theta'/dx^4 + (h^2\kappa^2 + \tfrac{4}{3}\mu' h^2 - ha^2 i) . d^2\Theta'/dx^2 - h^3 i\Theta' = 0.$$

Since this equation is linear, we set $\Theta' = e^{mx}$, and obtain

$$\text{(r)} \quad \kappa^2 m^4(b^2 + \tfrac{4}{3}\mu' hi) + m^2(h^2\kappa^2 + \tfrac{4}{3}\mu' h^2 - ha^2 i) - h^3 i = 0.$$

We will determine the exponent m only in the case in which the conductivity as well as the internal friction is very small. If $\kappa = 0$ and $\mu' = 0$, we have from (r) $m = -hi/a$. If therefore we set $m = (-hi + \delta)/a$, where δ is a small quantity whose higher powers may be disregarded, we have from (r), if the terms $\kappa^2\mu'$, $\kappa^2\delta$, etc., are neglected, (s) $\delta = -[\frac{4}{3}\mu' h^2 + (1 - b^2/a^2)\kappa^2 h^2]/2a^2$. But it follows from (n) and (q) that one value for Θ is $\Theta = e^{\delta x/a} \,.\, e^{hi(t - x/a)}$; the other is obtained by substituting $-i$ for i, which gives $\Theta = e^{\delta x/a} \,.\, e^{-hi(t - x/a)}$. Half the sum of the two values of Θ satisfies the conditions and is at the same time real, since

$$\text{(t)} \qquad \Theta = e^{-[4/3\mu' h^2 + (1 - b^2/a^2)\kappa^2 h^2] \,.\, x/2a^3} \,.\, \cos[h(t - x/a)].$$

From the exponent of e we see that the changes of temperature in the wave diminish the further it travels; at the same time u also diminishes. The sound, therefore, becomes weaker as the wave travels further. If T is the period of vibration and n the number of vibrations, we have $h = 2\pi/T = 2n\pi$.

By using this value of h it follows, from equation (t), that the higher tones lose their intensity more quickly than the lower ones.

The mathematical treatment of conduction is principally due to Fourier, who not only developed the partial differential equation which is at the foundation of the treatment of conduction, but also gave us methods for the solution of a great number of problems. His principal work is: *Théorie Analytique de la Chaleur*, Paris, 1822. Of the later works on this subject we mention Riemann, *Partielle Differentialgleichungen*, edited by Hattendorf, Braunschweig, 1876.

INDEX.

Acceleration, 2.
Centripetal, 11.
Resultant, 4.
Action and Reaction, 48.
Adiabatic Curves, 268.
Amplitude, 90.
Attraction, Universal, 30.

Biot and Savart's Law, 184.
Bodies, Structure of, 50.
Rigid, 58.
Equilibrium of, 59.
Motion of, 59.
Rotation of, 60.
Boyle's Law, 267.

Capacity, Electrical, 131, 149.
of Coaxial Cylinders, 153.
of Condenser, 141.
of Spherical Condenser, 140, 152.
of Parallel Plates, 151.
Specific Inductive.
(Dielectric Constant), 156.
Relation to Index of Refraction, 221.
Capillarity, 121.
Capillary Constant, 121.
Tubes, 125.
Carnot's Cycle, 272.
Theorem, 274.
Clausius's Equation of the State of a Gas, 290.
Theorem, 275, 277.
Collision, 49.
Condenser, Cylindrical, 153.
Parallel Plate, 151.
Spherical, 152.
Conductivity for Electricity, 197.
for Heat, 300.
Conductors, System of, 147.
Work done on, 150.
Contact Angle, 125.
Continuity, Equation of, 105.
Cooling of a Sphere by Conduction, 318.
by Radiation, 319.
Corresponding States, 289.
Critical Temperature, 284.
Current, Electrical, Continuity of, 191.
Force of, 184.
Force of Linear, 190.
Measurement of Constant, 194.
Measurement of Variable, 195.
Potential of, 185.
Potential Energy of, 193.
Currents, Electrical, Mutual Action of, 192.
Potential Energy of, 193.
Systems of, 186, 190.
Cycle (Cyclic Process), 268, 272.
Carnot's, 272.
Efficiency of Carnot's, 273.

Damping Action, 195.
Deformation, 74.
Relation of, to Stress, 79.
Density, 35.

Descartes's Explanation of Mutual Actions, 49.
Dielectric, 155.
  Equations of, 219.
  Plane Waves in, 221.
Dielectric Constant, 156.
  in Crystals, 246.
  Displacement, 156.
Dilatation, 76.
  Linear, 77.
  Principal Axes of, 78.
  Volume, 77.
Dirichlet's Principle, 136.
Dissociation, 295.
Dyne, 9.

Earth, Temperature of, 302.
Elastic Body, Equilibrium of, 82.
  Motion of, 89.
  Potential Energy of, 96.
Elasticity, Coefficient of, 79.
  Modulus of, 81.
Electrical Convection, 161.
  Displacement, 191.
  Distribution, 128.
    on a Conductor, 130.
    on Conductors, 139.
    on an Ellipsoid, 133.
    on a Plane, 135.
    on a Sphere, 132, 137.
  Double Sheets, 160.
  Energy, 145.
  Force, Law of, 128.
    Lines of, 143.
  Images, 135.
  Oscillations, 215, 223.
  Polarization, 191.
  Potential, 128.
    of a Conductor, 131.
    near a Surface, 129.
Electricity, Theories of, 127.
Electrified Body, Force on, 141.
Electro-kinetic Energy, 201, 210.
Electromagnetism, 184.
  Equations of, 188.
Electrometer, Quadrant, 154.
  Thomson's Absolute, 142.
Energy, Conservation of, 58.
  Kinetic, 14.
    of a System, 56.
  Potential, 57.
Entropy, 271, 272, 278, 292, 294.
Entropy Criterion of Equilibrium, 294.
Equilibrium, 58.
  Conditions of, 58.
  of Fluid Surfaces, 123.
Equipotential Surfaces, 23.
  Construction of, 132.
Equivalence of Heat and Energy, 268.
Equivalent, Mechanical, of Heat, 268.
Ether, 230, 231.
  Fresnel's Assumption Concerning, 233.
  Neumann's Assumption Concerning, 233.
Euler's Equations of Motion of Fluids, 103.

Falling Bodies, Laws of, 5.
Flexure, 87.
Flow of Fluid, 108.
  Through Tube, 119.
Flux of Force, 42.
Force, 8.
  Centripetal, 11.
  Components of, 9.
  Line of, 23.
  Measure of, 6, 9.
  Normal, 13.
  Tangential, 13.
  Tubes of, 145.
  Unit of, 9.
Forces, Conservative, 20.
  External, 53.
  Internal, 53, 62.
Fluid, Conduction of Heat in, 325.
  Elasticity of, 81.
  Equilibrium of, 62, 99.
  Motion of, 103.
    Viscous, 118.
  Motion of Sphere in, 109.
  Steady Motion of, 118.
Fourier's Theorem, 312, 315.
Fresnel's Formulas for Light, 231.
  Failure of, 235.
Friction of Fluids, 115.
  Coefficient of, 115.

Galileo's Laws of Falling Bodies, 5.
Gas, Elasticity of, 82.
  Ideal, 270.
    Specific Heats of, 270.

Gauss's Theorem, 41.
Gravity, Acceleration of, 5.
Centre of, 50.
Gyration, Radius of, 60.

Heat, Conduction of, 298.
in Fluids, 325.
Flow of, between two Bodies, 317.
from a Point, 204.
from a Surface, 303, 316.
in a Cylinder, 322.
in an Infinite Body, 305.
in a Plate, 308.
Fourier's Equation of, 298.
Steady Flow of, in a Cylinder, 323.
in a Plate, 300.
in a Sphere, 301.
in a Tube, 301.
Helmholtz's Transformation of Euler's Equations, 106.
Hertz's Apparatus, 216.
Form of Maxwell's Equations, 219.
Huygens's Construction, 262.

Ice, Formation of, 307.
Impulse, 8.
Measure of, 9.
Inclined Plane, 24.
Induction, Coefficients of Magnetic, 149.
of Electrical Currents, 199.
Coefficients of, 202.
Equations of, 208.
Law of, 200.
Measurement of Coefficients of, 203.
Mutual, 201.
Self-, 200.
Inertia, Moment of, 60.
Principle of, 6.
Isentropic Curves, 268.
Isothermal Curves, 267.

Joule's Law, 197.

Kepler's Laws, 27.

Lagrange's Equations of Motion of Fluids, 111.
Laplace's Equation, 45.
Application of, 46.
Lenz's Law, 200.
Light, Electromagnetic Theory of, 230, 235, 237.
Emission Theory of, 229.
Principal Laws of, 230.
Wave Theory of, 229.
Lines of Force, 23.
Electrical Force, 143.
Logarithmic Decrement, 196.

MacCullagh's Construction, 263.
Magnet, Constitution of, 163.
Forces acting on, 169.
Oscillation of, 170.
Potential Energy of, 171.
Potential of, 166.
Magnetic Axis, 166.
Induction, 173, 177, 179.
Coefficient of, 180.
Moment, 165.
Permeability, 180.
Poles, 163.
Shell, 180.
Strength of, 180.
Force, 166.
due to Electrical Current, 184.
Law of, 163.
Lines of, 174, 178.
Tubes of, 174.
Magnetism, 163.
Distribution of, 165.
Magnetization, Intensity of, 165.
Magnetized Sphere, Potential of, 168.
Material System, 53.
Maxwell's Electromagnetic Theory of Light, 230.
Theory of the Action of a Medium, 72.
Melting, 291.
Moment, of Force, 56.
of Inertia, 60.
of Momentum, 55.
Momentum, 49.
Moment of, 55.
Motion, Constrained, 24.
Curvilinear, 3.
Equations of, of a Particle, 10.
In a Circle, 11.
Oscillatory, 12.
Periodic, 1, 12.

Motion, Uniform, 1.
Variable, 1, 2.

Newton's Law of Attraction, 30.

Ohm's Law, 197.
Optic Axes, 249.
Optic Axes of Elasticity, 247.
Oscillatory Motion, 12.

Path, 1.
Equation of, 4.
Pendulum, 25, 61.
Period, 90.
Points, Electrical Action of, 144.
Poisson's Equation, 45.
Application, 46.
Polarization, Angle of, 234.
Potential, 20, 21.
Coefficients of, 148.
Difference of, 22.
of a Circular Plate, 38.
of a Cylinder, 39.
of a Solid Sphere, 37.
of a Spherical Shell, 36.
of a Straight Line, 39.
of a System, 34.
Poynting's Theorem, 224.
Pressure, 64.
Hydrostatic, 62, 63.
Principal Section, 260.
Projectiles, 7, 11.

Rays, Direction of, in Crystals, 256.
Velocity of Propagation of, 257.
Reflection of Polarized Light, 232, 237, 239.
Total, 240, 241.
Refraction, Double, 246.
at the Surface of a Crystal, 261.
in Uniaxial Crystals, 264.
In a Plate, 242.
Index of, 230.
of Polarized Light. 233, 237, 239.
Resistance, 198.
Measurement of, 205.
Resistance, Measurement of—
Lorenz's Method, 207.
Thomson's Method, 206.

Shear, 77.
Solenoid, 186.
Solid, Internal Forces in, 62.
Sound, Velocity of, in Gases, 330.
Spherical Shell under Pressure, 83.
Sphondyloid, 143.
State of a Body, 266.
Diagram Representing, 267.
Equation of, 267.
State of a Gas, Clausius's Equation of, 290.
Van der Waal's Equation of, 286.
Stokes's Theorem, 21.
Stress, 64.
Components of, 65.
Equilibrium under, 67.
Principal, 69.
Strings, Vibrating, 95.
Surface Tension, 122.

Tension, 64.
Thermodynamic Relations, 280.
Properties of Bodies, 281.
Thermodynamics, First Law of, 268.
Equation embodying, 269.
Second Law of, 170.
Application of, 279.
Torsion, 85.
Coefficient of, 86.
Trigonometrical Series Representing Arbitrary Functions, 309, 312.
Tubes of Force, 145.

Uniaxial Crystals, 259.
Wave Surface in, 259.
Units, Absolute, 2, 211.
Derived, 2.
Practical, 214.

Van der Waal's Equation of State, 283.
Vaporization, Heat of, 291.
Vapour, Saturated, 290.
Specific Heat of, 292.
Vector, 55.
Velocity, 1, 2.
Angular, 1.

Velocity, Components of, 3.
Resultant, 4.
Velocity Potential, 106.
Vibrations, Longitudinal, 91.
Transverse, 91.
Vortex Motion, 107.
Filament, 109.

Wave Surface, 251.
Reciprocal, 257.
Waves on the Surface of a Fluid, 112.
Waves, Plane, 90.
in Dielectric, 221.
Spherical, 93.
Stationary, 93.
Torsional, 93.
Velocity of, 91, 92.
in Crystals, 248, 249, 257.

Weight, 9.
Work, 14.
Done in Closed Path, 15, 16.

1897

UNIVERSITY·PRESS

ROBERT·MACLEHOSE·&·CO

GLASGOW

FLOREAT·GLASGUA·PRAEDICATIONE·EVANGELII

www.ingramcontent.com/pod-product-compliance
Lightning Source LLC
Chambersburg PA
CBHW021404210326
41599CB00011B/1007
* 9 7 8 3 3 3 7 2 7 7 5 4 3 *